Scientific Computing and Differential Equations

An Introduction to Numerical Methods

Scientific Computing and Differential Equations

An Introduction to Numerical Methods

Gene H. Golub

Computer Science Department
Stanford University
Stanford, California

James M. Ortega

Institute for Parallel Computation
School of Engineering and
Applied Science
University of Virginia
Charlottesville, Virginia

Academic Press
San Diego New York Boston
London Sydney Tokyo Toronto

Find Us on the Web! http: //www.apnet.com

This book is printed on acid-free paper. ∞

Academic Press
A Division of Harcourt Brace & Company
525 B Street, Suite 1900, San Diego, California 92101-4495

United Kingdom Edition published by
ACADEMIC PRESS LIMITED
24-28 Oval Road, London NW1 7DX

Library of Congress Cataloging-in-Publication Data

Golub, Gene H. (Gene Howard). 1932-
Scientific computing and differential equations : an introduction
to numerical methods / Gene H. Golub , James M. Ortega.
p. cm.
Rev. ed. of : Introduction to numerical methods for differential
equations / James M. Ortega. 1981.
Includes bibliographical references and index.
ISBN 0-12-289255-0
1. Differential equations — Numerical solutions — Data processing.
I. Ortega. James M., 1932- . II. Ortega, James M., 1932-
Introduction to numerical methods for differential equations.
III. Title.
QA371.G62 1992
515'.35—dc20 91-20106
CIP

Printed in the United States of America
98 99 QW 9 8 7 6

Dedicated to the Memory of

George Forsythe

who taught us to try to light the fire,

not fill the bucket

Table of Contents

Preface

This book is a revision of *Introduction to Numerical Methods for Differential Equations* by J. M. Ortega and W. G. Poole, Jr., published by Pitman Publishing, Inc., in 1981.

As discussed in Chapter 1, a large part of scientific computing is concerned with the solution of differential equations and, thus, differential equations is an appropriate focus for an introduction to scientific computing. The need to solve differential equations was one of the original and primary motivations for the development of both analog and digital computers, and the numerical solution of such problems still requires a substantial fraction of all available computing time. It is our goal in this book to introduce numerical methods for both ordinary and partial differential equations with concentration on ordinary differential equations, especially boundary-value problems. Although there are many existing packages for such problems, or at least for the main subproblems such as the solution of linear systems of equations, we believe that it is important for users of such packages to understand the underlying principles of the numerical methods. Moreover, it is even more important to understand the limitation of numerical methods: "Black Boxes" can't solve all problems. Indeed, it may be that one has several excellent black boxes for solving classes of problems, but the combination of such boxes may yield less than optimal results.

We treat initial-value problems for ordinary differential equations in Chapter 2 and introduce finite difference methods for linear boundary value problems in Chapter 3. The latter problems lead to the solution of systems of linear algebraic equations; Chapter 4 is devoted to the general treatment of direct methods for this problem, independently of any connection to differential equations. This chapter also considers the important problem of least-squares approximation. In Chapter 5 we return to boundary value problems, but now they are nonlinear. This motivates the treatment of the solution of nonlinear algebraic equations in the remainder of the chapter. The tool for discretizing differential equations to this point has been finite difference methods; in Chapter 6 we introduce Galerkin and collocation methods as alternatives. These methods lead to the important subtopics of numerical integration and spline approximation. Chapter 7 treats eigenvalue problems, motivated only partially by differential equations. Finally, in Chapters 8 and 9, we introduce some of the simplest methods for partial differential equations, first initial-boundary-value problems and then boundary-value problems.

As the previous paragraph indicates, the solution of differential equations requires techniques from a variety of other areas of numerical analysis: solution of linear and nonlinear algebraic equations, interpolation and approximation, integration, and so on. Thus, most of the usual topics of a first course in numerical methods are covered and the book can serve as text for such a course. Indeed, we view the book to be basically for this purpose but oriented toward students primarily interested in differential equations. We have found the organization of the book, which may seem rather unorthodox for a first course, to be more satisfactory and motivating than the usual one for a rather large number of students. Thus, the book can be used for a variety of different audiences, depending on the background of the students and the purposes of the course.

As a minimum background, the reader is assumed to have had calculus, first courses in computer programming and differential equations and at least some linear algebra. Some basic facts from calculus, differential equations and linear algebra are collected in two appendices and further background material appears in the text itself. Students with a minimum background and no prior numerical methods course will require a full year to cover the book completely; one semester or two quarter courses can easily be taught by eliminating some topics. On the other hand, the entire book is covered in a first-semester graduate course at the University of Virginia; in this case, much of the more elementary material is review and is covered quickly.

The style of the book is on the theoretical side, although rather few theorems are stated as such and many proofs are either omitted or partially sketched. On the other hand, enough mathematical argumentation is usually given to clarify what the mathematical properties of the methods are. In many cases, details of the proofs are left to exercises or to a supplementary discussion section, which also contains references to the literature. We have found this style quite satisfactory for most students, especially those outside of mathematics.

The development of numerical methods for solving differential equations has evolved to a state where accurate, efficient, and easy-to-use mathematical software exists for solving many of the basic problems. For example, excellent subroutine libraries are available for the initial-value problems of Chapter 2, linear equations of Chapter 4, and eigenvalue problems of Chapter 7. The supplementary discussions and references at the ends of sections give information on available software. On the other hand, some of the exercises require the reader to write programs to implement some of the basic algorithms for these same problems. The purpose of these exercises is not to develop high-grade software but, rather, to have the reader experience what is involved in coding the algorithms, thus leading to a deeper understanding of them.

We owe thanks to many colleagues and students for useful comments on the original version of this book as well as drafts of the current one. We are also indebted to Ms. Susanne Freund and Ms. Brenda Lynch for their expert LaTeXing of the manuscript.

Stanford, California
Charlottesville, Virginia

Chapter 1

The World of Scientific Computing

1.1 What is Scientific Computing?

The many thousands of computers now installed in this country and abroad are used for a bewildering – and increasing – variety of tasks: accounting and inventory control for industry and government, airline and other reservation systems, limited translation of natural languages such as Russian to English, monitoring of process control, and on and on. One of the earliest – and still one of the largest – uses of computers was to solve problems in science and engineering, and more specifically, to obtain solutions of mathematical models that represent some physical situation. The techniques used to obtain such solutions are part of the general area called *scientific computing*, and the use of these techniques to elicit insight into scientific or engineering problems is called *computational science* (or *computational engineering*).

There is now hardly an area of science or engineering that does not use computers for modeling. Trajectories for earth satellites and for planetary missions are routinely computed. Aerospace engineers also use computers to simulate the flow of air about an aircraft or other aerospace vehicle as it passes through the atmosphere, and to verify the structural integrity of aircraft. Such studies are of crucial importance to the aerospace industry in the design of safe and economical aircraft and spacecraft. Modeling new designs on a computer can save many millions of dollars compared to building a series of prototypes.

Electrical engineers use computers to design new computers, especially computer circuits and VLSI layouts. Civil engineers study the structural characteristics of large buildings, dams, and highways. Meteorologists use large amounts of computer time to predict tomorrow's weather as well as to make much longer range predictions, including the possible change of the earth's climate. Astronomers and astrophysicists have modeled the evolution of stars, and much of our basic knowledge about such phenomena as red giants and pulsating stars has come from such calculations coupled with observations. Ecologists and biologists are increasingly using the computer in such diverse areas as population dynamics (including the study of natural predator and prey relationships), the flow of blood in the human body, and the dispersion of pollutants in the oceans and atmosphere.

The mathematical models of all of these problems – and of most of the other problems in science and engineering – are systems of differential equations, either ordinary or partial. Thus, to a first approximation, scientific computing as currently practiced is the computer solution of differential equations. Even if this were strictly true, scientific computing would still be an intensely exciting discipline. Differential equations come in all "sizes and shapes," and even with the largest computers we are nowhere near being able to solve many of the problems posed by scientists and engineers.

But there is more to scientific computing, and the scope of the field is changing rapidly. There are many other mathematical models, each with its own challenges. In operations research and economics, large linear or nonlinear optimization problems need to be solved. Data reduction – the condensation of a large number of measurements into usable statistics – has always been an important, if somewhat mundane, part of scientific computing. But now we have tools (such as earth satellites) that have increased our ability to make measurements faster than our ability to assimilate them; fresh insights are needed into ways to preserve and use this irreplaceable information. In more developed areas of engineering, what formerly were difficult problems to solve even once on a computer are today's routine problems that are being solved over and over with changes in design parameters. This has given rise to an increasing number of computer-aided design systems. Similar considerations apply in a variety of other areas.

Although this discussion begins to delimit the area that we call scientific computing, it is difficult to define it exactly, especially the boundaries and overlaps with other areas.[1] We will accept as our working definition that *scientific computing is the collection of tools, techniques, and theories required to solve on a computer mathematical models of problems in science and engineering.*

A majority of these tools, techniques, and theories originally developed in mathematics, many of them having their genesis long before the advent of electronic computers. This set of mathematical theories and techniques is called numerical analysis (or numerical mathematics) and constitutes a major part of scientific computing. The development of the electronic computer, however, signaled a new era in the approach to the solution of scientific problems. Many

[1]Perhaps the only universally accepted definition of, say, mathematics is that it is what mathematicians do.

of the numerical methods that had been developed for the purpose of hand calculation (including the use of desk calculators for the actual arithmetic) had to be revised and sometimes abandoned. Considerations that were irrelevant or unimportant for hand calculation now became of utmost importance for the efficient and correct use of a large computer system. Many of these considerations – programming languages, operating systems, management of large quantities of data, correctness of programs – were subsumed under the new discipline of computer science, on which scientific computing now depends heavily. But mathematics itself continues to play a major role in scientific computing: it provides the language of the mathematical models that are to be solved and information about the suitability of a model (Does it have a solution? Is the solution unique?) and it provides the theoretical foundation for the numerical methods and, increasingly, many of the tools from computer science.

In summary, then, scientific computing draws on mathematics and computer science to develop the best ways to use computer systems to solve problems from science and engineering. This relationship is depicted schematically in Figure 1.1. In the remainder of this chapter, we will go a little deeper into these various areas.

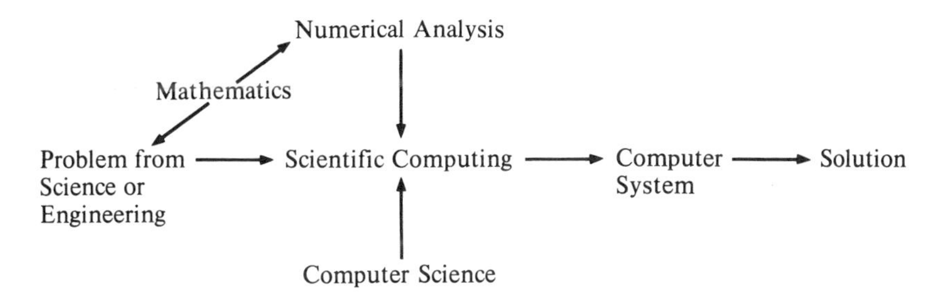

Figure 1.1: *Scientific Computing and Related Areas*

1.2 Mathematical Modeling

As was discussed in Section 1.1, we view scientific computing as the discipline that achieves a computer solution of mathematical models of problems from science and engineering. Hence, the first step in the overall solution process is the formulation of a suitable mathematical model of the problem at hand. This is a part of the discipline in which the problem arises: engineers devise models for engineering problems, and biologists for biological problems. Sometimes mathematicians and computer scientists are involved in this modeling process, at least as consultants.

Modeling

The formulation of a mathematical model begins with a statement of the factors to be considered. In many physical problems, these factors concern the balance of forces and other conservation laws of physics. For example, in the formulation of a model of a trajectory problem – which will be done in Section 2.1 – the basic physical law is Newton's second law of motion, which requires that the forces acting on a body equal the rate of change of momentum of the body. This general law must then be specialized to the particular problem by enumerating and quantifying the forces that will be of importance. For example, the gravitational attraction of Jupiter will exert a force on a rocket in Earth's atmosphere, but its effect will be so minute compared to the earth's gravitational force that it can usually be neglected. Other forces may also be small compared to the dominant ones but their effects not so easily dismissed, and the construction of the model will invariably be a compromise between retaining all factors that could likely have a bearing on the validity of the model and keeping the mathematical model sufficiently simple that it is solvable using the tools at hand. Classically, only very simple models of most phenomena were considered since the solutions had to be achieved by hand, either analytically or numerically. As the power of computers and numerical methods has developed, increasingly complicated models have become tractable.

In addition to the basic relations of the model – which in most situations in scientific computing take the form of differential equations – there usually will be a number of initial or boundary conditions. For example, in the predator-prey problem to be discussed in Chapter 2, the initial population of the two species being studied is specified. In studying the flow in a blood vessel, we may require a boundary condition that the flow cannot penetrate the walls of the vessel. In other cases, boundary conditions may not be so physically evident but are still required so that the mathematical problem has a unique solution. Or the mathematical model as first formulated may indeed have many solutions, the one of interest to be selected by some constraint such as a requirement that the solution be positive, or that it be the solution with minimum energy. In any case, it is usually assumed that the final mathematical model with all appropriate initial, boundary, and side conditions indeed has a unique solution. The next step, then, is to find this solution. For problems of current interest, such solutions rarely can be obtained in "closed form." The solution must be approximated by some method, and the methods to be considered in this book are numerical methods suitable for a computer. In the next section we will consider the general steps to be taken to achieve a numerical solution, and the remainder of the book will be devoted to a detailed discussion of these steps for a number of different problems.

Validation

Once we are able to compute solutions of the model, the next step usually is called the *validation of the model*. By this we mean a verification that the solution we compute is sufficiently accurate to serve the purposes for which the model was constructed. There are two main sources of possible error. First, there invariably are errors in the numerical solution. The general nature of these errors will be discussed in the next section, and one of the major themes in the remainder of the book will be a better understanding of the source and control of these numerical errors. But there is also invariably an error in the model itself. As mentioned previously, this is a necessary aspect of modeling: the modeler has attempted to take into account all the factors in the physical problem but then, in order to keep the model tractable, has neglected or approximated those factors that would seem to have a small effect on the solution. The question is whether neglecting these effects was justified. The first test of the validity of the model is whether the solution satisfies obvious physical and mathematical constraints. For example, if the problem is to compute a rocket trajectory where the expected maximum height is 100 kilometers and the computed solution shows heights of 200 kilometers, obviously some blunder has been committed. Or, it may be that we are solving a problem for which we know, mathematically, that the solution must be increasing but the computed solution is not increasing. Once such gross errors are eliminated – which is usually fairly easy – the next phase begins, which is, whenever possible, comparison of the computed results with whatever experimental or observational data are available. Many times this is a subtle undertaking, since even though the experimental results may have been obtained in a controlled setting, the physics of the experiment may differ from the mathematical model. For example, the mathematical model of airflow over an aircraft wing will usually assume the idealization of an aircraft flying in an infinite atmosphere, whereas the corresponding experimental results will be obtained from a wind tunnel where there will be effects from the walls of the enclosure. (Note that neither the experiment nor the mathematical model represents the true situation of an aircraft flying in our finite atmosphere.) The experience and intuition of the investigator are required to make a human judgement as to whether the results from the mathematical model are corresponding sufficiently well with observational data.

At the outset of an investigation this is quite often not the case, and the model must be modified. Usually this means that additional terms – which were thought negligible but may not be – are added to the model. Sometimes a complete revision of the model is required and the physical situation must be approached from an entirely different point of view. In any case, once the model is modified the cycle begins again: a new numerical solution, revalidation, additional modifications, and so on. This process is depicted schematically in Figure 1.2.

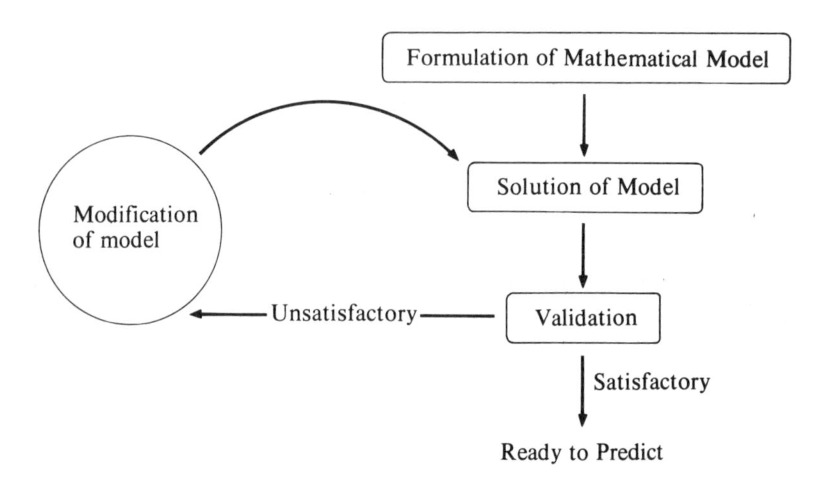

Figure 1.2: *The Mathematical Modeling and Solution Process*

Once the model is deemed adequate from the validation and modification process, it is ready to be used for prediction. This, of course, was the whole purpose. We should now be able to answer the questions that gave rise to the modeling effort: How high will the rocket go? Will the wolves eat all the rabbits? Of course, we must always take the answers with a healthy skepticism. Our physical world is simply too complicated and our knowledge of it too meager for us to be able to predict the future perfectly. Nevertheless, we hope that our computer solutions will give us increased insight into the problem being studied, be it a physical phenomenon or an engineering design.

1.3 The Process of Numerical Solution

We will discuss in this section the general considerations that arise in the computer solution of a mathematical model, and in the remainder of the book these matters will be discussed in more detail.

Once the mathematical model is given, our first thought typically is to try to obtain an explicit closed-form solution, but such a solution will usually only be possible for certain (perhaps drastic) simplifications of the problem. These simplified problems with known solutions may be of great utility in providing "check cases" for the more general problem.

After realizing that explicit solutions are not possible, we then turn to the task of developing a numerical method for the solution. Implicit in our thinking at the outset – and increasingly explicit as the development proceeds – will be the computing equipment as well as the software environment that is at our disposal. Our approach may be quite different for a microcomputer than for a very large computer. But certain general factors must be considered regardless of the computer to be used.

Rounding Errors

Perhaps the most important factor is that computers deal with a finite number of digits or characters. Because of this we cannot, in general, do arithmetic within the real number system as we do in pure mathematics. That is, the arithmetic done by a computer is restricted to finitely many digits, whereas the numerical representation of most real numbers requires infinitely many. For example, such fundamental constants as π and e require an infinite number of digits for a numerical representation and can never be entered exactly in a computer. Moreover, even if we could start with numbers that have an exact numerical representation in the computer, the processes of arithmetic require that eventually we make certain errors. For example, the quotient of two four-digit numbers may require infinitely many digits for its numerical representation. And even the product of two four-digit numbers will, in general, require eight digits. For example, assuming four-digit decimal arithmetic, $0.8132 \times 0.6135 = 0.49889820$ will be represented by 0.4988 or 0.4989, depending on the computer. Therefore, we resign ourselves at the outset to the fact that we cannot do arithmetic exactly on a computer. We shall make small errors, called *rounding errors*, on almost all arithmetic operations, and our task is to insure that these small errors do not accumulate to such an extent as to invalidate the computation.

The above example was given in terms of decimal arithmetic, but computers actually use the binary number system. Each machine has a word length consisting of the number of binary digits contained in each memory word, and this word length determines the number of digits that can be carried in the usual arithmetic, called *single-precision* arithmetic, of the machine. On most scientific computers, this is the equivalent of between 7 and 14 decimal digits. *Higher-precision* arithmetic can also be carried out. On many machines *double-precision* arithmetic, which essentially doubles the number of digits that are carried, is part of the hardware; in this case, programs with double-precision arithmetic usually require only modest, if any, increases in execution time compared to single-precision versions. On the other hand, some machines implement double precision by software, which may require several times as much time as single precision. Precision higher than double is always carried out by means of software and becomes increasingly inefficient as the precision increases. Higher-precision arithmetic is rarely used on practical problems, but it may be useful for generating "exact" solutions or other information for testing purposes.

Round-off errors can affect the final computed result in different ways. First, during a sequence of millions of operations, each subject to a small error, there is the danger that these small errors will accumulate so as to eliminate much of the accuracy of the computed result. If we round to the nearest digit, the individual errors will tend to cancel out, but the standard deviation of the accumulated error will tend to increase with the number of operations, leaving the possibility of a large final error. If chopping – that is, dropping the trailing digits rather than rounding – is used, there is a bias to errors in one direction, and the possibility of a large final error is increased. As an example of this phenomenon, consider the computation $0.8132 \times 0.6135 \times 0.2103 = 0.10491829$ correct to ten digits. Chopping the product of the first two numbers to four digits yields 0.4988, with an error of 0.9820×10^{-4}. Multiplying 0.4988 by 0.2103 gives 0.1048 after chopping, with an error of 0.9764×10^{-4}. The accumulated error is 0.1183×10^{-3}.

In addition to this possible accumulation of errors over a large number of operations, there is the danger of *catastrophic cancellation.* Suppose that two numbers a and b are equal to within their last digit. Then the difference $c = a - b$ will have only one significant digit of accuracy *even though no round-off error will be made in the subtraction.* Future calculations with c will then usually limit the final result to one correct digit. Whenever possible, one tries to eliminate the possibility of catastrophic cancellation by rearranging the operations. Catastrophic cancellation is one way in which an algorithm can be *numerically unstable,* although in exact arithmetic it may be a correct algorithm. Indeed, it is possible for the results of a computation to be completely erroneous because of round-off error even though only a small number of arithmetic operations have been performed. Examples of this will be given later.

Detailed round-off error analyses have now been completed for a number of the simpler and more basic algorithms such as those that occur in the solution of linear systems of equations; some of these results will be described in more detail in Chapter 4. A particular type of analysis that has proved to be very powerful is *backward error analysis.* In this approach the round-off errors are shown to have the same effect as that caused by changes in the original problem data. When this analysis is possible, it can be stated that the error in the solution caused by round off is no worse than that caused by certain errors in the original model. The question of errors in the solution is then equivalent to the study of the sensitivity of the solution to perturbations in the model. If the solution is highly sensitive, the problem is said to be *ill-posed* or *ill-conditioned,* and numerical solutions are apt to be meaningless.

Discretization Error

Another way that the finiteness of computers manifests itself in causing errors in numerical computation is due to the need to replace "continuous" problems by "discrete" ones. As a simple example, the integral of a continuous function requires knowledge of the integrand along the whole interval of integration, whereas a computer approximation to the integral can use values of the integrand at only finitely many points. Hence, even if the subsequent arithmetic were done exactly with no rounding errors, there would still be the error due to the discrete approximation to the integral. This type of error is usually called *discretization error* or *truncation error*, and it affects, except in trivial cases, all numerical solutions of differential equations and other "continuous" problems.

There is one more type of error which is somewhat akin to discretization error. Many numerical methods are based on the idea of an *iterative process.* In such a process, a sequence of approximations to a solution is generated with the hope that the approximations will converge to the solution; in many cases mathematical proofs of the convergence can be given. However, only finitely many such approximations can ever be generated on a computer, and, therefore, we must necessarily stop short of mathematical convergence. The error caused by such finite termination of an iterative process is sometimes called *convergence error*, although there is no generally accepted terminology here.

If we rule out trivial problems that are of no interest in scientific computing, we can summarize the situation with respect to computational errors as follows. Every calculation will be subject to rounding error. Whenever the mathematical model of the problem is a differential equation or other "continuous" problem, there also will be discretization error, and in many cases, especially when the problem is nonlinear, there will be convergence error. These types of errors and methods of analyzing and controlling them will be discussed more fully in concrete situations throughout the remainder of the book. But it is important to keep in mind that an acceptable error is very much dependent on the particular problem. Rarely is very high accuracy – say, 14 digits – needed in the final solution; indeed, for many problems arising in industry or other applications two or three digit accuracy is quite acceptable.

Efficiency

The other major consideration besides accuracy in the development of computer methods for the solution of mathematical models is *efficiency*. By this we will mean the amount of effort – both human and computer – required to solve a given problem. For most problems, such as solving a system of linear algebraic equations, there are a variety of possible methods, some going back many tens or even hundreds of years. Clearly, we would like to choose a method that minimizes the computing time yet retains suitable accuracy in the approximate solution. This turns out to be a surprisingly difficult problem which involves a number of considerations. Although it is frequently possible to estimate the computing time of an algorithm by counting the required arithmetic operations, the amount of computation necessary to solve a problem to a given tolerance is still an open question except in a few cases. Even if one ignores the effects of round-off error, surprisingly little is known. In the past several years these questions have spawned the subject of *computational complexity*. However, even if such theoretical results were known, they would still give only approximations to the actual computing time, which depends on a number of factors involving the computer system. And these factors change as the result of new systems and architectures. Indeed, the design and analysis of numerical algorithms should provide incentives and directions for such changes.

We give a simple example of the way a very inefficient method can arise. Many elementary textbooks on matrix theory or linear algebra present Cramer's rule for solving systems of linear equations. This rule involves quotients of certain determinants, and the definition of a determinant is usually given as the sum of all possible products (some with minus signs) of elements of the matrix, one element from each row and each column. There are $n!$ such products for an $n \times n$ matrix. Now, if we proceed to carry out the computation of a determinant based on a straightforward implementation of this definition, it would require about $n!$ multiplications and additions. For n very small, say $n = 2$ or $n = 3$, this is a small amount of work. Suppose, however, that we have a 20×20 matrix, a very small size in current scientific computing. If we assume that each arithmetic operation requires 1 microsecond (10^{-6} second), then the time required for this calculation – even ignoring all overhead operations in the computer program – will exceed one million years! On the other hand, the Gaussian elimination method, which will be discussed in Chapter 4, will do the arithmetic operations for the solution of a 20×20 linear system in less than 0.005 second, again assuming 1 microsecond per operation. Although this is an extreme example, it does illustrate the difficulties that can occur by naively following a mathematical prescription in order to solve a problem on a computer.

INTERESTING!!!

Good Programs

Even if a method is intrinsically "good," it is extremely important to implement the corresponding computer code in the best way possible, especially if other people are to use it. Some of the criteria for a good code are the following:

1. *Reliability* – the code does not have errors and can be trusted to compute what it is supposed to compute.

2. *Robustness*, which is closely related to reliability – the code has a wide range of applicability as well as the ability to detect bad data, "singular" or other problems that it cannot be expected to handle, and other abnormal situations, and deal with them in a way that is satisfactory to the user.

3. *Portability* – the code can be transferred from one computer to another with a minimum effort and without losing reliability. Usually this means that the code has been written in a general high-level language like FORTRAN and uses no "tricks" that are dependent on the characteristics of a particular computer. Any machine characteristics, such as word length, that must be used are clearly delineated.

4. *Maintainability* – any code will necessarily need to be changed from time to time, either to make corrections or to add enhancements, and this should be possible with minimum effort.

The code should be written in a clear and straightforward way so that such changes can be made easily and with a minimum likelihood of creating new errors. An important part of maintainability is that there be good *documentation* of the program so that it can be changed efficiently by individuals who did not write the code originally. Good documentation is also important so that the program user will understand not only how to use the code, but also its limitations. Finally, extensive *testing* of the program must be done to ensure that the preceding criteria have been met.

As examples of good software, LINPACK and EISPACK have been two standard packages for the solution of linear systems and eigenvalue problems, respectively. They are now being combined and revised into LAPACK, which is being designed to run on parallel and vector computers (see the next section). Another very useful system is MATLAB, which contains programs for linear systems, eigenvalues and many other mathematical problems and also allows for easy manipulation of matrices.

1.4 The Computational Environment

As indicated in the last section, there is usually a long road from a mathematical model to a successful computer program. Such programs are developed within the overall *computational environment*, which includes the computers to be used, the operating system and other systems software, the languages in which the program is to be written, techniques and software for data management and graphics output of the results, and programs that do symbolic computation. In addition, network facilities allow the use of computers at distant sites as well as the exchange of software and data.

Hardware

The computer hardware itself is of primary importance. Scientific computing is done on computers ranging from small PC's, which execute a few thousand floating point operations per second, to supercomputers capable of billions of such operations per second. Supercomputers that utilize hardware vector instructions are called *vector computers*, while those that incorporate multiple processors are called *parallel computers*. In the latter case, the computer system may contain a few, usually very powerful, processors or as many as several tens of thousands of relatively simple processors. Generally, algorithms designed for single processor "serial" computers will not be satisfactory, without modification, for parallel computers. Indeed, a very active area of research in scientific computing is the development of algorithms suitable for vector and parallel computers.

It is quite common to do program development on a workstation or PC prior to production runs on a larger computer. Unfortunately, a program will not always produce the same answers on two different machines due to different rounding errors. This, of course, will be the case if different precision arithmetic is used. For example, a machine using 48 digit binary arithmetic (14 decimal digits) can be expected to produce less rounding error then one using 24 binary digits (7 decimal digits). However, even when the precision is the same, two machines may produce slightly different results due to different conventions for handling rounding error. This is an unsatisfactory situation that has been addressed by the IEEE standard for floating point arithmetic. Although not all computers currently follow this standard, in the future they probably will, and then machines with the same precision will produce identical results on the same problem. On the other hand, algorithms for parallel computers often do the arithmetic operations in a different order than on a serial machine and this causes different errors to occur.

Systems and Languages

In order to be useful, computer hardware must be supplemented by systems software, including operating systems and compilers for high level languages. Although there are many operating systems, UNIX and its variants have increasingly become the standard for scientific computing and essentially all computer manufacturers now offer a version of UNIX for their machines. This is true for vector and parallel computers as well as more conventional ones. The use of a common operating system helps to make programs more portable. The same is true of programming languages. Since its inception in the mid 1950's, Fortran been the primary programming language for scientific computing. It has been continually modified and extended over the years, and now versions of Fortran also exist for parallel and vector computers. Other languages, especially the systems language "C," are sometimes used for scientific computing. However, it is expected that Fortran will continue to evolve and be the standard for the foreseeable future, at least in part because of the large investment in existing Fortran programs.

Data Management

Many of the problems in scientific computing require huge amounts of data, both input and output, as well as data generated during the course of the computation. The storing and retrieving of these data in an efficient manner is called *data management*. As an example of this in the area of computer-aided design, a data base containing all information relevant to a particular design application – which might be for an aircraft, an automobile, or a dam – may contain several billion characters. In an aircraft design this information would include everything relevant about the geometry of each part of the aircraft, the material properties of each part, and so on. An engineer may use this data base simply to find all the materials with a certain property. On the other hand, the data base will also be used in doing various analyses of the structural properties of the aircraft, which requires the solution of certain linear or nonlinear systems of equations. Large data management programs for use in business applications such as inventory control have been developed over many years, and some of the techniques used there are now being applied to the management of large data bases for scientific computation. It is interesting to note that in many scientific computing programs the number of lines of code to handle data management is far larger than that for the actual computation.

Visualization

The results of a scientific computation are numbers that may represent, for example, the solution of a differential equation at selected points. For large computations, such results may consist of the values of four or five functions at a million or more points. Such a volume of data cannot just be printed. *Scientific visualization* techniques allow the results of such computations to be represented pictorially. For example, the output of a fluid flow computation might be a movie which depicts the flow as a function of time in either two or three dimensions. The results of a calculation of the temperature distribution in a solid might be a color-coded representation in which regions of high temperatures are red and regions of low temperatures are blue, with a gradation of hues between the extremes. Or, a design model may be rotated in three-dimensional space to allow views from any angle. Such visual representations allow a quick understanding of the computation, although more detailed analysis of selected tables of numerical results may be needed for certain purposes, such as error checking.

Symbolic Computation

Another development which is having an increasing impact on scientific computing is *symbolic computation*. Systems such as MACSYMA, REDUCE, MAPLE, and MATHEMATICA allow the symbolic (as opposed to numerical) computation of derivatives, integrals and various algebraic quantities. For example, such systems can add, multiply and divide polynomials or rational expressions; differentiate expressions to obtain the same results that one would obtain using pencil and paper; and integrate expressions that have a "closed form" integral. This capability can relieve the drudgery of manipulating by hand lengthy algebraic expressions, perhaps as a prelude to a subsequent numerical computation. In this case, the output of the symbolic computation would ideally be a Fortran program. Symbolic computation systems can also solve certain mathematical problems, such as systems of linear equations, without rounding error. However, their use in this regard is limited since the size of the system must be small. In any case, symbolic computation is continuing to develop and can be expected to play an increasing role in scientific computation.

In this section we have discussed briefly some of the major components of the overall computing environment that pertain to scientific computing. In the remainder of the book we will point out in various places where these techniques can be used, although it is beyond the scope of this book to pursue their application in detail.

Supplementary Discussion and References: Chapter 1

For further reading on the computer science areas discussed in this chapter, see Hennessy and Patterson [1990] for computer architecture, Peterson and Silberschatz [1985] for operating systems, Pratt [1984] and Sethi [1989] for programming languages, Aho, Sethi, and Ullman [1988] and Fischer and LeBlanc [1988] for compilers, Elmasri and Navathe [1989] for data management, and Friedhoff and Benzon [1989] and Mendez [1990] for visualization. Another reference for computer graphics, which provides much of the technical foundation for visualization techniques is Newman and Sproul [1979]. The symbolic computation systems mentioned in the text are covered in Symbolics [1987] for MACSYMA, Rayna [1987] for REDUCE, Char et al. [1985] for MAPLE, and Wolfram [1988] for MATHEMATICA.

The packages EISPACK and LINPACK are discussed in Garbow et al. [1977] and Dongarra et al. [1979], respectively, and LAPACK in Dongarra and Anderson et al. [1990]. These and many other software packages are available on NETLIB; see, for example, Dongarra, Duff et al. [1990]. MATLAB can be obtained from The Math Works, Inc., South Natrick, MA 01760.

Chapter 2

Letting It Fly: Initial Value Problems

2.1 Examples of Initial Value Problems

In this section we shall derive the mathematical models for two initial-value problems, one from the field of ecology and the other with aerospace applications.

A Predator-Prey Problem

We consider the population dynamics of two interacting species that have a predator-prey relationship. That is, the prey is able to find sufficient food but is killed by the predator whenever they encounter each other. Examples of such species interaction are wolves and rabbits, and parasites and certain hosts. What we want to investigate is how the predator and prey populations vary with time.

Let $x = x(t)$ and $y = y(t)$ designate the number of prey and predators, respectively, at time t. To derive mathematical equations that approximate the population dynamics we make several simplifying assumptions. First, we assume that the prey population, if left alone, increases at a rate proportional to x. Second, we assume that the number of times that the predator kills the prey depends on the probability of the two coming together and is therefore proportional to xy. Combining these two assumptions, the prey population is governed by the ordinary differential equation

$$\frac{dx}{dt} = \alpha x + \beta xy, \tag{2.1.1}$$

where $\alpha > 0$ and $\beta < 0$.

For the predator equation we assume that the number of predators would decrease by natural causes if the prey were removed, contributing a γy term. However, the number of predators increases as a result of encounters with prey, leading to

$$\frac{dy}{dt} = \gamma y + \delta xy \tag{2.1.2}$$

with $\gamma < 0$ and $\delta > 0$. In summary, we have the system of two nonlinear ordinary differential equations

$$\frac{dx}{dt} = \alpha x + \beta xy, \qquad \frac{dy}{dt} = \gamma y + \delta xy, \tag{2.1.3}$$

with the assumptions $\alpha > 0$, $\beta < 0$, $\gamma < 0$, and $\delta > 0$. These equations were first formulated in 1925 and are known as the *Lotka-Volterra equations*. The problem statement is not complete; we must start the process at some time (for example, $t = 0$) with given values for the initial populations $x(0)$ and $y(0)$. Thus we supplement the differential equations by two *initial conditions*:

$$x(0) = x_0 \qquad y(0) = y_0. \tag{2.1.4}$$

Note that the model (2.1.3) -(2.1.4) gives continuous solutions although the number of predators and prey will always be an integer. This is typical of many mathematical models in which discrete quantities are approximated by continuous ones so as to obtain a differential equation.

A Trajectory Problem

Ballistics problems have a long history in scientific computing and were one of the motivations for the development of computers during World War II. We will consider a simple ballistics problem as a special case of the more general problem of rocket trajectories. Suppose that a rocket is launched at a given angle of inclination to the ground (the launch angle). How high will the rocket go? The answer depends on a number of factors: the characteristics of the rocket and its engine, the drag caused by air density, the gravitational forces, and so on. To set up a mathematical model for this problem, we make a number of simplifying assumptions. First, we consider only rockets going to a height and range of no more than 100 kilometers; in this case, we can assume that the earth is flat with little loss of accuracy. Second, we assume that the trajectory of the rocket lies entirely in a plane; for example, we assume no wind effects. With these two assumptions, we set up a two-dimensional coordinate system centered at the launching site, and in Figure 2.1 we depict a typical trajectory.

As shown in Figure 2.1, $x(t)$ and $y(t)$ denote the x and y coordinates of the rocket at time t, where we assume that launch occurs at $t = 0$ and, hence,

$$x(0) = y(0) = 0. \tag{2.1.5}$$

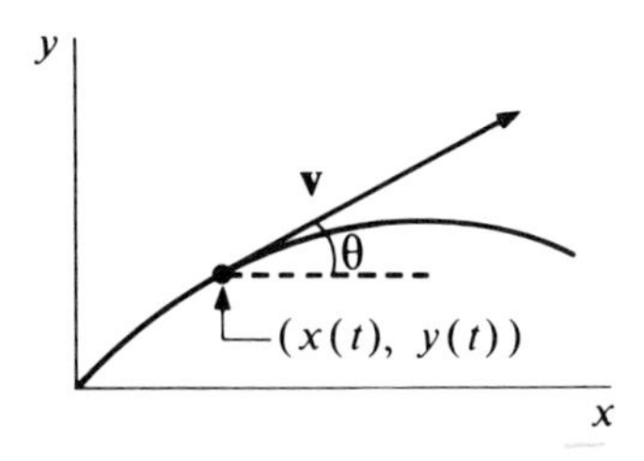

Figure 2.1: *A Typical Trajectory*

If we denote differentiation with respect to time by $\dot{x} = dx/dt$ and $\dot{y} = dy/dt$, then the velocity vector of the rocket at time t is $\mathbf{v}(t) = (\dot{x}(t), \dot{y}(t))$. We denote the magnitude of the velocity vector by $v(t)$ and its angle from the horizontal by $\theta(t)$, as shown in Figure 2.1. These quantities are then given by

$$v(t) = [(\dot{x}(t))^2 + (\dot{y}(t))^2]^{1/2}, \qquad \theta(t) = \tan^{-1}\frac{\dot{y}(t)}{\dot{x}(t)}. \tag{2.1.6}$$

The basic mathematical model of the trajectory is derived from Newton's laws of motion, which give

$$\frac{d}{dt}(m\mathbf{v}) = F. \tag{2.1.7}$$

Here $m(t)$ is the mass of the rocket, and F denotes the forces acting on the rocket and is composed of three terms: (1) the thrust, $T(t)$, when the rocket engine is firing; (2) the drag force

$$\tfrac{1}{2}c\rho s v^2, \tag{2.1.8}$$

where c is the coefficient of drag, ρ is air density, and s is the cross-sectional area of the rocket; and (3) the gravitational force, gm, where g is the acceleration of gravity.

To write (2.1.7) in terms of x and y, we note that the part of the force F that consists of the thrust and the drag acts along the axis of the rocket. If we call this part F_1, then

$$F_1 = T - \tfrac{1}{2}c\rho s v^2 \tag{2.1.9}$$

and (2.1.7) can be written

$$\dot{m}\dot{x} + m\ddot{x} = F_1\cos\theta, \qquad \dot{m}\dot{y} + m\ddot{y} = F_1\sin\theta - mg, \tag{2.1.10}$$

since the gravitational force acts only in the vertical direction. Using (2.1.9) and rearranging terms, we rewrite (2.1.10) as

$$\ddot{x} = \frac{1}{m}(T - \tfrac{1}{2}c\rho s v^2)\cos\theta - \frac{\dot{m}}{m}\dot{x}, \tag{2.1.11a}$$

$$\ddot{y} = \frac{1}{m}(T - \tfrac{1}{2}c\rho s v^2)\sin\theta - \frac{\dot{m}}{m}\dot{y} - g. \tag{2.1.11b}$$

This is a coupled system of two second-order nonlinear [recall equation (2.1.6)] differential equations. We are assuming that c and s are known constants, ρ is a known function of y (height above the surface), and T and m (and hence $\dot{m}$) are known functions of t. (The change in mass is caused by the expenditure of fuel.)

The solution of (2.1.11) must satisfy (2.1.5), and this gives two of the four initial conditions that are needed. The other two are

$$v(0) = 0, \qquad \theta(0) = \theta_0. \tag{2.1.12}$$

Thus, for a given rocket, the only "free parameter" is the launch angle θ_0, and changes in the launch angle obviously cause changes in the trajectory.

Equations (2.1.11) also serve as the mathematical model for the "projectile problem", examples of which are a shell being shot from a cannon or a rock launched from a slingshot. In this case we assume that the projectile starts with a given velocity v_0, and thus (2.1.12) is changed to

$$v(0) = v_0, \qquad \theta(0) = \theta_0. \tag{2.1.13}$$

There is now no thrust, and hence no change of mass, so (2.1.11) simplifies to

$$\ddot{x} = \frac{-c\rho s v^2}{2m}\cos\theta, \qquad \ddot{y} = \frac{-c\rho s v^2}{2m}\sin\theta - g, \tag{2.1.14}$$

which in the context of our simplified model shows that given the initial velocity and launch angle, the trajectory depends only on the drag and gravitational forces.

Our task, now, is to solve the equations (2.1.11) with the initial conditions (2.1.5) and (2.1.13). [Henceforth we shall use (2.1.13) since it includes the special case $v_0 = 0$ of (2.1.12).] In the trivial case in which there is neither thrust nor drag, the equations can be solved explicitly (Exercise 2.1.3). However, for any realistic specification of the air density ρ and the thrust this is not possible, and an approximate numerical solution is required.

For the numerical solutions to be discussed later, it will be convenient to reformulate the two second-order equations (2.1.11) as a system of four first-order equations. By differentiating the relations

$$\dot{x} = v\cos\theta, \qquad \dot{y} = v\sin\theta, \tag{2.1.15}$$

which are equivalent to (2.1.6), we have

$$\ddot{x} = \dot{v}\cos\theta - v\dot{\theta}\sin\theta, \qquad \ddot{y} = \dot{v}\sin\theta + v\dot{\theta}\cos\theta. \tag{2.1.16}$$

If we substitute (2.1.15) and (2.1.16) into (2.1.11) and solve for $\dot{v}$ and $\dot{\theta}$, we obtain

$$\dot{v} = \frac{1}{m}(T - \tfrac{1}{2}c\rho s v^2) - g\sin\theta - \frac{\dot{m}}{m}v \qquad (2.1.17)$$

$$\dot{\theta} = -\frac{g}{v}\cos\theta. \qquad (2.1.18)$$

Equations (2.1.17) and (2.1.18), together with (2.1.15), constitute a system of four first-order equations in the variables x, y, v, and θ. Again, the initial conditions are given by (2.1.5) and (2.1.13).

We shall return to the numerical solution of both the predator-prey problem and the trajectory problem after we have discussed the basic methods used for the solution.

Supplementary Discussion and References: 2.1

There is no known nontrivial analytical solution of the problem given by (2.1.3), (2.1.4), and we must use approximation methods. The primary concern of this book is with numerical methods that replace a continuous problem with a discrete problem that is solved on a computer. But we will consider here another approach to solving (2.1.3), (2.1.4). These *perturbation* methods replace the original continuous problem with a slightly different and simpler continuous problem which can be solved analytically.

The first step is to identify *stationary* or *equilibrium* states (x_s, y_s). In our case, the equations

$$x = x_s \equiv \frac{-\gamma}{\delta}, \qquad y = y_s \equiv \frac{-\alpha}{\beta}$$

represent stationary states because

$$\left.\frac{dx}{dt}\right|_{(x_s,y_s)} = x_s(\alpha + \beta y_s) = 0, \qquad \left.\frac{dx}{dt}\right|_{(x_s,y_s)} = y_s(\gamma + \delta x_s) = 0.$$

By expanding the right-hand sides of (2.1.3) in a Taylor series about (x_s, y_s), we obtain

$$x(\alpha + \beta y) = \beta x_s(y - y_s) + \cdots, \qquad y(\gamma + \delta x) = \delta y_s(x - x_s) + \cdots .$$

Thus in the neighborhood of (x_s, y_s) we approximate (2.1.3) by the linear equations

$$\dot{x} = \beta x_s(y - y_s), \qquad \dot{y} = \delta y_s(x - x_s). \qquad (2.1.19)$$

These equations may be solved in the form

$$\frac{(x - x_s)^2}{-\beta x_s} + \frac{(y - y_s)^2}{\delta y_s} = c, \qquad (2.1.20)$$

where c is some constant determined by the initial conditions. This is the equation of an ellipse whose center is at (x_s, y_s), and different starting values of $x(0)$ and $y(0)$ determine different ellipses. The figure that follows shows a family of ellipses about (x_s, y_s), with the arrows indicating the direction of increasing time. It can be seen that the populations are cyclical: after a certain time, they return to their original levels.

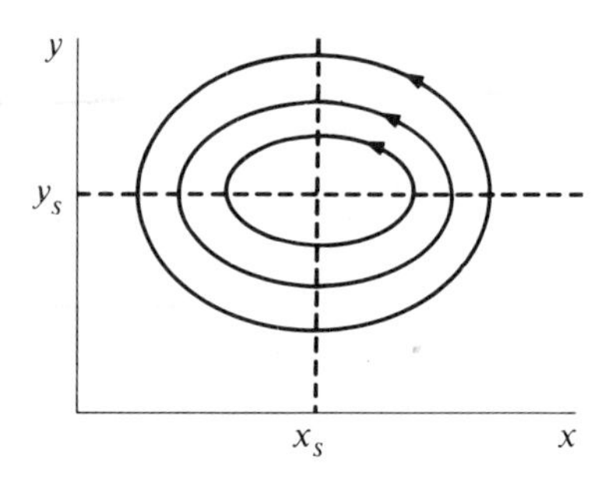

This type of perturbation analysis can provide useful information about the solution of (2.1.3) in the neighborhood of a stationary point. Because the equations (2.1.3) are approximated by equations (2.1.19), we might expect that the solutions to (2.1.3) would be close to the ellipses that solve (2.1.19). Such a relationship is verified by the numerical approximations described in the remainder of this chapter.

For further information on the predator-prey problem and other topics in mathematical biology, see Rubinow [1975], and for additional information on the theory of rocket trajectories, see, for example, Rosser, Newton, and Gross [1974].

EXERCISES 2.1

2.1.1. What relationships among the coefficients α, β, γ, and δ and the population levels x and y of (2.1.3) would guarantee stable populations for x and y (that is, $x(t+\Delta t) = x(t)$ and $y(t+\Delta t) = y(t)$ for all $\Delta t > 0$)?

2.1.2. Verify (2.1.6).

2.1.3. Show that the solution of $\ddot{x} = 0$, $\ddot{y} = -mg$, with initial conditions $x(0) = y(0) = 0$, $v(0) = v_0$, $\theta(0) = \theta_0$ is given by $x(t) = (v_0 \cos\theta_0)t$ and $y(t) = -mgt^2/2 + (v_0 \sin\theta_0)t$.

2.2 One-Step Methods

In the previous section we gave two examples of initial-value problems for systems of ordinary differential equations. We will now consider such problems in the general form

$$\frac{dy_i}{dx} = f_i(x, y_1(x), \ldots, y_n(x)), \qquad i = 1, \ldots, n, \quad a \le x, \tag{2.2.1}$$

with initial conditions

$$y_i(a) = \hat{y}_i, \quad i = 1, \ldots, n. \tag{2.2.2}$$

Here, the f_i are given functions, x is the independent variable and the $\hat{y}_i$ are given initial conditions. In the previous section the predator-prey problem gave rise to two equations, whereas in the trajectory problem there were four first-order equations (see Exercise 2.2.1). More generally, as shown in Appendix 1, a single higher-order equation or a system of higher-order equations may always be reduced to a system of first-order equations; thus, the problem (2.2.1), (2.2.2) is very general. For simplicity in the subsequent presentation, we shall restrict our attention to a single equation

$$\frac{dy}{dx} = f(x, y), \quad a \le x, \tag{2.2.3}$$

in the single unknown function y, and with the initial condition

$$y(a) = \hat{y}. \tag{2.2.4}$$

Later in the section we shall show how the methods extend easily to systems of the form (2.2.1).

As indicated in (2.2.3), we wish to find the solution for $x \ge a$. In some problems we will wish to find the solution on a prescribed interval $[a, b]$. In other problems we may wish to know the behavior of the solution as $x \to \infty$; of course, we can only compute the solution numerically in a finite interval, but we may wish to continue the computations until the behavior for large x becomes clear. And, in still other problems, we may wish to integrate until a prescribed condition is satisfied; for example, in the trajectory problem discussed in the previous section we may wish to know when the missile hits the ground.

Although some initial-value problems have solutions that can be obtained analytically, many problems, including most of those of practical interest, cannot be solved in this manner. The purpose of the chapter is to describe methods for approximating solutions by using numerical methods, particularly by what are known as *finite difference methods.*

The first step in the numerical solution is to introduce the *grid points* $a = x_0 < x_1 < \cdots < x_N$ as shown in Figure 2.2. Although unequal spacing of the grid points presents no particular difficulties, we shall assume that they are equally spaced in order to simplify the discussion and analysis. If we let h denote the spacing, then $x_k = a + kh$, $k = 0, 1, \ldots, N$. In what follows, $y(x_k)$ will denote the value of the exact solution of (2.2.3) at the point x_k, and

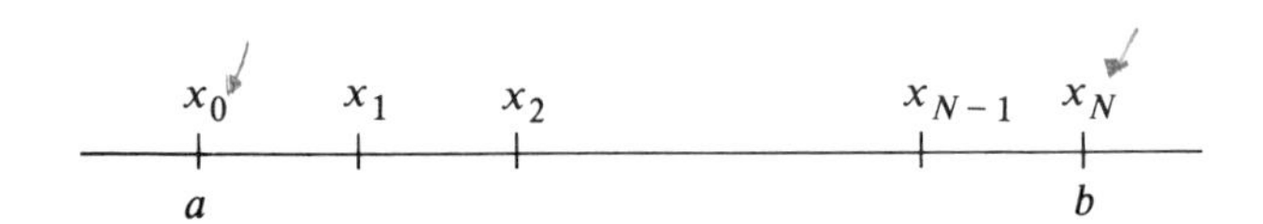

Figure 2.2: *Grid Points*

y_k will denote the approximation generated by the numerical method under consideration.

Euler's Method

Perhaps the simplest numerical scheme is *Euler's method*, which is defined by

$$y_0 = \hat{y}, \qquad y_{k+1} = y_k + hf(x_k, y_k), \quad k = 0, 1, \ldots, N-1. \tag{2.2.5}$$

The derivation of Euler's method is straightforward. By the Taylor expansion (see Appendix 1) of y about x_k, we have

$$\begin{aligned} y(x_{k+1}) &= y(x_k) + hy'(x_k) + \frac{h^2}{2}y''(z_k) \\ &= y(x_k) + hf(x_k, y(x_k)) + \frac{h^2}{2}y''(z_k), \end{aligned} \tag{2.2.6}$$

where z_k is some point in the interval (x_k, x_{k+1}). Here, and henceforth, we will always assume that all derivatives shown do exist. Now if y'' is bounded and h is small, we may ignore the last term and we have, using the notation $\doteq$ to mean "approximately equal to,"

$$y(x_{k+1}) \doteq y(x_k) + hf(x_k, y(x_k)).$$

This is the basis for (2.2.5). Geometrically, Euler's method consists of approximating the solution at x_{k+1} by following the tangent to the solution curve at x_k (see Figure 2.3).

Euler's method is very easy to carry out on a computer: at the k-th step we evaluate $f(x_k, y_k)$ and use this in (2.2.5). Hence, essentially all of the computation required is in the evaluation of $f(x_k, y_k)$. We now give a simple example of the use of the method. Consider the equation

$$y'(x) = y^2(x) + 2x - x^4, \quad y(0) = 0. \tag{2.2.7}$$

It is easily verified that the exact solution of this equation is $y(x) = x^2$. Here $f(x, y) = y^2 + 2x - x^4$, and therefore Euler's method for (2.2.7) becomes

$$y_{k+1} = y_k + h(y_k^2 + 2kh - k^4h^4), \quad k = 0, 1, \ldots, \qquad y_0 = 0, \tag{2.2.8}$$

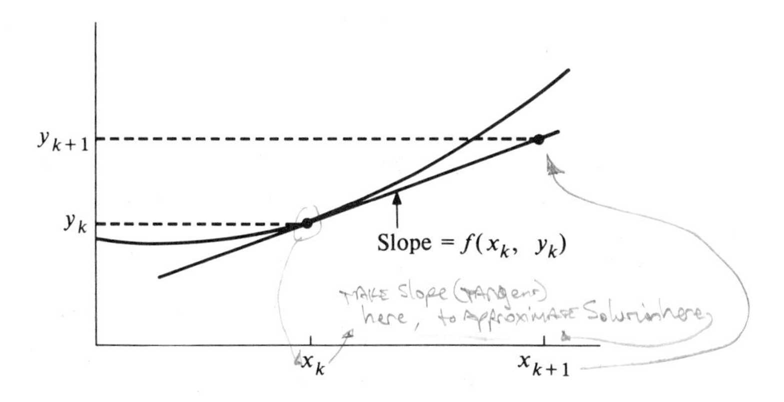

Figure 2.3: *One Step of Euler's Method*

since $x_k = kh$. In Table 2.1 we give some computed values for (2.2.8) for $h = 0.1$, as well as the corresponding values of the exact solution.

Table 2.1: *Computed and Exact Solutions for* (2.2.7) *by Euler's Method*

x	Computed Solution	Exact Solution
0.1	0.00	0.01
0.2	0.02	0.04
0.3	0.06	0.09
0.4	0.12	0.16
0.5	0.20	0.25
0.6	0.30	0.36

As Table 2.1 shows, the computed solution is in error, as is to be expected, and a major question regarding the use of Euler's method, or any other numerical method, is the accuracy of the approximations y_k. In general, the error in these approximations will come from two sources: (1) the discretization error that results from the replacement of the differential equation (2.2.3) by the approximation (2.2.5); and (2) the rounding error made in carrying out the arithmetic operations of the method (2.2.5). We shall consider the rounding error later, and for the moment we shall assume that the y_k of (2.2.5) are computed exactly so that the only error is the discretization error.

Discretization Error

We first consider the discretization error

$$E(h;b) = |y_N - y(b)| \tag{2.2.9a}$$

at a fixed point

$$b = a + hN. \tag{2.2.9b}$$

Note that N, and thus y_N, is function of h; in particular, for fixed b, as h decreases, then N increases and if $h \to 0$, then $N \to \infty$. We only allow h to vary, however, in such a way that (2.2.9b) is satisfied for an integer N. The maximum discretization error on the whole interval $[a, b]$,

$$E(h; [a,b]) = \max_{1 \le i \le N} |y_i - y(a+ih)|, \tag{2.2.9c}$$

is called the *global discretization error* (sometimes called the *global truncation error*) on the interval $[a, b]$. When the interval is clear, we will usually denote $E(h; [a, b])$ by $E(h)$. Intuitively, we expect – and certainly hope – that $E(h) \to 0$ as $h \to 0$.

We next show how the global discretization error can be bounded. First, we will assume that the exact solution y has a bounded second derivative y'' on the interval $[a, b]$ and set

$$\max_{a \le x \le b} |y''(x)| = M. \tag{2.2.10}$$

We then consider the expression

$$L(x, h) = \frac{1}{h}[y(x+h) - y(x)] - f(x, y(x)), \tag{2.2.11}$$

which is called the *local discretization error for Euler's method at point* x and is a measure of how much the difference quotient for $y'(x)$ differs from $f(x, y(x))$. Now suppose that y_k equals the exact solution $y(x_k)$. Then the difference between the Euler approximation y_{k+1} and the exact solution $y(x_{k+1})$ is simply

$$y(x_{k+1}) - y_{k+1} = y(x_{k+1}) - y(x_k) - hf(x_k, y(x_k)) = hL(x_k, h). \tag{2.2.12}$$

That is, h times the local discretization error is the error produced in a single step of Euler's method starting from the exact solution.

We shall be interested in the maximum size of $L(x, h)$ for any value of x, and we define the local discretization error for Euler's method by

$$L(h) = \max_{a \le x \le b-h} |L(x, h)|. \tag{2.2.13}$$

Note that $L(h)$ depends on the step length, h, as well as on the function f of the differential equation and the interval $[a, b]$. The only dependence we have explicitly delineated, however, is that on h, since under the assumption (2.2.10) and using a Taylor expansion analogous to (2.2.6) we obtain the bound

$$L(h) \le \frac{h}{2} M = 0(h). \tag{2.2.14}$$

Here we have used the standard notation $0(h)$ to denote a quantity that goes to zero as rapidly as h goes to zero. More generally, we will say that a function g of h is $0(h^p)$ if $g(h)/h^p$ is bounded as $h \to 0$ but $g(h)/h^q$ is unbounded if $q > p$.

The problem now is to relate the local discretization error to the global discretization error. If we denote the error $y(x_k) - y_k$ by e_k, then we have, by using (2.2.5) and (2.2.11),

$$\begin{aligned} e_{k+1} &= y(x_{k+1}) - y_{k+1} \\ &= y(x_k) + hf(x_k, y(x_k)) + hL(x_k, h) - y_k - hf(x_k, y_k) \\ &= e_k + h[f(x_k, y(x_k)) - f(x_k, y_k)] + hL(x_k, h). \end{aligned} \tag{2.2.15}$$

Now assume that the function f has a bounded partial derivative with respect to its second variable:

$$\left|\frac{\partial f}{\partial y}(x,y)\right| \le M_1, \quad a \le x \le b, \quad |y| < \infty. \tag{2.2.16}$$

Then by the mean-value theorem (Appendix 1), we have for some $0 < \theta < 1$,

$$\begin{aligned} |f(x_k, y(x_k)) - f(x_k, y_k)| &= \left|\frac{\partial f}{\partial y}(x_k, \theta y(x_k) + (1-\theta)y_k)(y(x_k) - y_k)\right| \\ &\le M_1|e_k|. \end{aligned}$$

Substituting this into (2.2.15) and bounding $L(x_k, h)$ by $L(h)$ gives

$$|e_{k+1}| \le (1 + hM_1)|e_k| + h|L(h)|. \tag{2.2.17}$$

This is an inequality of the form

$$\bar{e}_{k+1} \le c\bar{e}_k + d,$$

where $\bar{e}_k = |e_k|$, $c = 1 + hM_1$, $d = hL(h)$ and $\bar{e}_0 = 0$. Thus,

$$\begin{aligned} \bar{e}_n &\le c(c\bar{e}_{n-2} + d) + d \le \cdots \le (1 + c + \cdots c^{n-1})d \\ &= \left(\frac{c^n - 1}{c - 1}\right) d = \frac{[(1 + hM_1)^n - 1]}{hM_1} hL(h) \le \frac{(1 + hM_1)^n}{M_1} L(h) \\ &\le \frac{e^{hnM_1}}{M_1} L(h), \end{aligned} \tag{2.2.18}$$

where the last inequality follows from $1 + x \le e^x$ for $x \ge 0$. Now $x_n = x_0 + nh$ and if we keep x_n fixed at some point x^+ in the interval $[a, b]$ as $h \to 0$, then (2.2.18) becomes

$$\bar{e}_n \le \frac{L(h)}{M_1} e^{M_1(x^+ - x_0)}.$$

This estimate is an upper bound on the absolute value of the error and may be rather pessimistic. Nevertheless, it does show that $e_n \to 0$ since $L(h) = 0(h)$. Moreover, $E(h) = 0(h)$. Thus, we have proved the following result.

THEOREM 2.2.1 (Euler Discretization Error) *If the function f has a bounded partial derivative with respect to its second variable, and if the solution of* (2.2.3), (2.2.4) *has a bounded second derivative, then the Euler approximations converge to the exact solution as $h \to 0$, and the global discretization error of Euler's method satisfies $E(h) = 0(h)$.*

The fact that the global discretization error is $0(h)$ is usually expressed by saying that Euler's method is *first order*. The practical consequence of this is that as we decrease h, we expect that the approximate solution will become more accurate and converge to the exact solution at a linear rate in h as h tends to zero; that is, if we halve the step size, h, we expect that the error will decrease by about a factor of 2. This error behavior is shown in the following example.

Consider the equation $y' = y,\ y(0) = 1$, for which the exact solution is $y(x) = e^x$. We compute the solution at $x = 1$ by Euler's method using various values of h (see Table 2.2). The exact solution at $x = 1$ is $e = 2.718\ldots$; the errors for the different step sizes are given in the middle column. The ratios of the errors for successive halvings of h are given in the right-hand column, and it is seen that these ratios tend to $\frac{1}{2}$, as expected.

Table 2.2: *Error in Euler's Method*

h	Computed Value	Error	Error Ratio
1	2.000	0.718	
1/2	2.250	0.468	0.65
1/4	2.441	0.277	0.59
1/8	2.566	0.152	0.55
1/16	2.638	0.080	0.53

Runge-Kutta Methods

The very slow rate of convergence shown in Table 2.2 as h decreases is typical of first-order methods and militates against their use. Much of the rest of this chapter will be devoted to studying other methods for which the error tends to zero at a faster rate as h tends to zero. As an example of one of the approaches to such methods, we next discuss the *Heun method*, which is given by

$$y_{k+1} = y_k + \frac{h}{2}[f(x_k, y_k) + f(x_{k+1}, y_k + hf(x_k, y_k))]. \tag{2.2.19}$$

Note that we have just replaced $f(x_k, y_k)$ in Euler's method by an average of f evaluated at two different places. This is illustrated in Figure 2.4. The Heun method is also known as a second-order *Runge-Kutta method* and has a local discretization error that is $0(h^2)$, as we will show shortly.

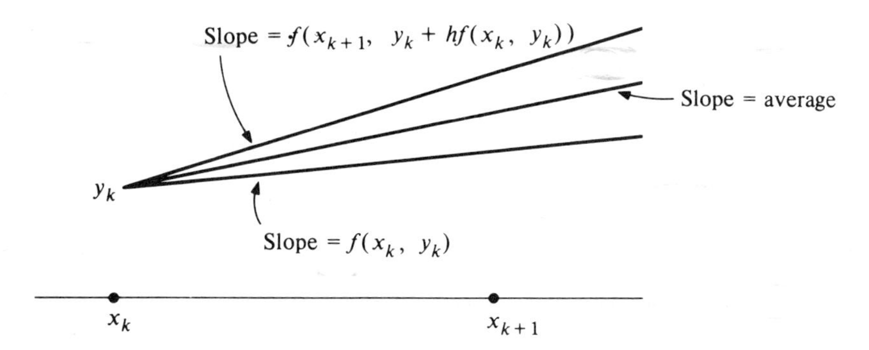

Figure 2.4: *The Heun Method*

The most famous of the Runge-Kutta methods is the classical fourth-order method, given by

$$y_{k+1} = y_k + \frac{h}{6}(F_1 + 2F_2 + 2F_3 + F_4) \tag{2.2.20}$$

where

$$\begin{aligned} F_1 &= f(x_k, y_k), \qquad F_2 = f\left(x_k + \frac{h}{2}, y_k + \frac{h}{2}F_1\right), \\ F_3 &= f\left(x_k + \frac{h}{2}, y_k + \frac{h}{2}F_2\right), \qquad F_4 = f(x_{k+1}, y_k + hF_3). \end{aligned}$$

Here the $f(x_k, y_k)$ in Euler's method has been replaced by a weighted average of f evaluated at four different points. It is instructive to draw the figure corresponding to Figure 2.4; this is left to Exercise 2.2.9.

One-Step Methods

In Section 2.4 we will consider methods based on using information from prior steps so that y_{k+1} will be a function not only of y_k but also of y_{k-1} and, perhaps, other prior values. The present section deals with methods that depend only on y_k. Such methods are called *one step methods* and can be written in the general form

$$y_{k+1} = y_k + h\phi(x_k, y_k) \tag{2.2.21}$$

for some suitable function ϕ. In the case of Euler's method ϕ is just f itself, whereas for the Heun method

$$\phi(x, y) = \tfrac{1}{2}[f(x, y) + f(x + h, y + hf(x, y))]. \tag{2.2.22}$$

The fourth-order Runge-Kutta method (2.2.20) is also a one-step method, and the corresponding function ϕ can be written in a manner similar to (2.2.22) (see Exercise 2.2.5).

For any one-step method (2.2.21), we define the local discretization error in a manner analogous to that for Euler's method by

$$L(h) = \max_{a \le x \le b-h} |L(x,h)|, \ L(x,h) = \frac{1}{h}[y(x+h) - y(x)] - \phi(x, y(x)) \quad (2.2.23)$$

where, again, $y(x)$ is the exact solution of the differential equation. If, for a given ϕ, $L(h) = 0(h^p)$ for some integer p, then it is possible to show, under suitable assumptions on ϕ and f, that the global discretization error will also be of order p in h:

$$E(h) \equiv \max_{1 \le k \le N} |y(x_k) - y_k| = 0(h^p). \quad (2.2.24)$$

The *order* of the method (2.2.21) is defined to be the integer p for which $L(h) = 0(h^p)$. This definition of order is a statement about the method and assumes that the solution y of the differential equation has bounded derivatives of suitably high order. For example, we showed that $p = 1$ for Euler's method under the assumption (2.2.10), and Table 2.2 illustrated that the error decreased by a factor of about 2^{-1} when the step length h was halved. For a pth order method, we expect the error to decrease by a factor of about 2^{-p} when we halve h, at least for h sufficiently small.

It is a relatively simple matter to show that the local discretization error for Heun's method is $0(h^2)$, but this will be a consequence of the following more general analysis. Consider a function ϕ defined by

$$\phi(x, y) = c_2 f(x, y) + c_3 f(x + c_1 h, y + c_1 h f(x, y)),$$

where we wish to determine the constants c_1, c_2, and c_3 so as to maximize the order of the one-step method (2.2.21); that is, we wish the best linear combination, as determined by c_2 and c_3, of two values of f, and how far along the interval the second evaluation of f should be done, as determined by c_1.

We expand f in a Taylor series in two variables about the point (x, y). First, in the x variable, we have

$$\phi = c_2 f + c_3[f(x, y + c_1 hf) + c_1 h f_x(x, y + c_1 hf) + 0(h^2)],$$

where we have denoted $f(x, y)$ simply by f and the partial derivative of f with respect to x by f_x. Next, expand in y where all partial derivatives shown are evaluated at (x, y):

$$\begin{aligned}\phi &= c_2 f + c_3[f + c_1 h f f_y + 0(h^2) + c_1 h f_x + 0(h^2)] \\ &= (c_2 + c_3) f + c_1 c_3 h (f f_y + f_x) + 0(h^2).\end{aligned} \tag{2.2.25}$$

On the other hand, the exact solution $y(x)$ of the differential equation satisfies

$$\begin{aligned}\frac{1}{h}[y(x+h) - y(x)] &= y'(x) + \tfrac{1}{2} y''(x) h + 0(h^2) \\ &= f + \tfrac{1}{2} h \frac{df}{dx} + 0(h^2) \\ &= f + \tfrac{1}{2} h (f f_y + f_x) + 0(h^2).\end{aligned} \tag{2.2.26}$$

Therefore, (2.2.25) and (2.2.26) combine to yield

$$\begin{aligned}&\frac{1}{h}[y(x+h) - y(x)] - \phi(x, y(x)) \\ &\quad = (1 - c_2 - c_3) f + h(\tfrac{1}{2} - c_1 c_3)(f f_y + f_x) + 0(h^2).\end{aligned} \tag{2.2.27}$$

If we require that

$$c_2 + c_3 = 1, \qquad c_1 c_3 = \frac{1}{2}, \tag{2.2.28}$$

then the first two terms of (2.2.27) vanish for any f. Therefore $L(h) = 0(h^2)$. Moreover, by carrying out the Taylor expansions one more term, it can be shown that, in general, we cannot achieve $L(h) = 0(h^3)$ no matter what choice of the constants c_1, c_2, c_3 is made. Hence we have

$$\frac{1}{h}[y(x+h) - y(x)] - \phi(x, y(x)) = 0(h^2), \tag{2.2.29}$$

which will hold whenever (2.2.28) is satisfied and the various derivatives we have used are bounded. Therefore, the methods delineated by (2.2.28) are all second order so that there is not a unique second-order method of this type.

If we set $c_1 = \gamma/2$ and solve the two equations of (2.2.28) in terms of γ, we obtain a function that will always satisfy (2.2.28). Therefore, the method

$$y_{k+1} = y_k + h\left[\left(1 - \frac{1}{\gamma}\right) f(x_k, y_k) + \frac{1}{\gamma} f\left(x_k + \frac{\gamma}{2} h, y_k + \frac{\gamma h}{2} f(x_k, y_k)\right)\right]$$

is second-order accurate for any $\gamma \neq 0$. The special choice $\gamma = 2$ gives the second-order Runge-Kutta method (2.2.19). The derivation of higher-order Runge-Kutta methods, and in particular the fourth-order method (2.2.20), can proceed in an analogous, but more complicated, manner.

Systems of Equations

We next indicate how the above methods can be used for systems of equations. Consider the system (2.2.1), which we will write in the vector form

$$\mathbf{y}'(x) = \mathbf{f}(x, \mathbf{y}(x)). \tag{2.2.30}$$

Here $\mathbf{y}(x)$ denotes the vector with components $y_1(x), \ldots, y_n(x)$, and $\mathbf{f}$ is the vector with components $f_1, \ldots, f_n$. The vector $\hat{\mathbf{y}}$ will denote the initial values (2.2.2). Then Euler's method (2.2.5) can be written for the system (2.2.30) as

$$\mathbf{y}_0 = \hat{\mathbf{y}} \qquad \mathbf{y}_{k+1} = \mathbf{y}_k + h\mathbf{f}(x_k, \mathbf{y}_k), \quad k = 0, 1, \ldots \tag{2.2.31}$$

where $\mathbf{y}_1, \mathbf{y}_2, \ldots$ are vector approximations to the solution $\mathbf{y}$. We could, of course, write out (2.2.31) in component form; for $n = 2$, this would be

$$y_{1,0} = \hat{y}_1 \qquad y_{2,0} = \hat{y}_2$$

$$\left.\begin{aligned} y_{1,k+1} &= y_{1,k} + hf_1(x_k, y_{1,k}, y_{2,k}) \\ y_{2,k+1} &= y_{2,k} + hf_2(x_k, y_{1,k}, y_{2,k}) \end{aligned}\right\}, \quad k = 0, 1 \ldots .$$

Clearly, the succinct vector notation (2.2.31) is advantageous.

Similarly, Heun's method (2.2.19) can be written in vector form for (2.2.30) by

$$\mathbf{y}_{k+1} = \mathbf{y}_k + \frac{h}{2}[\mathbf{f}(x_k, \mathbf{y}_k) + \mathbf{f}(x_{k+1}, \mathbf{y}_k + h\mathbf{f}(x_k, \mathbf{y}_k))]. \tag{2.2.32}$$

It is left to Exercise 2.2.7 to write the fourth-order Runge-Kutta method (2.2.20) in vector form.

Rounding Error

We now turn to a brief discussion of the rounding error in the methods of this section. Consider Euler's method, in which there are two sources of rounding error. The first is the error that occurs in the evaluation of $f(x_k, y_k)$; we will denote this error by ε_k. The second error, η_k, is the error made in the Euler formula. Thus, the computed approximations y_k satisfy

$$y_{k+1} = y_k + h[f(x_k, y_k) + \varepsilon_k] + \eta_k, \quad k = 0, 1, \ldots . \tag{2.2.33}$$

It is possible to bound the effects of these errors in terms of bounds on the ε_k and η_k. However, we will content ourselves with the following intuitive discussion. As we have seen, the global discretization error in Euler's method goes to zero as h goes to zero. Hence we can make the discretization error as small as we like by making h sufficiently small. However, the smaller h is, the more steps of Euler's method that will be required and, in general, the larger the effect of the rounding error on the computed solution. In practice, for a fixed word length in the computer arithmetic there will be a size of h below which the rounding error will become the dominant contribution to the overall error. The situation is depicted schematically in Figure 2.5, in which the step-size h_0 is the practical minimum that can be used. This minimum step-size is very difficult to ascertain in advance, but for problems for which only a moderate accuracy is required, the step size used will be far larger than this minimum, and the discretization error will be the dominant contributor to the error. The same general behavior occurs in all the methods, although the minimum step size, h_0, will depend on the method and the problem.

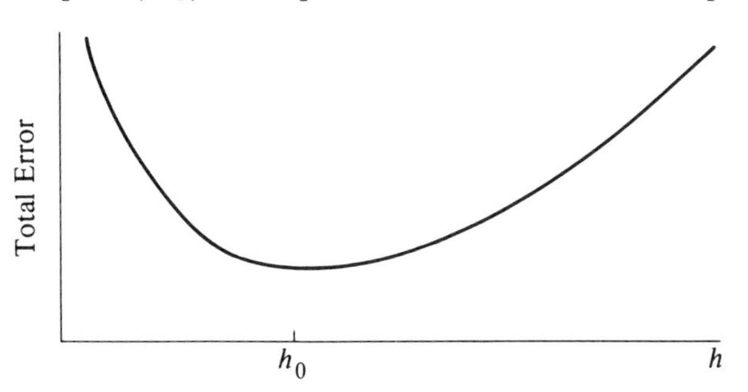

Figure 2.5: *Error in Euler's and Other Methods*

Change in Step Size

So far in this section we have considered a fixed step-size h. However, in practice it is usually beneficial to allow the step to vary. By the bound (2.2.14) on the local discretization error, it follows that the error is small if the second derivative of the solution is small. In regions where this is the case, suitable accuracy may be obtained by using a relatively large step-size. On the other hand, if the second derivative is large a smaller step-size will be necessary to control the discretization error. For the higher order Runge-Kutta methods, the same idea will hold but higher-order derivatives of the solution will be the determining factor. Generally, one will not know much about these higher derivatives, but many times it will be known in advance that the solution is varying slowly in some region and rapidly in another, and the step-size can be changed accordingly. For further discussion on ways to estimate the discretization error so as to ascertain the proper step-size, see the Supplementary Discussion.

Numerical Examples

We complete this section with some simple calculations for the predator-prey equations introduced in Section 2.1. Recall that these equations are

$$\frac{dx}{dt} = \alpha x + \beta xy, \qquad \frac{dy}{dt} = \gamma y + \delta xy, \tag{2.2.34}$$

with the initial conditions

$$x(0) = x_0, \qquad y(0) = y_0. \tag{2.2.35}$$

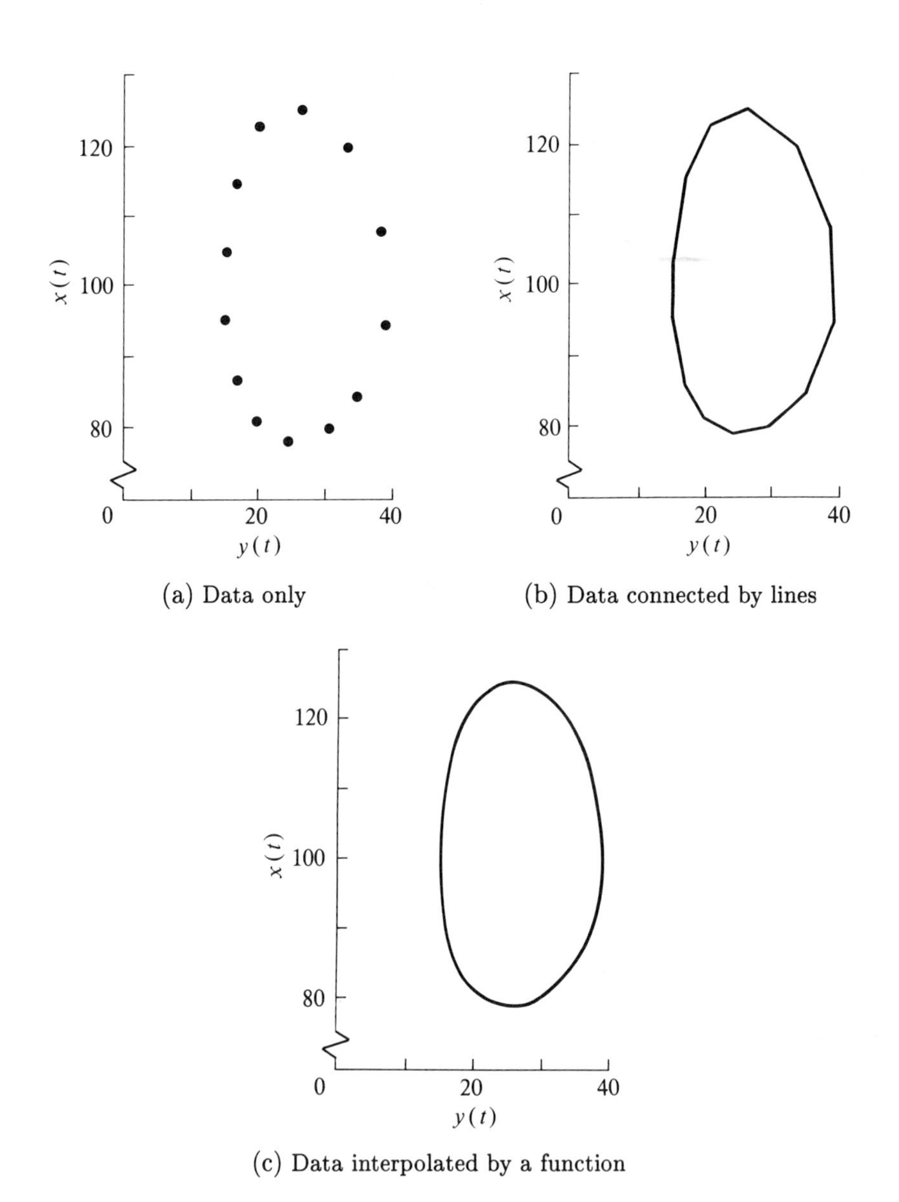

(a) Data only

(b) Data connected by lines

(c) Data interpolated by a function

Figure 2.6: *Options for Graphical Output. Solution to* (2.2.34), (2.2.35)

The Supplementary Discussion of Section 2.1 showed that

$$x_s = \frac{-\gamma}{\delta}, \qquad y_s = \frac{-\alpha}{\beta}$$

is a stationary point of (2.2.34), and that in the neighborhood of (x_s, y_s) the path traced out by $(x(t),\ y(t))$ for $t > 0$ is approximately an ellipse. For illustration purposes, we have chosen initial values x_0 and y_0 that are near stationary points. We have used the following values for the parameters: $\alpha = 0.25$, $\beta = -0.01$, $\gamma = -1.00$, and $\delta = 0.01$. For these parameter values, there is a stationary point at $(x_s, y_s) = (100, 25)$. Initial values of $x_0 = 80$ and $y_0 = 30$ were used in all cases.

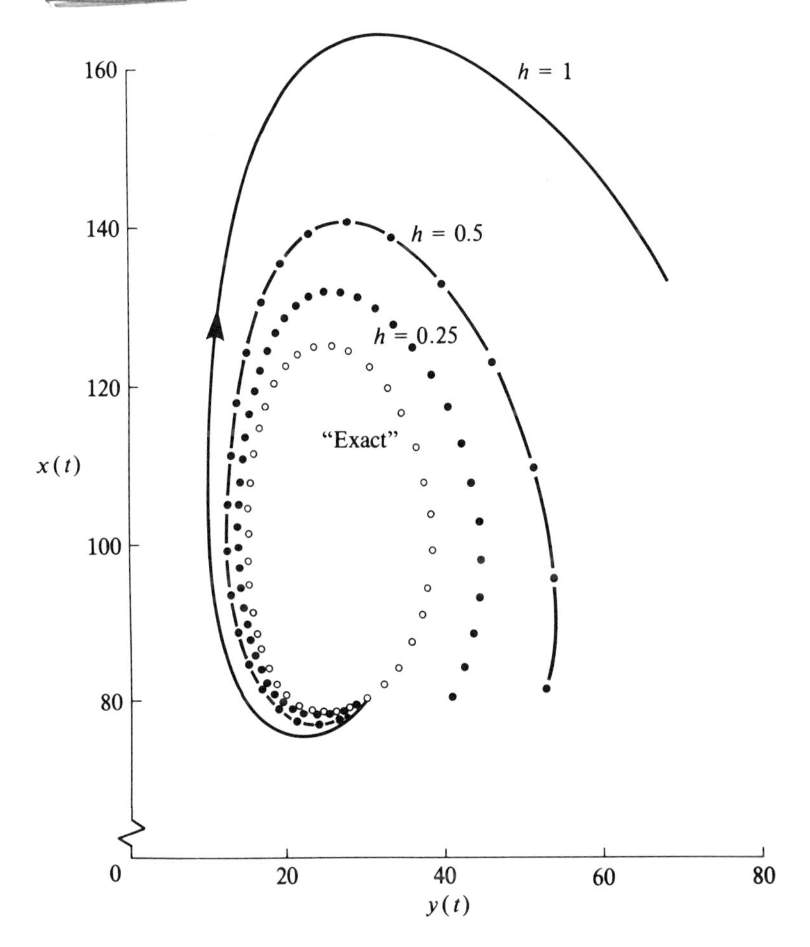

Figure 2.7: *Euler's Method for* (2.2.34), (2.2.35) *Using Different Step Sizes*

Figures 2.6 – 2.8 are the plotted approximations to solutions of (2.2.34), (2.2.35) generated by several of the numerical methods in this section. In all cases we have plotted x (the prey) versus y (the predator), both as functions of time, t. The motion is in a clockwise direction as t increases. Figure 2.6 demonstrates three options that are usually available for graphics. Part (a) plots just the discrete values (x_i, y_i) generated by the numerical method, emphasizing the discrete nature of the methods. In part (b) the points are connected by straight lines, giving a polygonal shape to the approximation. Part (c) shows the dots connected by smooth curves. This option requires some special software based on approximation methods like those discussed in Section 6.2.

Figure 2.7 demonstrates the dependency of the approximate solution on the value of the step size, h. The numerical method used for this figure is Euler's method defined by (2.2.5). The values of h are 1, 0.5, and 0.25. One sees that as the step size is halved the error is roughly halved also, suggesting $0(h)$ convergence. Clearly, the errors are rather large even for $h = 0.25$. The "exact" solution used for comparison was obtained by a higher-order Runge-Kutta method, and the solution so obtained may be considered to be exact for the purpose of comparing with the lower-order methods.

Figure 2.8 shows the effect of using a second-order method rather than the first-order Euler method. Here the error for a step size of $h = 1$ is less than that for Euler's method with a step size of $h = 0.25$. Note that, as with Euler's method, the approximate solution is spiraling out away from the exact solution.

Supplementary Discussion and References: 2.2

Perhaps the conceptually simplest approach to higher-order one-step methods is Taylor-series expansion of the solution. Consider the "method"

$$y_{k+1} = y_k + hy'(x_k) + \tfrac{1}{2}h^2 y''(x_k) + \cdots + \frac{h^p}{p!}y^{(p)}(x_k), \tag{2.2.36}$$

where y is the exact solution. It is easy to see that the order of this method is p. The higher derivatives of the solution can be obtained in principle from the differential equation itself. Thus $y'(x) = f(x, y(x))$, and

$$y''(x) = \frac{d}{dx}f(x, y(x)) = f_x(x, y(x)) + f_y(x, y(x))y'(x). \tag{2.2.37}$$

We then approximate $y'(x_k)$ by $f(x_k, y_k)$, and proceed similarly for higher derivatives. Thus the method for $p = 1$ is simply Euler's method, whereas for $p = 2$ it becomes

$$y_{k+1} = y_k + hf(x_k, y_k) + \frac{h^2}{2}[f_x(x_k, y_k) + f_y(x_k, y_k)f(x_k, y_k)],$$

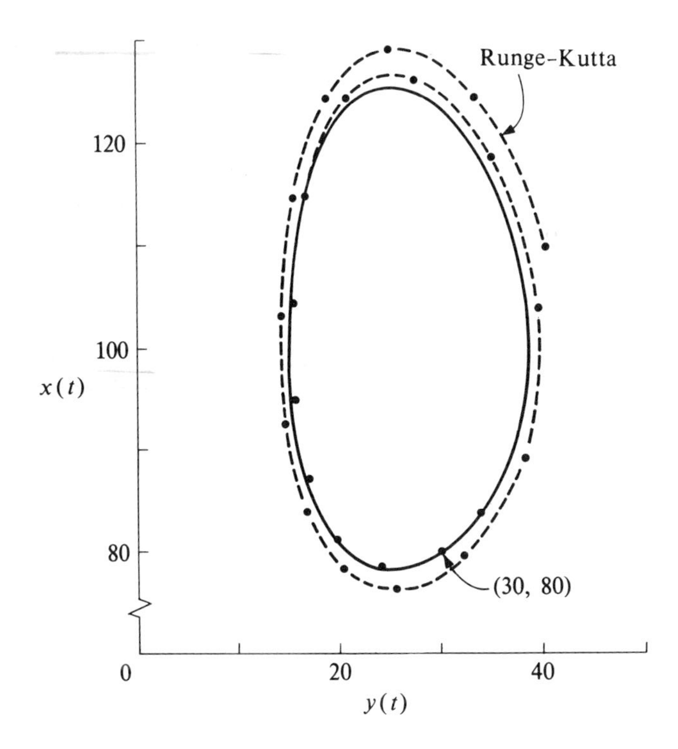

Figure 2.8: *Second-Order Runge-Kutta (Spiraling Outward) with Step-Size* $h = 1$, *Compared to the "Exact" Solution of* (2.2.34), (2.2.35)

which is a second-order method. One can continue to differentiate (2.2.37) to obtain higher derivatives of y in terms of higher partial derivatives of f, but the methods become exceedingly cumbersome. Symbol manipulation techniques have proved somewhat useful in generating the derivatives. Note that Taylor series methods achieve higher order by using derivatives of f evaluated at a single point, whereas Runge-Kutta methods evaluate only f at different points. For further discussion of Taylor-series methods, see Daniel and Moore [1970].

Runge-Kutta methods of order higher than four may be obtained but at still additional costs in evaluations of the function f. Runge-Kutta methods of order p require p evaluations of f for $2 \leq p \leq 4$, $p+1$ evaluations for $5 \leq p \leq 7$, and $p + 2$ evaluations for $p \geq 8$. For a thorough discussion of Runge-Kutta methods, see, for example, Henrici [1962], Butcher [1987], or Hairer, Norsett, and Wanner [1987].

All good computer codes using Runge-Kutta methods employ some mechanism for automatically changing the step size h as the integration proceeds. The problem is to ascertain the proper step size before the start of the next integration step. The usual approach is to estimate the local discretization error and adjust the step size accordingly. There are several ways to estimate the local error; two simple approaches are to repeat the last step of the integration with a step size half as large and then to compare the two results, or to use two Runge-Kutta formulas of different order. Both of these ways are costly in evaluations of f, and an alternative approach is by means of the Runge-Kutta-Fehlberg formulas (see Butcher [1987] for further discussion). Here one can use, for example, a Runge-Kutta method of order five to estimate the error in a fourth-order Runge-Kutta method in such a way that only six evaluations of f are needed, as opposed to ten if the usual Runge-Kutta formulas were used. One of the best and most used Runge-Kutta codes is RKF45; see Shampine, Watts, and Davenport [1976].

As we have shown, in the absence of rounding error Euler's method will yield the true solution of the differential equation as $h \to 0$. But, as Table 2.2 illustrates, the rate of convergence can be rather slow. An important technique to accelerate the convergence is *Richardson extrapolation to the limit*. For some fixed point x^*, denote the approximation to the solution by $y(x^*; h)$. Under certain assumptions, it can be shown that

$$y(x^*; h) = y(x^*) + c_1 h + c_2 h^2 + \cdots + c_p h^p + 0(h^{p+1}), \tag{2.2.38}$$

where the c_i are functions of x^*. Now suppose that we also compute the approximate solution with step length $h/2$. Then we can combine $y(x^*; h)$ and $y(x^*; h/2)$ to obtain a better approximation. In particular,

$$\bar{y}(x^*; h) \equiv 2y(x^*; \frac{h}{2}) - y(x^*; h) = y(x^*) + d_2 h^2 + \cdots + 0(h^{p+1}),$$

so that $\bar{y}(x^*; h)$ is a second order approximation. (Note that the more accurate solution is obtained only at the points with spacing h, and not at the intermediate points.) The process can then be repeated, if desired, to eliminate the coefficient d_2 and obtain a third order approximation, and so on. The same idea can be applied to other problems such as numerical integration or differentiation, or other types of differential equations, provided that an "asymptotic expansion" of the form (2.2.38) holds for the approximate solution.

An important modification of a system of differential equations occurs when there are side conditions. For example, a system of the form

$$\begin{aligned} y_i' &= f_i(x, y,(x), \ldots, y_n(x)), \qquad i = 1, \ldots, m, \\ 0 &= g_i(x, y,(x), \ldots, y_n(x)), \qquad i = m+1, \ldots, n, \end{aligned}$$

consists of m differential equations and $n - m$ non-differential equations. Such a system is called *differential-algebraic*. For further discussion, see, for example, Hairer, Lubich, and Roche [1989], and Brenan, Campbell, and Petzold [1989].

EXERCISES 2.2

2.2.1. Rewrite the predator-prey equations (2.1.3) in the form (2.2.1); that is, give the functions f_1 and f_2. Do the same for the trajectory equations (2.1.15), (2.1.17), and (2.1.18).

2.2.2. Apply Euler's method (2.2.5) to the initial-value problem $y' = -y$, $0 \leq x \leq 1$, $y(0) = 1$, with $h = 0.25$. Compare your answers to the exact solution $y(x) = e^{-x}$. Repeat for $h/2$ and $h/4$.

2.2.3. Verify the calculations of Tables 2.1 and 2.2.

2.2.4. Apply the Heun method to the problem of Exercise 2.2.2. Compare your results with Euler's method.

2.2.5. Give the function ϕ of (2.2.21) for the fourth-order Runge-Kutta method (2.2.20).

2.2.6. The method $y_{k+1} = y_k + hf(x_k + (h/2), y_k + (h/2)f(x_k, y_k))$ is known as the *midpoint rule.* Show that it is second-order accurate.

2.2.7. Write the fourth-order Runge-Kutta method (2.2.20) in vector form for the system (2.2.30).

2.2.8. Apply Euler's method and the Heun method to the problem $y'(x) = x^2 + [y(x)]^2, y(0) = 1$, $x \geq 0$, and compute y_2 for $h = 0.1$.

2.2.9. For the fourth-order Runge-Kutta method (2.2.20), draw the figure corresponding to Figure 2.4.

2.2.10. Repeat the calculations of Figure 2.7 using Euler's method with step sizes 0.5 and 0.25. How small a step size do you have to use for the graph of the solution to close back on itself to visual accuracy?

2.2.11. Test the stability of the solution of the predator-prey equations (2.2.34) with respect to changes in the initial conditions by changing $x_0 = 80$, $y_0 = 30$ by a unit amount in each direction (four different cases) and repeating the calculation using the second-order Runge-Kutta method.

2.2.12. Suppose that the initial condition of (2.2.5) is in error. Modify the analysis of the discretization error in Euler's method to obtain a bound on the error caused by using an inexact initial condition $y_0 = \bar{y} \neq \hat{y}$.

2.3 Polynomial Interpolation

The methods described in Section 2.2 were all one-step methods – methods that estimate y at x_{k+1} using information only at the previous point, x_k. In Section 2.4 methods that use information at several previous points will be described. In order to develop such methods, however, we must first describe polynomial interpolation, the subject of this section.

Suppose that one is given a set of points or *nodes* $x_0, x_1, \ldots, x_n$ and a set of corresponding numbers $y_0, y_1, \ldots, y_n$. The *interpolation problem* is to find a function g that satisfies

$$g(x_i) = y_i, \qquad i = 0, 1, \ldots, n; \tag{2.3.1}$$

if (2.3.1) holds we say that g interpolates the data. There are many types of possible functions g that might be used, but the functions of interest in this section will be polynomials.

It is clear that a polynomial of a given degree can not always be found so that (2.3.1) is satisfied. For example, if the data y_i are given at three distinct nodes, no polynomial of degree 1 (a linear function) can satisfy (2.3.1) unless the data lie on a straight line. On the other hand, there is a polynomial of degree 2 and infinitely many polynomials of degree 3 which satisfy (2.3.1). The basic result for polynomial interpolation is given in the following theorem.

THEOREM 2.3.1 (Existence and Uniqueness for Polynomial Interpolation) *If $x_0, x_1, \ldots, x_n$ are distinct nodes, then for any $y_0, y_1, \ldots, y_n$ there exists a unique polynomial $p(x)$ of degree n or less, such that*

$$p(x_i) = y_i, \qquad i = 0, 1, \ldots, n. \tag{2.3.2}$$

Proof: The existence can be proved by constructing the *Lagrange polynomials* defined by

$$\begin{aligned} l_j(x) &= \frac{(x - x_0)(x - x_1)\cdots(x - x_{j-1})(x - x_{j+1})\cdots(x - x_n)}{(x_j - x_0)(x_j - x_1)\cdots(x_j - x_{j-1})(x_j - x_{j+1})\cdots(x_j - x_n)} \\ &= \prod_{\substack{k=0 \\ k\neq j}}^{n} \left(\frac{x - x_k}{x_j - x_k}\right), \qquad j = 0, 1, \ldots, n. \end{aligned} \tag{2.3.3}$$

It is easy to verify that these polynomials, which are all of degree n, satisfy

$$l_j(x_i) = \begin{cases} 1 & \text{if } i = j \\ 0 & \text{if } i \neq j. \end{cases} \tag{2.3.4}$$

Therefore $l_j(x)y_j$ has the value 0 at all nodes x_i, $i = 0, 1, \ldots, n$, except for x_j, where $l_j(x_j)y_j = y_j$. Thus, by defining

$$p(x) = \sum_{j=0}^{n} l_j(x)y_j, \tag{2.3.5}$$

we have a polynomial of degree n or less that interpolates the data.

To prove uniqueness, suppose, on the contrary, that there is another interpolating polynomial of degree n or less, say $q(x)$. By defining

$$r(x) = p(x) - q(x)$$

we obtain a polynomial, r, of degree n or less that is equal to zero at the $n+1$ distinct points $x_0, x_1, \ldots, x_n$. By the fundamental theorem of algebra, such a polynomial must be identically equal to zero, and it follows that $p(x) = q(x)$. Thus, uniqueness is proved.

As an example of polynomial interpolation, let us determine the polynomial $p(x)$ of degree 2 or less that satisfies $p(-1) = 4$, $p(0) = 1$, and $p(1) = 0$. The interpolating polynomial (2.3.5) is

$$\begin{aligned} p(x) &= \frac{(x-0)(x-1)}{(-1-0)(-1-1)}4 + \frac{(x-(-1))(x-1)}{(0-(-1))(0-1)}1 + \frac{(x-(-1))(x-0)}{(1-(-1))(1-0)}0 \\ &= 2x^2 - 2x + 1 - x^2 + 0 = x^2 - 2x + 1. \end{aligned}$$

One can easily verify that this $p(x)$ does interpolate the given data.

Error Estimates

We next consider the question of accuracy in polynomial interpolation. In many applications of interpolation there is some function f defined over the entire interval of interest, even though values are used only at discrete points to determine an interpolating polynomial p. Thus it is of interest to discuss the discrepancy between $p(x)$ and $f(x)$ for values of x that lie between the nodes. The following theorem gives an expression for the error in terms of higher derivatives of f:

THEOREM 2.3.2 (Polynomial Interpolation Error) *Let $f(x)$ be a function with $n+1$ continuous derivatives on an interval containing the distinct nodes $x_0 < x_1 < \cdots < x_n$. If $p(x)$ is the unique polynomial of degree n or less satisfying*

$$p(x_i) = f(x_i), \qquad i = 0, 1, \ldots, n,$$

then for any $x \in [x_0, x_n]$,

$$f(x) - p(x) = \frac{(x-x_0)(x-x_1)\cdots(x-x_n)}{(n+1)!} f^{(n+1)}(z) \tag{2.3.6}$$

for some z, depending on x, in the interval (x_0, x_n).

We indicate a proof of this theorem in the Supplementary Discussion of this section; here we only discuss some of its ramifications. First of all, if n is at all large (even 4 or 5), it will probably be difficult, if not impossible, to compute the $(n+1)$th derivative of f. Even if n is only 1 (linear interpolation) and only the second derivative of f is needed, this also may be impossible if f is an unknown function for which values are known only at some discrete points; at best, we might be able to estimate some bound for the second derivative on the basis of our assumed knowledge of f. In any case, it will almost never be the case that (2.3.6) can be used to give a very precise bound on the error. It can, however, be useful in providing insight into the errors that are produced. As an example of this, suppose that the points x_i are equally spaced with spacing h. Then it is easy to see (Exercise 2.3.2.) that

$$|(x-x_0)(x-x_1)\cdots(x-x_n)| \le n!\ h^{n+1}$$

for any x in the interval $[x_0, x_n]$. Thus (2.3.6) can be bounded by

$$|f(x)-p(x)| \le \frac{Mh^{n+1}}{n+1}, \tag{2.3.7}$$

where

$$M = \max_{x_0 \le x \le x_n} |f^{(n+1)}(x)|.$$

Piecewise Polynomials

The bound (2.3.7) is, of course, still difficult to compute because of the quantity M. But it is useful in the following way. Suppose that we wish to approximate the function f over a given interval $[a, b]$ by means of *piecewise polynomials*, that is, functions that are polynomials on given subintervals of $[a, b]$. For example, if $a = \gamma_0 < \gamma_1 < \cdots < \gamma_p < \gamma_{p+1} = b$ is a partitioning of the interval $[a, b]$, and g is a function that is continuous on $[a, b]$ and is a polynomial on each of the intervals (γ_i, γ_{i+1}), $i = 0, 1, \cdots, p$, then g is called a piecewise polynomial function on $[a, b]$.

As an example of a piecewise quadratic function, suppose that the values of the function f on the interval $[0, 1]$ are given by

x	0	1/6	1/3	1/2	2/3	5/6	1
f	1	3	2	1	0	2	1

Then, the function g defined by

$$\begin{aligned} g(x) &= -54x^2 + 21x + 1, && 0 \le x \le \tfrac{1}{3} \\ &= -6x + 4, && \tfrac{1}{3} \le x \le \tfrac{2}{3} \\ &= -54x^2 + 93x - 38, && \tfrac{2}{3} \le x \le 1 \end{aligned} \tag{2.3.8}$$

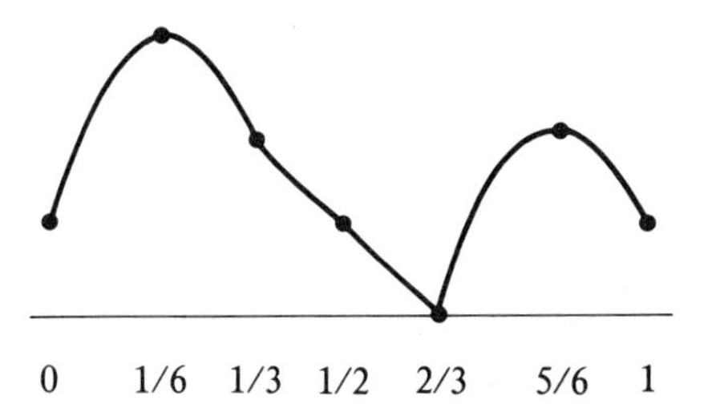

Figure 2.9: *A Piecewise Quadratic Function*

is the piecewise quadratic function on $[0,1]$ that agrees with f at the given nodes, is continuous on the whole interval, and is a quadratic on each of the subintervals $[0,\frac{1}{3}]$, $[\frac{1}{3},\frac{2}{3}]$, $[\frac{2}{3},1]$. This function is shown in Figure 2.9.

Consider now the error in approximating the function f by the function g of (2.3.8). Suppose that M is a bound for the third derivative of f on the entire interval $[0,1]$. Then on each of the intervals $[0,\frac{1}{3}]$, $[\frac{1}{3},\frac{2}{3}]$, and $[\frac{2}{3},1]$, the error bound (2.3.7) can be applied; here $h=\frac{1}{6}$, and $n=2$. Therefore

$$|f(x)-g(x)| \le \frac{h^3 M}{3} = \frac{M}{3\cdot 6^3}, \qquad 0 \le x \le 1 \tag{2.3.9}$$

Without further information on M, this estimate does not furnish a quantitative bound. It does, however, show how the spacing h enters the error estimate. In particular, approximation by piecewise quadratics has an error estimate that is $0(h^3)$, so that it is third-order. Higher order approximation will result from using piecewise cubics or higher degree polynomials.

The Vandermonde Matrix

Even though the interpolating polynomial is unique, as shown by Theorem 2.3.1, there are several alternative ways to obtain or represent the polynomial, other than by the Lagrange polynomials. Perhaps the most basic approach is the following. Suppose that the interpolation polynomial p is

$$p(x) = a_0 + a_1 x + \cdots + a_n x^n.$$

Then we want

$$a_0 + a_1 x_i + \cdots + a_n x_i^n = y_i, \qquad i = 0,1,\ldots,n. \tag{2.3.10}$$

Since the x_i's and the y_i's are known, this is a system of $n+1$ linear equations in the $n+1$ unknowns $a_0, a_1, \ldots, a_n$. We write this system in the matrix-vector form

$$\begin{bmatrix} 1 & x_0 & x_0^2 & \cdots & x_0^n \\ 1 & x_1 & x_1^2 & \cdots & x_1^n \\ \vdots & \vdots & \vdots & & \\ 1 & x_n & x_n^2 & \cdots & x_n^n \end{bmatrix} \begin{bmatrix} a_0 \\ a_1 \\ \vdots \\ a_n \end{bmatrix} = \begin{bmatrix} y_0 \\ y_1 \\ \vdots \\ y_n \end{bmatrix}. \tag{2.3.11}$$

The coefficient matrix of (2.3.11), which we denote by V, is called the *Vandermonde matrix* and is nonsingular if the x's are distinct. (This statement can be proved directly rather easily, but note that we already have proved it indirectly by means of Theorem 2.3.1, which showed the existence and uniqueness of the interpolating polynomial. For if V were singular, this would imply that either no interpolating polynomial exists for the given data or infinitely many exist.)

The Vandermonde matrix approach is sometimes useful for theoretical purposes, but less so for computation of the polynomial. For the latter the Lagrange polynomials are usually better, but they are not convenient if a node is added or dropped from the data. For example, if (x_{n+1}, y_{n+1}) were added to the set of data (x_i, y_i), $i = 0, 1, \ldots, n$, and we wished to compute the polynomial of degree $n+1$ that interpolated this data, then the Lagrange polynomials would all have to be recomputed. There is another representation of the interpolating polynomial that is very useful in this context; this is the *Newton form*, which we now describe.

The Newton Form

We assume now that the points x_i are equally spaced with spacing h. We define differences of the data y_i by means of $\Delta y_i = y_{i+1} - y_i$, and higher differences by repeated application of this:

$$\begin{aligned} &\Delta^2 y_0 = \Delta y_1 - \Delta y_0 = y_2 - 2y_1 + y_0 \\ &\Delta^3 y_0 = \Delta^2 y_1 - \Delta^2 y_0 = y_3 - 3y_2 + 3y_1 - y_0 \\ &\vdots \\ &\Delta^n y_0 = y_n - \binom{n}{1} y_{n-1} + \binom{n}{2} y_{n-2} - \cdots + (-1)^n y_0, \end{aligned} \tag{2.3.12}$$

where the binomial coefficients are given by

$$\binom{n}{i} = \frac{n(n-1)\cdots(n-i+1)}{i!}.$$

In terms of the differences (2.3.12), we define a polynomial of degree n by

$$\begin{aligned} p_n(x) &= y_0 + \frac{(x-x_0)}{h}\Delta y_0 + \frac{(x-x_0)(x-x_1)}{2h^2}\Delta^2 y_0 \\ &\quad + \cdots + \frac{(x-x_0)(x-x_1)\cdots(x-x_{n-1})}{n!\, h^n}\Delta^n y_0. \end{aligned} \tag{2.3.13}$$

Clearly, $p_n(x_0) = y_0$ since all remaining terms in (2.3.13) vanish. Similarly,

$$p_n(x_1) = y_0 + \frac{(x_1 - x_0)}{h}(y_1 - y_0) = y_1,$$

and

$$\begin{aligned} p_n(x_2) &= y_0 + \frac{(x_2 - x_0)}{h}(y_1 - y_0) + \frac{(x_2 - x_0)(x_2 - x_1)}{2h^2}(y_2 - 2y_1 + y_0) \\ &= y_0 + 2(y_1 - y_0) + (y_2 - 2y_1 + y_0) = y_2. \end{aligned}$$

It is easy to verify in an analogous way that $p_n(x_i) = y_i$, $i = 3, \cdots, n$, although the computations become increasingly tedious.

It is of interest to note that the polynomial p_n of (2.3.13) is analogous to the first $n + 1$ terms of a Taylor expansion about x_0. Now suppose that we add (x_{n+1}, y_{n+1}) to the data set. Then the polynomial p_{n+1} that satisfies $p_{n+1}(x_i) = y_i$, $i = 0, 1, \ldots, n + 1$, is given by

$$p_{n+1}(x) = p_n(x) + \frac{(x - x_0)(x - x_1) \cdots (x - x_n)}{(n + 1)!\, h^{n+1}} \Delta^{n+1} y_0,$$

and it is this feature of the Newton form of the interpolating polynomial that is sometimes useful in practice. This is similar to taking one more term in a Taylor expansion.

In the next section we will use interpolating polynomials to derive other methods for the solution of differential equations. This will be done by integrating an interpolating polynomial, but once the polynomial has been determined other manipulations, such as differentiation, can also be performed. Moreover, interpolation is useful for approximating the solution between grid points, if that should be desired. However, it should be pointed out that interpolation may fail to preserve such desirable properties as monotonicity and convexity. It is a more difficult and subtle problem to obtain approximating functions that do maintain such properties.

Supplementary Discussion and References: 2.3

We will indicate the proof of the basic error theorem 2.3.2. Assume that $x \neq x_j$, $j = 0, 1, \ldots, n$; otherwise, both sides of (2.3.6) are zero, and the result is trivially true. Now, for x held fixed, define the function

$$\phi(s) = f(s) - p(s) - q(x)\psi(s),$$

where

$$\psi(s) = (s - x_0)(s - x_1) \cdots (s - x_n), \qquad q(x) = \frac{f(x) - p(x)}{\psi(x)}.$$

It is clear that $\psi(x_i) = 0$, $i = 0, 1, \ldots, n$, and $\phi(x) = 0$; hence ϕ has at least $n+2$ distinct roots $x_0, x_1, \ldots, x_n, x$. It follows by repeated application of Rolle's theorem that ϕ' has at least $n+1$ distinct roots, ϕ'' has at least n distinct roots, and so on. In particular, $\phi^{(n+1)}$ has at least one root z in the interval spanned by $x_0, x_1, \ldots, x_n, x$. But

$$\begin{aligned}\phi^{(n+1)}(s) &= f^{(n+1)}(s) - p^{(n+1)}(s) - q(x)\psi^{(n+1)}(s) \\ &= f^{(n+1)}(s) - (n+1)!\, q(x),\end{aligned}$$

since p is a polynomial of degree n and ψ is a polynomial of degree $n+1$. Thus,

$$0 = \phi^{(n+1)}(z) = f^{(n+1)}(z) - (n+1)!\, q(x),$$

and solving for $q(x)$ gives (2.3.6).

For further discussion of interpolation, see, for example, Young and Gregory [1990].

EXERCISES 2.3

2.3.1. Compute the polynomial p of degree 2 that satisfies $p(0) = 0$, $p(1) = 1$, $p(2) = 0$ by all three methods, that is, by using Lagrange polynomials, the Vandermonde matrix, and the Newton representation. Conclude that the polynomial is the same in all three cases.

2.3.2. Verify the bound (2.3.7). Then let $f(x) = \sin \pi x/2$, and let p be the polynomial of Exercise 2.3.1 that agrees with f at the points $x = 0, 1, 2$. Use (2.3.7) to compute a bound for $|f(x) - p(x)|$ on the interval $[0, 2]$. Compare this bound with the actual error at selected points in the interval, and in particular at $x = \frac{1}{4}$ and $\frac{3}{4}$.

2.3.3. Find the piecewise linear and quadratic functions that agree with the following data:

x	0	1/6	1/3	1/2	2/3	5/6	1
f	1	4	1	−1	2	4	0

Compute error bounds for these functions on the interval $[0, 1]$, assuming that the function f satisfies $|f''(z)| \leq 4$, $|f'''(z)| \leq 10, 0 \leq z \leq 1$.

2.3.4. Let $f(x) = \sin x$, and let p and q be two polynomials of degree 3 that satisfy $p(k/3) = q(k/3) = f(k/3)$, $k = 0, 1, 2, 3$. Compute a bound for $|p(x) - q(x)|$ that holds on the whole interval $[0, 1]$.

2.3.5. For given $y_0, \ldots, y_n$ and distinct $x_0, \ldots, x_n$, let p be the polynomial of degree n that satisfies $p(x_i) = y_i$, $i = 0, \ldots, n$. Suppose that we wish to write p in the form $p(x) = c_0 q_0(x) + \cdots + c_n q_n(x)$, where $q_0(x) \equiv 1$ and $q_i(x) = (x - x_0) \cdots (x - x_i)$. Give an algorithm for finding $c_0, \ldots, c_n$.

2.3.6. Find a polynomial of degree 3 that agrees with $\sqrt{x}$ at $0, 1, 3, 4$. Compare the approximation $p(2)$ with $\sqrt{2} = 1.414216$.

2.3.7. Let p be a polynomial that satisfies $p(-2) = -5$, $p(-1) = 1$, $p(0) = 1$, $p(1) = 1$, $p(2) = 7$, $p(3) = 25$. What can you say about the degree of p?

2.4 Multistep Methods

We return now to the initial-value problem

$$y' = f(x, y), \qquad a \leq x, \qquad y(a) = \hat{y}. \tag{2.4.1}$$

In the methods of Section 2.2, the value of y_{k+1} depended only on information at the previous point, x_k. It seems plausible that more accuracy might be gained if information at several previous points, $x_k, x_{k-1}, \ldots$, were used. Multistep methods do just that.

A large and important class of multistep methods arises from the following approach. If we integrate (2.4.1) for the exact solution $y(x)$ over the interval $[x_k, x_{k+1}]$, we have

$$\begin{aligned} y(x_{k+1}) - y(x_k) &= \int_{x_k}^{x_{k+1}} y'(x)dx = \int_{x_k}^{x_{k+1}} f(x, y(x))dx \qquad (2.4.2) \\ &\doteq \int_{x_k}^{x_{k+1}} p(x)dx, \end{aligned}$$

where in the last term we assume that $p(x)$ is a polynomial that approximates $f(x, y(x))$. To obtain this polynomial, suppose that, as in Section 2.2, $y_k, y_{k-1}, \ldots, y_{k-N}$ are approximations to the solution at $x_k, x_{k-1}, \ldots, x_{k-N}$, where we assume that the x_i are equally spaced with spacing h. Then $f_i \equiv f(x_i, y_i)$, $i = k, k-1, \ldots, k-N$, are approximations to $f(x, y(x))$ at $x_k, x_{k-1}, \ldots$, x_{k-N}, and we take p to be the interpolating polynomial for the data set (x_i, f_i), $i = k, k-1, \ldots, k-N$. Thus p is the polynomial of degree N that satisfies $p(x_i) = f_i$, $i = k, k-1, \ldots, k-N$. In principle, we can integrate this polynomial explicitly to give the method

$$y_{k+1} = y_k + \int_{x_k}^{x_{k+1}} p(x)dx. \tag{2.4.3}$$

Adams-Bashforth Methods

As the simplest example, if $N = 0$, then p is the constant f_k and (2.4.3) is simply Euler's method. If $N = 1$, then p is the linear function that interpolates

(x_{k-1}, f_{k-1}) and (x_k, f_k). We may obtain this and the subsequent formulas by applying (2.3.13) backwards from (x_k, f_k); that is, with $\Delta f_k = f_{k-1} - f_k$

$$p(x) = p_1(x) = f_k - \frac{(x - x_k)}{h} \Delta f_k, \tag{2.4.4}$$

where the minus sign occurs since $h = |x_{k-1} - x_k|$. If we integrate (2.4.4) from x_k to x_{k+1}, we obtain the method

$$y_{k+1} = y_k + h f_k - \frac{h}{2} \Delta f_k = y_k + \frac{h}{2}(3f_k - f_{k-1}), \tag{2.4.5}$$

which is a two-step method since it uses information at the two points x_k and x_{k-1}. Note that the first form of (2.4.5) shows how Euler's method is modified to obtain the new method.

Similarly, if $N = 2$, then p is the interpolating quadratic polynomial for (x_{k-2}, f_{k-2}), (x_{k-1}, f_{k-1}), and (x_k, f_k). If we again use (2.3.13), this polynomial may be written as

$$p_2(x) = p_1(x) + \frac{(x - x_k)(x - x_{k-1})}{2h^2} \Delta^2 f_k, \tag{2.4.6}$$

where $\Delta^2 f_k = f_k - 2f_{k-1} + f_{k-2}$. Thus, by (2.4.3), the method is

$$y_{k+1} = y_k + h f_k - \frac{h}{2} \Delta f_k + \frac{5}{6} h \Delta^2 f_k. \tag{2.4.7}$$

This exhibits how the two-step formula (2.4.5) has been modified. We can also collect terms in (2.4.7) and write it as

$$y_{k+1} = y_k + \frac{h}{12}(23 f_k - 16 f_{k-1} + 5 f_{k-2}). \tag{2.4.8}$$

If $N = 3$, the interpolating polynomial is a cubic, and the method is

$$y_{k+1} = y_k + \frac{h}{24}(55 f_k - 59 f_{k-1} + 37 f_{k-2} - 9 f_{k-3}). \tag{2.4.9}$$

Note that (2.4.8) is a three-step method, whereas (2.4.9) is a four-step method. It is left to Exercise 2.4.1 to give a detailed verification of the formulas (2.4.5), (2.4.8), and (2.4.9).

The formulas (2.4.5), (2.4.8), and (2.4.9) are known as *Adams-Bashforth methods.* As we shall see later, (2.4.5) is second-order accurate and hence is known as the *second-order Adams-Bashforth method.* Similarly, (2.4.8) and (2.4.9) are the *third-* and *fourth-order Adams-Bashforth methods*, respectively. We can, in principle, continue the preceding process to obtain Adams-Bashforth methods of arbitrarily high order by increasing the number of prior points and the degree of the interpolating polynomial p. The formulas become increasingly complex as N increases, but the principle is still the same.

The Starting Problem

Multistep methods suffer from a problem not encountered with one-step methods. Consider the fourth-order Adams-Bashforth method of (2.4.9). The initial value y_0 is given, but for $k = 0$ in (2.4.9), information is needed at x_{-1}, x_{-2}, and x_{-3}, which doesn't exist. The problem is that multistep methods need "help" getting started. We cannot use (2.4.9) until $k \geq 3$, nor can we use (2.4.8) until $k \geq 2$. The usual tactic is to use a one-step method, such as Runge-Kutta, of the same order of accuracy until enough values have been computed so that the multistep method is usable. Alternatively, one may use a one-step method at the first step, a two-step method at the second, and so on until enough starting values have been built up. However, it is important that the starting values obtained in this fashion be as accurate as those to be produced by the final method, and if the starting methods are of lower order, this will necessitate using a smaller step size and generating more intermediate points at the outset. The same problem arises if the step length is changed during the calculation. If, for example, at x_k the step length is changed from h to $\frac{h}{2}$, then we will need values of y and f at $x_k - \frac{h}{2}$, which we do not have. Again, a Runge-Kutta method starting at x_{k-1} with step length $\frac{h}{2}$ could be used. An attractive alternative is to use an interpolation formula, but care must be exercised to ensure that suitable accuracy is maintained. In general, it is much easier to change step-size with a one-step method than with a multistep method.

Adams-Moulton Methods

The Adams-Bashforth methods were obtained by using information already computed at x_k and prior points. In principle, we can form the interpolating polynomial by using forward points as well. The simplest situation is to use the points $x_{k+1}, x_k, \ldots, x_{k-N}$ and form the interpolating polynomial of degree $N+1$ that satisfies $p(x_i) = f_i$, $i = k+1, k, \ldots, k-N$. This generates a class of methods known as *Adams-Moulton methods.* If $N = 0$, then p is the linear function that interpolates (x_k, f_k) and (x_{k+1}, f_{k+1}), and the corresponding method is

$$y_{k+1} = y_k + \frac{h}{2}(f_{k+1} + f_k), \tag{2.4.10}$$

which is the *second-order Adams-Moulton method.* If $N = 2$, then p is the cubic polynomial that interpolates (x_{k+1}, f_{k+1}), (x_k, f_k), (x_{k-1}, f_{k-1}), and (x_{k-2}, f_{k-2}); in this case the corresponding method is

$$y_{k+1} = y_k + \frac{h}{24}(9f_{k+1} + 19f_k - 5f_{k-1} + f_{k-2}), \tag{2.4.11}$$

which is the *fourth-order Adams-Moulton method.*

Note that in the formulas (2.4.10) and (2.4.11) f_{k+1} is not known, since we need y_{k+1} to evaluate $f(x_{k+1}, y_{k+1}) = f_{k+1}$, but y_{k+1} is not yet known. Hence the Adams-Moulton methods define y_{k+1} only implicitly. For example, (2.4.10) is really an equation,

$$y_{k+1} = y_k + \frac{h}{2}[f(x_{k+1}, y_{k+1}) + f_k], \tag{2.4.12}$$

for the unknown value y_{k+1}, and similarly for (2.4.11). Thus the Adams-Moulton methods are called *implicit*, whereas the Adams-Bashforth methods are called *explicit* since no equation needs to be solved in order to obtain y_{k+1}.

Predictor-Corrector Methods

Implicit methods are useful for so-called stiff equations, to be discussed in the next section. However, another use of implicit methods is to combine an explicit with an implicit formula to form a *predictor-corrector method.* A commonly used predictor-corrector method is the combination of the fourth-order Adams methods (2.4.9) and (2.4.11):

$$\begin{aligned} y_{k+1}^{(p)} &= y_k + \frac{h}{24}(55f_k - 59f_{k-1} + 37f_{k-2} - 9f_{k-3}), \\ f_{k+1}^{(p)} &= f(x_{k+1}, y_{k+1}^{(p)}), \\ y_{k+1} &= y_k + \frac{h}{24}(9f_{k+1}^{(p)} + 19f_k - 5f_{k-1} + f_{k-2}),. \end{aligned} \tag{2.4.13}$$

Note that this method is entirely explicit. First a "predicted" value $y_{k+1}^{(p)}$ of y_{k+1} is computed by the Adams-Bashforth formula, then $y_{k+1}^{(p)}$ is used to give an approximate value of f_{k+1}, which is used in the Adams-Moulton formula. The Adams-Moulton formula "corrects" the approximation given by the Adams-Bashforth formula. We could also take additional corrector steps in (2.4.13). In fact, repeated use of the corrector formula gives an iterative method to solve the nonlinear equation (2.4.11). We will return to this topic in Chapter 5.

Discretization Error

We turn now to the question of the discretization error and, for simplicity, we will consider in detail only the Adams-Bashforth method (2.4.5). In a manner analogous to (2.2.23) for one-step methods, we define the local discretization error at x by

$$hL(x, h) = y(x + h) - y(x) - \frac{h}{2}[3f(x, y(x)) - f(x - h, y(x - h))], \tag{2.4.14}$$

where $y(x)$ is the exact solution of the differential equation. Since $y'(x) = f(x, y(x))$, the term in brackets in (2.4.14) can be expanded in a Taylor series as

$$3y'(x) - y'(x-h) = 2y'(x) + hy''(x) + 0(h^2).$$

Combining this with

$$y(x+h) - y(x) = hy'(x) + \frac{h^2}{2}y''(x) + 0(h^3)$$

yields

$$L(x,h) = 0(h^2). \tag{2.4.15}$$

Therefore, assuming that suitably high-order derivatives of the solution are bounded, we have that the local discretization error $L(h)$ on the interval $[a, b]$ satisfies

$$L(h) = \max_{a \le x \le b-h} |L(x,h)| = 0(h^2), \tag{2.4.16}$$

which shows that the method is second order.

We could define the local discretization error separately for each of the other methods of this section. However, all of these methods are special cases of what are called *linear multistep* methods of the form

$$y_{k+1} = \sum_{i=1}^{m} \alpha_i y_{k+1-i} + h \sum_{i=0}^{m} \beta_i f_{k+1-i}, \tag{2.4.17}$$

where, as usual, $f_j = f(x_j, y_j)$, and m is some fixed integer. The method (2.4.17) is called linear since y_{k+1} is a linear combination of the y_i and f_i. If $\beta_0 = 0$, the method is explicit, and if $\beta_0 \neq 0$, then the method is implicit. In all of the Adams methods $\alpha_1 = 1$ and $\alpha_i = 0$, $i > 1$; in the Adams-Bashforth methods $\beta_0 = 0$ and for Adams-Moulton $\beta_0 \neq 0$.

For the general linear multistep method (2.4.17), we define the local discretization error $L(x, h)$ at x by

$$L(x,h) = \frac{1}{h}[y(x+h) - \sum_{i=1}^{m} \alpha_i y(x-(i-1)h)] - \sum_{i=0}^{m} \beta_i y'(x-(i-1)h).$$

For any given method, that is, for any given choice of m and the constants α_i and β_i, one can compute the local discretization error by expansion of y and y' in Taylor series about x. In particular, under suitable assumptions on the differentiability of the solution, one can show that the Adams-Bashforth methods (2.4.8) and (2.4.9) are third and fourth order, respectively, whereas the Adams-Moulton methods (2.4.10) and (2.4.11) are second and fourth order. The verification of these statements is left to Exercise 2.4.6.

Once the local discretization error is known, there remains the problem of bounding the global discretization error, which is defined by

$$E(h) = \max\{|y(x_k) - y_k| : 1 \leq k \leq N\}.$$

In general, this is a difficult problem, but under suitable assumptions on f and the solution y, it can be shown for all of the methods of this section that $E(h) = 0(h^p)$ when $L(h) = 0(h^p)$.

Multistep methods constitute an attractive alternative to the one-step methods of Section 2.2. High-order methods can be constructed that require only one evaluation of f at each step, but at the price that the methods are not self-starting. Indeed, high-order Adams methods are the basis of the most efficient computer codes available today (see the Supplementary Discussion).

Numerical Examples

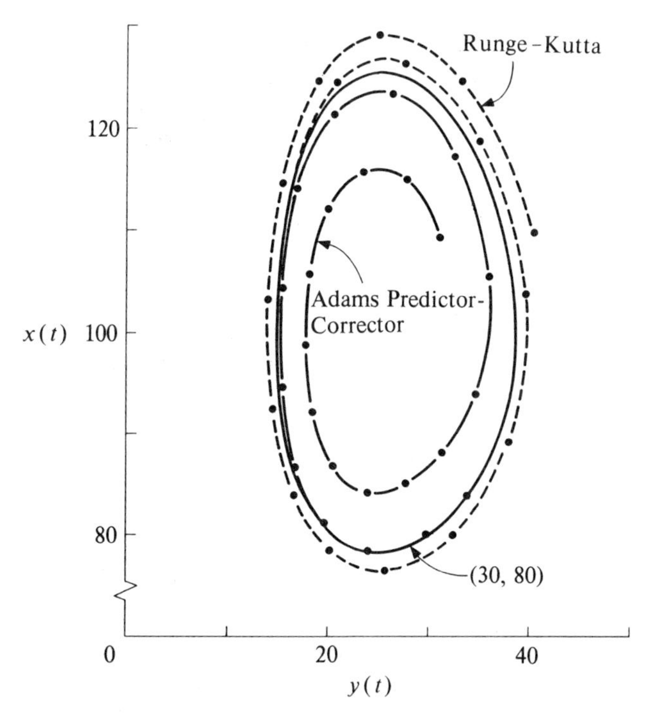

Figure 2.10: *Second-Order Runge-Kutta and Adams Predictor-Corrector*

We conclude this section by returning to the sample problems of Section 2.1. In Figure 2.8 (Section 2.2) we gave an approximate solution of the predator-prey equations (2.2.34), (2.2.35) obtained by the second-order Runge-Kutta method. In Figure 2.10 we superimpose on Figure 2.8 an approximate solution obtained by a second-order Adams-Bashforth/Adams-Moulton predictor-corrector method, which is based on (2.4.5) and (2.4.10) and given explicitly in Exercise 2.4.5. Note that the two methods in Figure 2.10 are in error by a comparable amount, although one approximate solution spirals in from the exact solution whereas the other spirals out. The step size used for both methods was $h = 1$. Both of these methods require two evaluations of f per step. However, the corresponding fourth-order predictor-corrector method still requires only two evaluations of f, whereas the fourth-order Runge-Kutta method requires four.

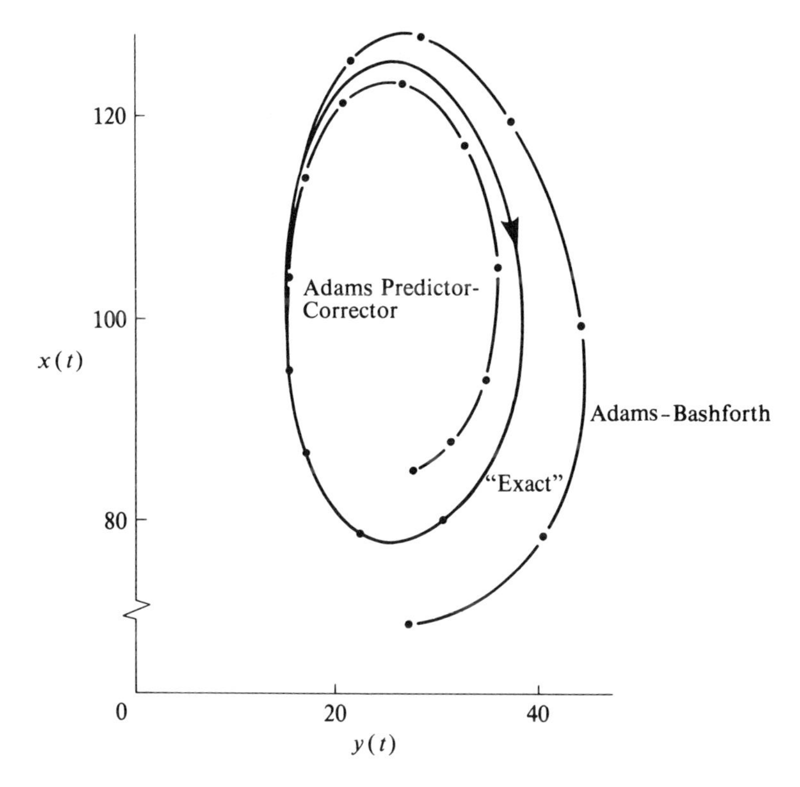

Figure 2.11: *Second-Order Adams-Bashforth and Adams Predictor-Corrector*

Figure 2.11 compares the second-order predictor-corrector method of Figure 2.10 with the second-order Adams-Bashforth method. Note the strong effect that the correction step has on the Adams-Bashforth method; the accuracy is improved somewhat but, more noticeably, the approximate solution now spirals in rather than out. Again, the step size for both methods was $h = 1$.

We now apply the second-order Adams predictor-corrector method to the trajectory problem discussed in Section 2.1. The system of ordinary differential equations used for the projectile problem is given by (2.1.15), (2.1.17), and (2.1.18) with $T = 0$ and $\dot{m} = 0$:

$$\begin{aligned} \dot{x} &= v\cos\theta, \qquad \dot{y} = v\sin\theta, \\ \dot{v} &= -\frac{1}{2m}c\rho s v^2 - g\sin\theta, \qquad \dot{\theta} = -\frac{g}{v}\cos\theta, \end{aligned} \tag{2.4.18}$$

with initial conditions

$$x(0) = 0, \qquad y(0) = 0, \qquad v(0) = v_0, \qquad \theta(0) = \theta_0. \tag{2.4.19}$$

Values for the parameters in (2.4.20) are $m = 15$ kg, $c = 0.2$, $\rho = 1.29$ kg/m, $s = 0.25$ m^2, and $g = 9.81$ m/s^2. The initial value for v is $v_0 = 50$ m/s, and two different initial angles were used: $\theta_0 = 0.6$ and 1.2 radians. Figure 2.12 is a plot of the height versus range of the projectile.

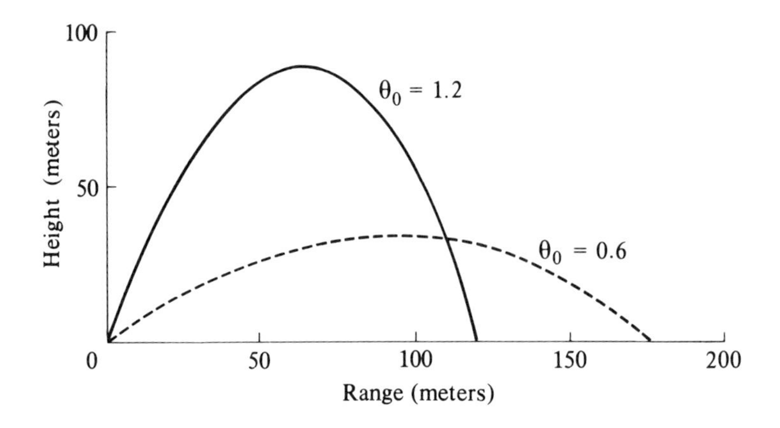

Figure 2.12: *Adams Predictor-Corrector Method* (2.4.18), (2.4.19) *Applied to the Trajectory Problem with Different Initial Angles*

Supplementary Discussion and References: 2.4

The Adams methods form the basis for a number of highly sophisticated computer codes. In these codes the Adams methods are implemented with the capability of changing not only the step size – as we discussed for the Runge-Kutta methods – but also the order of the method. For details on one particular collection of codes – ODEPACK – see Hindmarsh [1983]. For more discussion of the theory and practice of Adams methods, see, for example, Gear [1971] and Butcher [1987]. For an excellent account of how good computer codes are developed, see Shampine, Watts, and Davenport [1976]; this article makes the important point that the way methods are implemented on a computer can be more important than the intrinsic difference between methods.

Another approach to the derivation of multistep methods starts with the general linear method (2.4.17) and requires that it be exact when the solution y of the differential equation is a polynomial of degree q. This then implies that the method is order q. For example, if $q = 1$, then (2.4.17) must be exact whenever the solution is a constant; in this case the f_i all vanish since $f(x, y(x)) = y'(x) = 0$, and we are left with the condition

$$1 = \sum_{i=1}^{m} \alpha_i. \tag{2.4.20}$$

Similarly, the requirement that (2.4.17) be exact when the solution is $y(x) = x$ leads to the condition

$$m + 1 = \alpha_1 m + \alpha_2(m-1) + \cdots + \alpha_m + \sum_{i=0}^{m} \beta_i. \tag{2.4.21}$$

The relations (2.4.20) and (2.4.21) for the coefficients α_i and β_i are known as the *consistency conditions* for the multistep method and are necessary and sufficient conditions that the method be first order. One can continue this process to obtain relations on the α_i and β_i that are necessary and sufficient that the method be of any given order. For further discussions of multistep methods, see, for example, Henrici [1962] and Butcher [1987].

We can combine the one-step methods of Section 2.2 with the multistep methods of this section into the same general formulation:

$$y_{k+1} = \sum_{i=1}^{m} \alpha_i y_{k+1-i} + h\phi(x_{k+1}, \ldots, x_{k+1-m}; y_{k+1}, \ldots, y_{k+1-m}). \tag{2.4.22}$$

For one-step methods $m = 1$, and if $\alpha_1 = 1$ and ϕ is independent of x_{k+1} and y_{k+1}, then (2.4.22) reduces to the one-step method (2.2.24). On the other hand, if ϕ is the function

$$\phi = \sum_{i=0}^{m} \beta_i f(x_{k+1-i}, y_{k+1-i}),$$

then (2.4.22) reduces to the linear multistep method (2.4.17). The formulation (2.4.22) contains virtually all methods in current use.

EXERCISES 2.4

2.4.1. Verify that p_1 of (2.4.4) is the linear interpolating polynomial for (x_k, f_k) and (x_{k-1}, f_{k-1}), and that (2.4.5) follows from (2.4.3) for this p_1. Similarly, verify that (2.4.6) is the quadratic interpolating polynomial for (x_k, f_k), (x_{k-1}, f_{k-1}), and (x_{k-2}, f_{k-2}). Then carry out the integration of p_2 to verify the formula (2.4.7). Finally, give the cubic interpolating polynomial p_3 by adding the appropriate cubic term (see (2.3.13)) to p_2. Then integrate p_3 in (2.4.3) to obtain the formula (2.4.9).

2.4.2. Write a computer program to carry out the second-order Adams-Bashforth method (2.4.5). Use the second-order Runge-Kutta method to supply the missing starting value y_1. Apply your program to the problems of Exercises 2.2.2 and 2.2.8 and compare your results with the Euler and Heun methods.

2.4.3. Repeat Exercise 2.4.2 using the fourth-order Adams-Bashforth method (2.4.9).

2.4.4. Carry out in detail the derivation of the Adams-Moulton method (2.4.10). Do the same for the method (2.4.11).

2.4.5. Use as much as possible of your program of Exercise 2.4.2 to write a computer program to carry out the predictor-corrector method

$$\begin{aligned} y_{k+1}^{(p)} &= y_k + \frac{h}{2}(3f_k - f_{k-1}), \\ f_{k+1}^{(p)} &= f(x_{k+1}, y_{k+1}^{(p)}), \\ y_{k+1} &= y_k + \frac{h}{2}(f_{k+1}^{(p)} + f_k). \end{aligned}$$

Apply this to the problem $y' = -y$, $y(0) = 1$ and compare your results with the methods of Exercise 2.4.2. Similarly, write a program to carry out (2.4.13).

2.4.6. Compute the local discretization errors for the Adams-Bashforth methods (2.4.8) and (2.4.9) and show that they are third and fourth order, respectively. (Assume that the solution is sufficiently differentiable.) Do the same for the Adams-Moulton methods (2.4.10) and (2.4.11) and verify that they are second and fourth order, respectively.

2.4.7. Give the coefficients α_i and β_i in the linear multistep formulation (2.4.17) for the Adams-Bashforth and Adams-Moulton methods of second, third, and fourth order.

2.4.8. Consider the method $y_{k+1} = y_{k-1} + \frac{h}{2}(f_{k+1} + 2f_k + f_{k-1})$.

a. Find the order of the method.

b. Discuss how to apply this method to the *system* of equations $\mathbf{y}' = \mathbf{f}(x, \mathbf{y})$. What difficulties do you expect to encounter in carrying out the method?

2.4.9. Repeat the calculations of Figure 2.11 using the second-order Adams-Bashforth method and the predictor-corrector method of Exercise 2.4.5.

2.4.10. Repeat the calculations of Figure 2.12. Find the value of θ_0 such that the range of the rocket is 150 m.

2.4.11. Consider the problem of evaluating the "normal density"

$$p(x) = \frac{1}{\sqrt{2\pi}} \int_0^x e^{-t^2/2} dt + \frac{1}{2}$$

by solving the differential equation

$$p'(x) = \frac{1}{\sqrt{2\pi}} e^{-x^2/2}, \qquad p(0) = \frac{1}{2}.$$

Use the second-order Runge-Kutta method and the second-order predictor-corrector method of Exercise 2.4.5 to solve this differential equation. Compare your results.

2.5 Stability, Instability, and Stiff Equations

One of the pervading concerns of scientific computing is that of *stability*, a much overused word that tends to have somewhat different meanings depending on the context. In this section we will discuss several aspects of stability as it pertains to the numerical solution of ordinary differential equations.

Unstable Solutions

Consider the second-order differential equation

$$y'' - 10y' - 11y = 0 \tag{2.5.1}$$

with the initial conditions

$$y(0) = 1, \qquad y'(0) = -1. \tag{2.5.2}$$

The solution of (2.5.1), (2.5.2) is $y(x) = e^{-x}$, as is easily verified. Now suppose we change the first initial condition by a small quantity ε, so that the initial conditions are

$$y(0) = 1 + \varepsilon, \qquad y'(0) = -1. \tag{2.5.3}$$

Then, as is again easily verified, the solution of (2.5.1) with the initial conditions (2.5.3) is

$$y(x) = (1 + \tfrac{11}{12}\varepsilon)e^{-x} + \frac{\varepsilon}{12}e^{11x}. \tag{2.5.4}$$

Therefore for any $\varepsilon > 0$, no matter how small, the second term in (2.5.4) causes the solution to tend to infinity as $x \to \infty$. The two solutions are shown in Figure 2.13. We say that the solution $y(x) = e^{-x}$ of the problem (2.5.1), (2.5.2) is *unstable*, since arbitrarily small changes in the initial conditions can produce arbitrarily large changes in the solution as $x \to \infty$. In the parlance of numerical analysis, one would also say that this problem is *ill-conditioned*; it is extremely difficult to obtain the solution numerically because rounding and discretization error will cause the same effect as changing the initial conditions, and the approximate solution will tend to diverge to infinity (see Exercise 2.5.1).

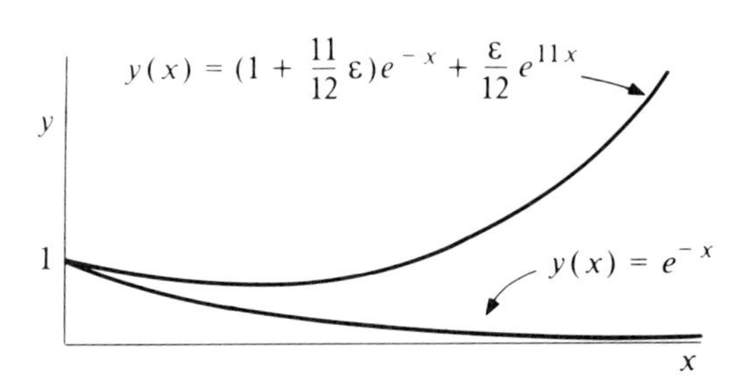

Figure 2.13: *Solutions of Slightly Different Problems*

Even more pronounced instabilities can occur with nonlinear equations. For example, the problem

$$y' = xy(y-2), \qquad y(0) = 2, \tag{2.5.5}$$

has the solution $y(x) \equiv 2$, which is unstable. To see this, note that for the initial condition $y(0) = y_0$ the solution is

$$y(x) = \frac{2y_0}{y_0 + (2 - y_0)e^{x^2}} \, .$$

Thus if $y_0 < 2$, then $y(x) \to 0$ as $x \to \infty$, and if $y_0 > 2$, the solution increases and has a singularity when $y_0 + (2 - y_0)e^{x^2} = 0$. Typical solutions are shown in Figure 2.14.

Unstable Methods

The two previous examples illustrated instabilities of solutions of the differential equation itself. We now turn to possible instabilities in the numerical method. Let us consider the method

$$y_{n+1} = y_{n-1} + 2hf_n, \tag{2.5.6}$$

which is similar to Euler's method but is a multistep method and is second-order accurate, as is easy to verify (Exercise 2.5.3).

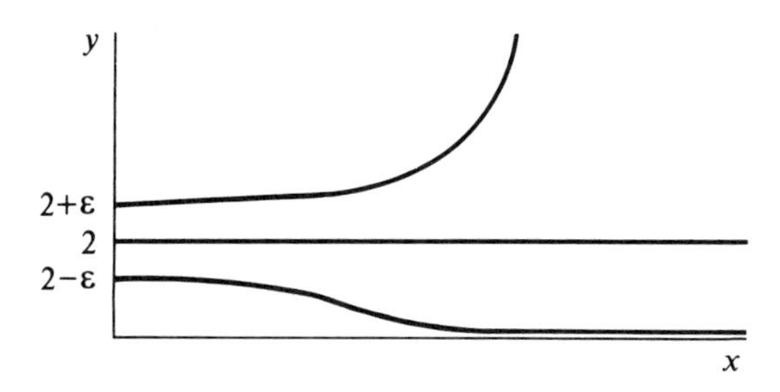

Figure 2.14: *Solutions for Three Slightly Different Initial Conditions*

We now apply (2.5.6) to the problem

$$y' = -2y + 1, \qquad y(0) = 1, \tag{2.5.7}$$

whose exact solution is

$$y(x) = \tfrac{1}{2}e^{-2x} + \tfrac{1}{2}. \tag{2.5.8}$$

This solution is stable, since if the initial condition is changed to $y(0) = 1 + \varepsilon$ the solution becomes

$$y(x) = (\tfrac{1}{2} + \varepsilon)e^{-2x} + \tfrac{1}{2},$$

and the change in (2.5.8) is only εe^{-2x}. The method (2.5.6) applied to (2.5.7) is

$$y_{n+1} = y_{n-1} + 2h(-2y_n + 1) = -4hy_n + y_{n-1} + 2h, \quad y_0 = 1, \tag{2.5.9}$$

with y_0 taken as the initial condition. However, since (2.5.6) is a two-step method, we also need to supply y_1 in order to start the process, and we will take y_1 to be the exact solution (2.5.8) at $x = h$:

$$y_1 = \tfrac{1}{2}e^{-2h} + \tfrac{1}{2}. \tag{2.5.10}$$

If, for any fixed $h > 0$, we now generate the sequence $\{y_n\}$ by (2.5.9) and (2.5.10), we will see that $|y_n| \to \infty$ as $n \to \infty$, rather than mirroring the behavior as $x \to \infty$ of the solution (2.5.8) of the differential equation. Thus the method (2.5.6), although second-order accurate, exhibits unstable behavior and we next wish to address why this is so.

Difference Equations

It is relatively easy to analyze the behavior of the sequence $\{y_n\}$ generated by (2.5.9) by viewing (2.5.9) as a *difference equation.* The theory of difference equations parallels that of differential equations, and we will sketch the basic parts of this theory in the case of *linear difference equations of order m with constant coefficients*; such an equation is of the form

$$y_{n+1} = a_m y_n + \cdots + a_1 y_{n-m+1} + a_0, \qquad n = m-1, m, m+1, \ldots, \tag{2.5.11}$$

with given constants $a_0, a_1, \ldots, a_m$. The *homogeneous part* of (2.5.11) is

$$y_{n+1} = a_m y_n + \cdots + a_1 y_{n-m+1}. \tag{2.5.12}$$

By analogy with differential equations, we attempt to find an exponential type solution of (2.5.12), only now the exponential takes the form $y_k = \lambda^k$ for some unknown constant λ. We see that $y_k = \lambda^k$ indeed is a solution of (2.5.12) provided that λ satisfies

$$\lambda^m - a_m \lambda^{m-1} - \cdots - a_1 = 0, \tag{2.5.13}$$

which is the *characteristic equation* of (2.5.12). If we assume that the m roots $\lambda_1, \ldots, \lambda_m$ of (2.5.13) are distinct, then the *fundamental solutions* of (2.5.12) are $\lambda_1^k, \ldots, \lambda_m^k$, and the general solution of (2.5.12) is

$$y_k = \sum_{i=1}^{m} c_i \lambda_i^k, \qquad k = 0, 1, \ldots, \tag{2.5.14}$$

where the c_i are arbitrary constants. A *particular solution* of (2.5.11) is given by

$$y_k = \frac{a_0}{1 - a_1 - \cdots - a_m}, \tag{2.5.15}$$

provided that the denominator is not zero, as is easily verified. Therefore the general solution of (2.5.11) is the sum of (2.5.14) and (2.5.15):

$$y_k = \sum_{i=1}^{m} c_i \lambda_i^k + \frac{a_0}{1 - a_1 - \cdots - a_m}, \qquad k = 0, 1 \ldots . \tag{2.5.16}$$

The arbitrary constants in (2.5.16) can be determined – just as for differential equations – by imposing additional conditions on the solution. In particular, suppose we are given the initial conditions

$$y_0, y_1, \ldots, y_{m-1}. \tag{2.5.17}$$

Then (2.5.16) requires that

$$\sum_{i=1}^{m} c_i \lambda_i^k + \frac{a_0}{1 - a_1 - \cdots - a_m} = y_k, \quad k = 0, 1, \ldots m-1, \tag{2.5.18}$$

which is a system of m linear equations in the m unknowns $c_1, \ldots, c_m$, and can be used to determine the c_i.

We now apply this theory to the difference equation (2.5.9), which we write in the form

$$y_{n+1} = -4hy_n + y_{n-1} + 2h, \quad y_0 = 1, \ y_1 = \tfrac{1}{2}e^{-2h} + \tfrac{1}{2}. \tag{2.5.19}$$

The characteristic equation (2.5.13) is $\lambda^2 + 4h\lambda - 1 = 0$ with roots

$$\lambda_1 = -2h + \sqrt{1+4h^2}, \quad \lambda_2 = -2h - \sqrt{1+4h^2}. \tag{2.5.20}$$

The conditions (2.5.18) then become

$$c_1 + c_2 + \tfrac{1}{2} = y_0 = 1, \qquad c_1\lambda_1 + c_2\lambda_2 + \tfrac{1}{2} = y_1 = \tfrac{1}{2}e^{-2h} + \tfrac{1}{2},$$

which can be solved for c_1 and c_2 to give

$$c_1 = \tfrac{1}{4} + \frac{y_1 - \frac{1}{2} + h}{2\sqrt{1+4h^2}}, \qquad c_2 = \tfrac{1}{4} - \frac{(y_1 - \frac{1}{2} + h)}{2\sqrt{1+4h^2}}. \tag{2.5.21}$$

Thus the solution of (2.5.19) is

$$y_n = c_1(-2h + \sqrt{1+4h^2})^n + c_2(-2h - \sqrt{1+4h^2})^n + \tfrac{1}{2}. \tag{2.5.22}$$

Although this representation of the solution is perhaps a little formidable, it allows us to see very easily the behavior of y_n as $n \to \infty$. In particular, for any fixed step size $h > 0$, it is evident that

$$0 < -2h + \sqrt{1+4h^2} < 1, \qquad 2h + \sqrt{1+4h^2} > 1.$$

Therefore the first term in (2.5.22) tends to zero, while the second tends to infinity, in an oscillatory way, as n tends to infinity. Since the exact solution (2.5.8) of the differential equation tends to $1/2$ as x tends to infinity, we see that the error in the approximate solution $\{y_n\}$ diverges to infinity, and the method (2.5.9) is unstable applied to the problem (2.5.7). Note that this divergence of the error has nothing to do with rounding error; (2.5.22) is the exact mathematical representation of y_n, and if the sequence (2.5.19) were computed in exact arithmetic it would correspond precisely with that given by (2.5.22).

Stability of Methods

From the preceding example it is clear that an important property of a method is that it be stable in some sense. The most basic definition of stability may be given in terms of the general method (2.4.22):

$$y_{n+1} = \sum_{i=1}^{m} \alpha_i y_{n+1-i} + h\phi(x_{n+1}, \ldots, x_{n+1-m}, y_{n+1}, \ldots, y_{n+1-m}). \tag{2.5.23}$$

The method (2.5.23) is *stable* provided that all roots λ_i of the polynomial

$$\rho(\lambda) \equiv \lambda^m - \alpha_1\lambda^{m-1} - \cdots - \alpha_m \tag{2.5.24}$$

satisfy $|\lambda_i| \leq 1$, and any root for which $|\lambda_i| = 1$ is simple. The method is *strongly stable* if, in addition, $m-1$ roots of (2.5.24) satisfy $|\lambda_i| < 1$. (We note that some authors use the terms *weakly stable* and *stable* in place of stable and strongly stable.)

Any method that is at least first-order accurate must satisfy the condition $\sum_{i=1}^{m}\alpha_i = 1$, so that 1 is a root of (2.5.24). In this case, a strongly stable method will then have one root of (2.5.24) equal to 1 and all the rest strictly less than 1 in absolute value. For Runge-Kutta methods, $\rho(\lambda) = \lambda - 1$ since the method is one-step; hence there are no roots of (2.5.24) besides the root $\lambda = 1$, and these methods are always strongly stable. For an m-step Adams method, $\rho(\lambda) = \lambda^m - \lambda^{m-1}$, so the other $m-1$ roots of (2.5.24) are zero, and these methods also are strongly stable.

For the method (2.5.6), the polynomial (2.5.24) is $\rho(\lambda) = \lambda^2 - 1$ with roots ± 1; hence this method is stable but not strongly stable, and it is this lack of strong stability that gives rise to the unstable behavior of the sequence $\{y_k\}$ defined by (2.5.19). The reason for this is as follows. The difference equation (2.5.19) is second order (since y_{n+1}, y_n, and y_{n-1} appear in the equation) and has two fundamental solutions, λ_1^n and λ_2^n, where λ_1 and λ_2 are the roots (2.5.20). The sequence $\{y_k\}$ generated by (2.5.19) is meant to approximate the solution of the differential equation (2.5.7), which is a first-order equation with only one fundamental solution. This fundamental solution is approximated by λ_1^n; λ_2^n is spurious and should rapidly go to zero. However, for any $h > 0$, $|\lambda_2| > 1$, and hence λ_2^n tends to infinity and not zero; it is this that causes the instability. Now, note that λ_1 and λ_2 converge to the roots of the stability polynomial (2.5.24) as $h \to 0$; indeed, this polynomial is just the limit, as $h \to 0$, of the characteristic polynomial $\lambda^2 + 4h\lambda - 1$ of (2.5.19). The idea of strong stability now becomes more evident. If all the roots except one of the stability polynomial are less than 1 in magnitude, then all but one of the roots of the characteristic equation of the method must be less than 1 for sufficiently small h; hence powers of these roots – the spurious fundamental solutions of the difference equation – tend to zero and cause no instability.

The stability theory that we have just discussed is essentially stability in the limit as $h \to 0$, and the example of instability that we gave shows what can happen for arbitrarily small h if the method is stable but not strongly stable and the interval is infinite. On a finite interval, a stable method will give accurate results for sufficiently small h. On the other hand, even strongly stable methods can exhibit unstable behavior if h is too large. Although in principle h can be taken sufficiently small to overcome this difficulty, it may be that the computing time then becomes prohibitive. This is the situation with differential equations that are known as *stiff*, and we shall conclude this section with a short discussion of such problems.

Stiff Equations

Consider the equation

$$y' = -100y + 100, \qquad y(0) = y_0. \tag{2.5.25}$$

The exact solution of this problem is

$$y(x) = (y_0 - 1)e^{-100x} + 1. \tag{2.5.26}$$

It is clear that the solution is stable, since if we change the initial condition to $y_0 + \varepsilon$ the solution changes by εe^{-100x}. Euler's method applied to (2.5.25) is

$$y_{n+1} = y_n + h(-100y_n + 100) = (1 - 100h)y_n + 100h, \tag{2.5.27}$$

and the exact solution of this first-order difference equation is

$$y_n = (y_0 - 1)(1 - 100h)^n + 1. \tag{2.5.28}$$

For concreteness, suppose that $y_0 = 2$ so that the exact solutions (2.5.26) and (2.5.28) become

$$y(x) = e^{-100x} + 1, \tag{2.5.29}$$

$$y_n = (1 - 100h)^n + 1. \tag{2.5.30}$$

Now, $y(x)$ decreases very rapidly from $y_0 = 2$ to its limiting value of 1; for example, $y(0.1) \doteq 1 + 5 \times 10^{-5}$. Initially, therefore, we expect to require a small step size h to compute the solution accurately. However, beyond, say, $x = 0.1$, the solution varies slowly and is essentially equal to 1; intuitively we would expect to obtain sufficient accuracy with Euler's method using a relatively large h. However, we see from (2.5.30) that if $h > 0.02$, then $|1 - 100h| > 1$ and the approximation y_n grows rapidly at each step and shows an unstable behavior. If we compare the exact solutions (2.5.29) and (2.5.30), we see that the particular solutions of (2.5.25) and (2.5.27) are identical (and equal to 1). The quantity $(1 - 100h)^n$ is an approximation to the exponential term e^{-100x} and is, indeed, a good approximation for small h but rapidly becomes a poor approximation as h becomes as large as 0.02. Even though this exponential term contributes virtually nothing to the solution after $x = 0.1$, Euler's method still requires that we approximate it with sufficient accuracy to maintain stability. This is the typical problem with stiff equations: the solution contains a component that contributes very little to the solution, but the usual methods require that it be approximated accurately in order to maintain stability.

This problem occurs very frequently in systems of equations. For example, consider the second-order equation

$$y'' + 101y' + 100y = 0. \tag{2.5.31}$$

As discussed in Appendix 1, we can convert (2.5.31) to an equivalent system of two first-order equations, but it is sufficient for our purposes to treat it in its second-order form. The general solution of (2.5.31) is

$$y(x) = c_1 e^{-100x} + c_2 e^{-x},$$

and if we impose the initial conditions

$$y(0) = 1.01, \qquad y'(0) = -2,$$

the solution is

$$y(x) = \tfrac{1}{100} e^{-100x} + e^{-x}. \tag{2.5.32}$$

Clearly, the first term of this solution contributes very little after x reaches a value such as $x = 0.1$. Yet we will have the same problem as in the previous example if we apply Euler's method to the first-order system corresponding to (2.5.31); that is, we will need to make the step size sufficiently small to approximate e^{-100x} accurately even though this term contributes very little to the solution. This example illustrates the essence of the problem of stiffness in systems of equations. Usually the independent variable in such problems is time and the physical problem that is being modeled has transients that decay to zero very rapidly, but the numerical scheme must cope with them even after they no longer contribute to the solution.

The general approach to the problem of stiffness is to use implicit methods. It is beyond the scope of this book to discuss this in detail, and we will only give an indication of the value of implicit methods in this context by applying one of the simplest such methods to the problem (2.5.25). For the general equation $y' = f(x, y)$, the method

$$y_{n+1} = y_n + hf(x_{n+1}, y_{n+1}) \tag{2.5.33}$$

is known as the *backward Euler method.* It is of the same form as Euler's method except that f is evaluated at (x_{n+1}, y_{n+1}) rather than at (x_n, y_n); hence the method is implicit. If we apply (2.5.33) to (2.5.25), we obtain

$$y_{n+1} = y_n + h(-100y_{n+1} + 100), \tag{2.5.34}$$

which can be put in the form

$$y_{n+1} = (1 + 100h)^{-1}(y_n + 100h). \tag{2.5.35}$$

The exact solution of the difference equation (2.5.35) is

$$y_n = (y_0 - 1)(1 + 100h)^{-n} + 1, \tag{2.5.36}$$

as is easily verified (Exercise 2.5.10). In particular, for the initial condition $y_0 = 2$, which was treated previously, (2.5.36) becomes

$$y_n = \frac{1}{(1 + 100h)^n} + 1, \tag{2.5.37}$$

and we see that there is no unstable behavior regardless of the size of h. Note that with Euler's method we are attempting to approximate the solution by a polynomial, and no polynomial (except 0) can approximate e^{-x} as $x \to \infty$. With the backward Euler's method we are approximating the solution by a rational function, and such functions can indeed go to zero as $x \to \infty$.

The backward Euler method, like Euler's method itself, is only first-order accurate, and a better choice would be the second-order Adams-Moulton method (2.4.10):

$$y_{n+1} = y_n + \frac{h}{2}[f_n + f(x_{n+1}, y_{n+1})], \tag{2.5.38}$$

which is also known as the *trapezoid rule.* The use of this method on (2.5.25) is left to Exercise 2.5.13.

The application of an implicit method to (2.5.25) was deceptively simple since the differential equation is linear and hence we could easily solve for y_{n+1} in (2.5.34). If the differential equation had been nonlinear, however, the method would have required the solution of a nonlinear equation for y_{n+1} at each step. More generally, for a system of differential equations the solution of a system of equations (linear or nonlinear, depending on the differential equations) would be needed at each step. This is costly in computer time, but the effective handling of stiff equations requires that some kind of implicitness be brought into the numerical method.

Supplementary Discussion and References: 2.5

There is a vast literature on the theory of stability of solutions of differential equations. For a readable introduction, see LaSalle and Lefschetz [1961].

We have given the basic result for linear difference equations with constant coefficients only for the case where the roots of the characteristic equation are distinct. If there are multiple roots, then polynomial terms in n enter the solution in a manner entirely analogous to that for differential equations. For more discussion of the theory of linear difference equations, see, for example, Ortega [1987].

The method (2.5.6) arises in a natural way by differentiation of an interpolating polynomial; for the derivation, see Henrici [1962, p. 219].

The basic results of the theory of stability of multistep methods were developed by G. Dahlquist in the 1950s; for a detailed treatment of this theory, see Henrici [1962]. Since then there have been a number of refined definitions of stability; in particular, the terms stiffly-stable and A-stable deal with types of stability needed for methods to handle stiff equations. For more on different definitions of stability, see, for example, Gear [1971], Lambert [1973], and Butcher [1987].

There are a number of codes available for solving stiff equations. One of the best is the VODE package of Brown, Byrne, and Hindmarsh [1989], which may also be used for problems which are not stiff. For stiff equations a so-called Backward Differention Formula (BDF) implicit method is used, which is of the general form (2.4.17) with $\beta_i = 0$ for $i > 0$.

EXERCISES 2.5

2.5.1. By letting $z = y'$, show that the problem (2.5.1), (2.5.2) is equivalent to the first-order system $y' = z$, $z' = 10z + 11y$, with initial conditions $y(0) = 1$ and $z(0) = -1$. Attempt to solve this system numerically by any of the methods of this chapter and discuss your results.

2.5.2. Attempt to solve the problem (2.5.5) numerically by any of the methods of Sections 2.2 or 2.4 and discuss your results.

2.5.3. Verify that the method (2.5.6) is second-order accurate.

2.5.4. Carry out the algorithm (2.5.9), (2.5.10) numerically for various values of h. Discuss your results.

2.5.5. Solve the difference equation $y_{n+1} = \frac{5}{2}y_n + y_{n-1}$, $y_0 = y_1 = 1$, in terms of the roots of its characteristic equation. Discuss the behavior of the sequence $\{y_n\}$ as $n \to \infty$.

2.5.6. Find a value of y_1 such that the resulting solution of (2.5.9) with $y_0 = 1$ tends to zero as n tends to infinity. Write a program to carry out (2.5.9) with y_1 given in this way as well as by (2.5.10). Discuss your results.

2.5.7. Consider the method $y_{n+1} = y_{n-3} + (4h/3)(2f_n - f_{n-1} + 2f_{n-2})$, which is known as *Milne's method.* Ascertain whether this method is stable and strongly stable.

2.5.8. Write a program to carry out Euler's method (2.5.27) for different values of h both less than and greater than 0.02. Discuss your results.

2.5.9. The system $y' = x$, $z' = -100y - 101z$ is the first-order system equivalent to the second-order equation (2.5.31). Using the initial conditions $y(0) = 2$ and $z(0) = -2$, apply Euler's method to this system and determine experimentally how small the step size h must be to maintain stability. Attempt to verify analytically your conclusion about the size of h.

2.5.10. Find the solution of the difference equation $y_{n+1} = cy_n + d$ as a function of the initial condition y_0. Apply your result to verify that (2.5.36) is the solution of (2.5.35).

2.5.11. The function $y(x) = e^{-x}$ is the solution of the problem $y'' = y$, $y(0) = 1$, $y'(0) = -1$. Is this solution stable? Prove your assertion.

2.5.12. Ascertain which of the following methods are stable and which are strongly stable, and also find their order.

a. $y_{k+1} = y_k + \frac{h}{6}(6f_k - 3f_{k-1} + 3f_{k-2})$
b. $y_{k+1} = y_{k-1} + \frac{h}{2}(f_{k+1} + 2f_k + f_{k-1})$
c. $y_{k+1} = 3y_k - 2y_{k-1} + \frac{h}{2}(f_{k+1} + 2f_k + f_{k-1})$
d. $y_{k+1} = \frac{1}{2}(y_k + y_{k-1}) + \frac{3h}{4}(3f_k - f_{k-1})$

2.5.13. Apply the trapezoid rule (2.5.38) to the equation (2.5.25).

2.5.14. Show that for any constants c_1 and c_2, $c_1 k a^{k-1} + c_2 a^k$ is a solution of the difference equation $y_{k+1} - 2ay_k + a^2 y_{k-1} = 0$.

Chapter 3

Pinning It Down: Boundary Value Problems

3.1 The Finite Difference Method for Linear Problems

In the previous section we considered initial value problems for ordinary differential equations. In many problems, however, there will be conditions on the solution given at more than one point. For a single first order equation $y' = f(x, y)$, data at one point completely determines the solution so that if conditions at more than one point are given, either higher order equations or systems of equations must be treated. Consider the second-order equation

$$v''(x) = f(x, v(x), v'(x)), \qquad 0 \le x \le 1. \tag{3.1.1}$$

Because it is second-order, such an equation requires two additional conditions, and the simplest possibility is to prescribe the solution values at the end-points:

$$v(0) = \alpha, \qquad v(1) = \beta. \tag{3.1.2}$$

Equations (3.1.1) and (3.1.2) define a *two-point boundary value* problem. We note that the restriction of the interval in (3.1.1) to [0, 1] is no loss of generality since a problem on any finite interval may be converted to one on the interval [0, 1] (Exercise 3.1.1). Alternatively, the following development is easily done directly on any finite interval (Exercise 3.1.6).

If the function f of (3.1.1) is nonlinear in either $v(x)$ or $v'(x)$, the boundary value problem is *nonlinear*. Nonlinear boundary value problems are intrinsically more difficult to solve, and we will consider them in Chapter 5. In this chapter we treat only linear problems, in which case (3.1.1) may be written in the form

$$v''(x) = b(x)v'(x) + c(x)v(x) + d(x), \qquad 0 \le x \le 1, \tag{3.1.3}$$

where b, c, and d are given functions of x. The boundary conditions that we consider first will be (3.1.2); later, we shall treat other types of boundary conditions. Equations (3.1.3) and (3.1.2) define a linear two-point boundary-value problem for the unknown function v, and our task is to develop procedures to approximate the solution. We will assume that the problem has a unique solution that is at least two times continuously differentiable.

We first consider the special case of (3.1.3) in which $b(x) \equiv 0$, so that the equation is

$$v''(x) = c(x)v(x) + d(x), \qquad 0 \le x \le 1. \tag{3.1.4}$$

We will assume that $c(x) \ge 0$ for $0 \le x \le 1$; this is a sufficient condition for the problem (3.1.4), (3.1.2) to have a unique solution. To begin the numerical solution we divide the interval $[0, 1]$ into a number of equal subintervals of length h, as shown in Figure 3.1. As in Chapter 2, in Figure 3.1 the points x_i are called the grid points, or nodes, and h is the grid spacing; x_0 and x_{n+1} are the *boundary* points, and $x_1, \ldots, x_n$ are the *interior* grid points.

Figure 3.1: *Grid Points*

We now need to approximate $v''(x)$ in (3.1.4), and we do this by *finite differences.* Let x_i be any interior grid point and approximate the first derivatives at the points $x_i \pm \frac{h}{2}$ by

$$v'(x_i - \tfrac{h}{2}) \doteq \frac{[v(x_i) - v(x_{i-1})]}{h}, \qquad v'(x_i + \tfrac{h}{2}) \doteq \frac{[v(x_{i+1}) - v(x_i)]}{h}.$$

These relations are then used to approximate the second derivative:

$$v''(x_i) \doteq \frac{v'(x_i + \frac{h}{2}) - v'(x_i - \frac{h}{2})}{h} \doteq \frac{v(x_{i+1}) - 2v(x_i) + v(x_{i-1})}{h^2}. \tag{3.1.5}$$

If we now put this approximation into equation (3.1.4) and denote the functions c and d evaluated at x_i by c_i and d_i, we obtain

$$\frac{1}{h^2}[v(x_{i+1}) - 2v(x_i) + v(x_{i-1})] - c_i v(x_i) \doteq d_i, \quad i = l, \ldots, n. \tag{3.1.6}$$

What we have shown so far is that if we replace the second derivatives of the solution v by finite difference approximations and put these approximations into the differential equation, we obtain the approximate relations (3.1.6) that the solution must satisfy. We now turn this procedure around. Suppose that we can find numbers $v_1, \ldots, v_n$ that satisfy the equations

$$\frac{1}{h^2}(v_{i+1} - 2v_i + v_{i-1}) - c_i v_i = d_i, \quad i = 1, \ldots, n, \tag{3.1.7}$$

or

$$-v_{i+1} + 2v_i - v_{i-1} + c_i h^2 v_i = -h^2 d_i, \quad i = 1, \ldots, n, \tag{3.1.8}$$

with $v_0 = \alpha$ and $v_{n+1} = \beta$. Then we can consider $v_1, \ldots, v_n$ to be approximations at the grid points $x_1, \ldots, x_n$ to the solution v of the boundary-value problem (3.1.4), (3.1.2). We shall return shortly to the question of the accuracy of these approximations.

The equations (3.1.8) form a system of n linear equations in the n unknowns $v_1, \ldots, v_n$ and can be written in matrix-vector form as

$$\begin{bmatrix} 2+c_1h^2 & -1 & & & \\ -1 & 2+c_2h^2 & \ddots & & \\ & \ddots & \ddots & & \\ & & & & -1 \\ & & & -1 & 2+c_nh^2 \end{bmatrix} \begin{bmatrix} v_1 \\ v_2 \\ \vdots \\ \\ v_n \end{bmatrix} = \begin{bmatrix} -h^2d_1+\alpha \\ -h^2d_2 \\ \vdots \\ -h^2d_{n-1} \\ -h^2d_n+\beta \end{bmatrix}. \tag{3.1.9}$$

Thus, to obtain the approximate solution $v_1, \ldots, v_n$, we need to solve this system of linear equations. Techniques for this will be discussed in the next section.

The coefficient matrix of (3.1.9) in the case that the c_i are all zero is

$$\begin{bmatrix} 2 & -1 & & & \\ -1 & 2 & \ddots & & \\ & \ddots & \ddots & & \\ & & & & -1 \\ & & & -1 & 2 \end{bmatrix}. \tag{3.1.10}$$

This is an important matrix which arises in many contexts, as we shall see. Matrices of the form (3.1.9) or (3.1.10) are called *tridiagonal* since only the three main diagonals of the matrix have non-zero elements. Tridiagonal matrices arise in a variety of applications in addition to the two-point boundary value problems of this chapter.

Discretization Error

We next consider the important question of the error in the approximations $v_1, \ldots, v_n$. Since the linear system (3.1.9) which determines these quantities will be solved numerically, the computed v_i will be in error because of rounding; this will be discussed in more detail in Chapter 4. For the present we assume that the v_i are computed with no rounding error so that $v_1, \ldots, v_n$ is the exact solution of the system (3.1.9). Let $v(x_i)$ again be the exact solution of the boundary-value problem at x_i. Then, analogous to the definition for initial-value problems in Chapter 2,

$$\max_{1 \le i \le n} |v_i - v(x_i)| \tag{3.1.11}$$

is the *(global) discretization error*. We now indicate how an analysis of the discretization error proceeds, and give results in a particular case. Again, we restrict our attention to the equation (3.1.4).

We first define the *local discretization error*, in a manner analogous to that for initial-value problems, by

$$L(x,h) = \frac{1}{h^2}[v(x+h) - 2v(x) + v(x-h)] - c(x)v(x) - d(x), \tag{3.1.12}$$

where v is the exact solution of the differential equation (3.1.4) and h is the grid spacing. By means of (3.1.4), we can replace $cv + d$ in (3.1.12) with v'' so that

$$L(x,h) = \frac{1}{h^2}[v(x+h) - 2v(x) + v(x-h)] - v''(x). \tag{3.1.13}$$

Thus the local discretization error is just the error in approximating v''. To estimate this error, we assume that v is four times continuously differentiable and expand $v(x+h)$ and $v(x-h)$ in Taylor series. After we collect terms (the details of which are left to Exercise 3.1.2), we obtain

$$L(x,h) = \frac{1}{12}v^{(4)}(x)h^2 + 0(h^4) = 0(h^2). \tag{3.1.14}$$

The problem now is to relate this local discretization error to the global error (3.1.11). To do this we evaluate (3.1.12) at the grid points x_i, set $\sigma_i = L(x_i, h)$, and then subtract (3.1.7) from (3.1.12). Setting $e_i = v(x_i) - v_i$, this gives

$$\sigma_i = \frac{1}{h^2}[e_{i+1} - 2e_i + e_{i-1}] - c_i e_i, \quad i = 1, \cdots, n,$$

or

$$(2 + c_i h^2)e_i - e_{i+1} - e_{i-1} = -h^2 \sigma_i, \quad i = l, \ldots, n, \tag{3.1.15}$$

where $e_0 = e_{n+1} = 0$. If A is the coefficient matrix of (3.1.9) and $\mathbf{e}$ and $\boldsymbol{\sigma}$ are vectors with components $e_1, \ldots, e_n$ and $\sigma_1, \ldots, \sigma_n$, we can write (3.1.15) as

$$A\mathbf{e} = -h^2 \boldsymbol{\sigma}, \tag{3.1.16}$$

or, assuming that A^{-1} exists,

$$\mathbf{e} = -h^2 A^{-1}\boldsymbol{\sigma}. \tag{3.1.17}$$

This is the basic relationship between the global and local discretization errors. Note that the global discretization errors $e_1, \ldots, e_n$ and the approximate solutions $v_1, \ldots, v_n$ satisfy systems of equations with exactly the same coefficient matrix, but different right hand sides.

The problem now is to study the behavior of A^{-1} as $h \to 0$. This is made more difficult by the fact that n, the order of A, tends to infinity as $h \to 0$. It is beyond the scope of this book to pursue this problem in any generality, but we can give a relatively simple analysis of the discretization error in the case where

$$c(x) \geq \gamma > 0, \qquad x \in [0, 1], \tag{3.1.18}$$

so that $c_i \geq \gamma$, $i = 1, \ldots, n$. If we set $e = \max |e_i|$ and $\sigma = \max |\sigma_i| = 0(h^2)$, we obtain from (3.1.15) and (3.1.18) that

$$(2 + \gamma h^2)|e_i| \leq 2e + h^2\sigma, \qquad i = 1, \ldots, n. \tag{3.1.19}$$

Since (3.1.19) holds for all i, we must have

$$(2 + \gamma h^2)e \leq 2e + h^2\sigma.$$

Therefore, since by (3.1.14) $\sigma = 0(h^2)$, we conclude that

$$e \leq \frac{\sigma}{\gamma} = 0(h^2), \tag{3.1.20}$$

which shows that the global discretization error is $0(h^2)$ provided that the local discretization is $0(h^2)$. It can be shown that the same result holds more generally, particularly for the important special case in which $c(x) \equiv 0$, but a more difficult analysis is required.

More General Equations

We next consider the more general equation (3.1.3), in which v' is present. The standard centered difference approximation to $v'(x)$ is

$$v'(x) \doteq \frac{1}{2h}[v(x+h) - v(x-h)], \tag{3.1.21}$$

and it is easy to show (Exercise 3.1.3) that the error in this approximation is $0(h^2)$. If we replace $v'(x)$ in (3.1.3) at the grid points and proceed as before, the equations corresponding to (3.1.8) now become

$$\left(-1 - \frac{b_i h}{2}\right) v_{i-1} + (2 + c_i h^2)v_i + \left(-1 + \frac{b_i h}{2}\right) v_{i+1} = -h^2 d_i, \quad i = 1, \ldots, n,$$

or

$$r_i v_{i-1} + p_i v_i + q_i v_{i+1} = -h^2 d_i, \quad i = 1, \dots, n,$$

where we have set

$$p_i = 2 + c_i h^2, \qquad q_i = -1 + \frac{b_i h}{2}, \qquad r_i = -1 - \frac{b_i h}{2}. \tag{3.1.22}$$

Then we can write these equations in matrix form as

$$\begin{bmatrix} p_1 & q_1 & & & \\ r_2 & p_2 & q_2 & & \\ & \ddots & \ddots & \ddots & \\ & & & & q_{n-1} \\ & & & r_n & p_n \end{bmatrix} \begin{bmatrix} v_1 \\ \\ \vdots \\ \\ v_n \end{bmatrix} = -h^2 \begin{bmatrix} d_1 + r_1\alpha/h^2 \\ d_2 \\ \vdots \\ d_{n-1} \\ d_n + q_n\beta/h^2 \end{bmatrix}. \tag{3.1.23}$$

Diagonal Dominance

One desirable property of the coefficient matrix of (3.1.23) is diagonal dominance. A general $n \times n$ matrix $A = (a_{ij})$ is *row diagonally dominant* if

$$|a_{ii}| \geq \sum_{j \neq i} |a_{ij}|, \quad i = 1, \dots, n; \tag{3.1.24}$$

that is, the absolute value of the diagonal element of each row is at least as large as the sum of the absolute values of all the off-diagonal elements in that row. The matrix is *column diagonally dominant* if

$$|a_{ii}| \geq \sum_{j \neq i} |a_{ji}|, \quad i = 1, \dots, n. \tag{3.1.25}$$

Diagonal dominance is important for a number of reasons, one of which is that it is an approach to showing non-singularity of the matrix. Diagonal dominance by itself is not sufficient for this; for example, a 2×2 matrix with all elements equal to 1 is diagonally dominant but singular. However, the following strengthening of diagonal dominance guarantees non-singularity. The matrix A is *strictly* row (column) diagonally dominant if strict inequality holds in (3.1.24) for all i (or (3.1.25) for column). We then have the following result.

THEOREM 3.1.1. *If the $n \times n$ matrix A is strictly row or column diagonally dominant, then it is non-singular.*

To prove this theorem, assume that A is singular so that $A\mathbf{x} = 0$ for some non-zero $\mathbf{x}$ (see Theorem A.2.1 in Appendix 2). Let $|x_k| = \max\{|x_i| : i = 1, \dots, n\}$. The kth equation of $A\mathbf{x} = 0$ is

$$a_{kk} x_k = -\sum_{j \neq k} a_{kj} x_j$$

so that

$$|a_{kk}||x_k| = |a_{kk}x_k| = |\sum_{j\neq k} a_{kj}x_j| \leq |x_k| \sum_{j\neq k} |a_{kj}|.$$

By assumption, $|x_k| > 0$ so it can be divided out and the resulting inequality contradicts strict row diagonal dominance. This contradiction shows that A is non-singular. If A is strictly column diagonally dominant, then A^T, the transpose of A, is strictly row diagonally dominant, so it is non-singular. Hence A is non-singular since $\det A = \det A^T$. This completes the proof.

For the matrix of (3.1.23) to be row diagonally dominant we need that

$$|p_i| \geq |r_i| + |q_i|, \qquad i = 1, \ldots, n,$$

or, using (3.1.22),

$$|2 + c_i h^2| \geq |1 + \frac{b_i h}{2}| + |1 - \frac{b_i h}{2}|, \qquad i = 1, \ldots, n. \tag{3.1.26}$$

If we assume that $c_i \geq 0$, then (3.1.26) holds if h is sufficiently small. In particular, if

$$|b_i h| \leq 2, \qquad i = 1, \ldots, n, \tag{3.1.27}$$

then the absolute values of the quantities on the right side of (3.1.26) are the quantities themselves, so that

$$|2 + c_i h^2| \geq 1 + \frac{b_i h}{2} + 1 - \frac{b_i h}{2} = 2.$$

The condition (3.1.27) on h, which also ensures column diagonal dominance, is a rather stringent one and can be avoided by using one-sided differences in place of the central difference (3.1.21) to approximate the first derivative. More precisely, we use the approximations

$$v'(x_i) \doteq \begin{cases} \frac{1}{h}(v_{i+1} - v_i) & \text{if} \quad b_i < 0, \\ \frac{1}{h}(v_i - v_{i-1}) & \text{if} \quad b_i \geq 0, \end{cases} \tag{3.1.28}$$

so that the direction of the one-sided difference is determined by the sign of b_i. Such differences are quite commonly used in fluid dynamics problems and in that context are called *upwind* (or *upstream*) differences. With (3.1.28) the ith row of the coefficient matrix (3.1.23) becomes

$$\begin{array}{llll} -1 & 2 + c_i h^2 - b_i h & -1 + b_i h, & b_i \leq 0, \\ -(1 + b_i h) & 2 + c_i h^2 + b_i h & -1, & b_i > 0, \end{array} \tag{3.1.29}$$

and it is easy to verify that row diagonal dominance holds, independent of the size of h, assuming again that $c_i \geq 0$. We note, however, that although the centered difference approximation (3.1.21) is second-order accurate, the one-sided approximations (3.1.28) are only first-order accurate, and this increase in the discretization error must be weighed against the better properties of the coefficient matrix.

Symmetry

Another desirable property of the coefficient matrix of (3.1.23) is symmetry. A matrix $A = (a_{ij})$ is *symmetric* if $A = A^T$; that is, if

$$a_{ij} = a_{ji}, \quad i, j = 1, \ldots, n. \tag{3.1.30}$$

The matrix of (3.1.9) is symmetric, but for the matrix of (3.1.23) to be symmetric we need that $q_i = r_{i+1}$, or using (3.1.22),

$$-1 + \tfrac{1}{2}b_i h = -1 - \tfrac{1}{2}b_{i+1}h, \quad i = 1, \ldots, n-1.$$

Clearly, these relations will not usually hold. In many situations, however, the differential equation is of the form

$$[a(x)v']' = d(x). \tag{3.1.31}$$

In this case we can obtain a symmetric coefficient matrix by "symmetric differencing," as described in the Supplementary Discussion.

Other Boundary Conditions

The previous discussion has used the boundary conditions (3.1.2). In many problems, however, boundary conditions on the derivative rather than the function itself may be given, and we now consider the modifications that this requires.

Suppose, for example, that we have the boundary conditions

$$v'(0) = \alpha \qquad v'(1) = \beta \tag{3.1.32}$$

in place of (3.1.2); that is, we specify derivative values rather than function values. Consider now the difference equations (3.1.8). In the equation for $i = 1$, the value of v_0 is no longer known from the boundary condition at $x = 0$. Instead v_0 will be an additional unknown and we will need another equation that can be derived as follows. We approximate v'' at $x = 0$ by

$$v''(0) \doteq \frac{1}{h^2}[v_{-1} - 2v_0 + v_1], \tag{3.1.33}$$

using a grid point $-h$ outside the interval. Then by the boundary condition

$$\alpha = v'(0) \doteq \frac{1}{2h}[v_1 - v_{-1}] \tag{3.1.34}$$

we have $v_{-1} \doteq v_1 - 2\alpha h$, and we can use this to eliminate v_{-1} in (3.1.33):

$$v''(0) \doteq \frac{1}{h^2}[2v_1 - 2v_0 - 2\alpha h]. \tag{3.1.35}$$

In general this approximation is only first-order accurate, but in the important special case that $\alpha = 0$ it is second-order (Exercise 3.1.7). We can now use (3.1.35) to obtain an additional equation and we have $n+1$ equations in the $n+1$ unknowns $v_0, \ldots, v_n$. If $v(1) = \beta$ is specified as before, then this is the system of equations to be solved. If $v'(1)$ is specified, as in (3.1.32), then another equation would arise from the approximation

$$v''(1) \doteq \frac{1}{h^2}[2v_n - 2v_{n+1} + 2\beta h], \tag{3.1.36}$$

and v_{n+1} would be an additional unknown.

If only the function value is specified, as in (3.1.2), the boundary conditions are called *Dirichlet*, whereas if only the derivatives are specified, as in (3.1.32), the boundary conditions are called *Neumann*. More generally, mixed boundary conditions may be given as linear combinations of both function and derivative values at both end points:

$$\eta_1 v(0) + \eta_2 v'(0) = \alpha, \qquad \gamma_1 v(1) + \gamma_2 v'(1) = \beta.$$

In this case approximations analogous to those discussed previously would be used.

We return to the boundary conditions (3.1.32), and consider the differential equation (3.1.4). The approximations (3.1.35) and (3.1.36) give the two equations

$$(2 + c_0 h^2)v_0 - 2v_1 = 2\alpha h - h^2 d_0,$$
$$(2 + c_{n+1}h^2)v_{n+1} - 2v_n = 2\beta h - h^2 d_{n+1},$$

which are added to the system (3.1.8) to give $n+2$ equations in the $n+2$ unknowns $v_0, \ldots, v_{n+1}$. The coefficient matrix of this system is

$$A = \begin{bmatrix} 2 + c_0 h^2 & -2 & & & \\ -1 & 2 + c_1 h^2 & -1 & & \\ & \ddots & \ddots & \ddots & \\ & & -1 & & -1 \\ & & & -2 & 2 + c_n h^2 \end{bmatrix}. \tag{3.1.37}$$

This matrix is no longer symmetric, but it can be easily symmetrized if desired (Exercise 3.1.8). If $c_i > 0$, $i = 0, \ldots, n+1$, Theorem 3.1.1 shows that A is non-singular. But if the c_i are all zero it is singular, since $A\mathbf{e} = 0$, where

$\mathbf{e} = (1, 1, \ldots, 1)^T$. This singularity of A mirrors the non-uniqueness of solutions of the differential equation itself: if $c(x) \equiv 0$ in (3.1.4) and v is a solution satisfying the boundary conditions (3.1.32), then $v + \gamma$ is also a solution for any constant γ.

Another type of boundary condition that leads to non-unique solutions is called *periodic*:

$$v(0) = v(1). \tag{3.1.38}$$

In this case v_0 and v_{n+1} are both unknowns but since they are equal only one needs to be added to the system of equations. Again, we can use (3.1.33) as an additional equation. By (3.1.38), we can assume that the solution v is extended periodically outside $[0, 1]$; in particular, we can take $v_{-1} = v_n$ in (3.1.33) so that it becomes

$$v''(0) \doteq \frac{1}{h^2}[v_n - 2v_0 + v_1]. \tag{3.1.39}$$

Similarly, in the approximation at x_n we can take $v_{n+1} = v_0$:

$$v''(x_n) \doteq \frac{1}{h^2}[v_{n-1} - 2v_n + v_{n+1}] = \frac{1}{h^2}[v_{n-1} - 2v_n + v_0]. \tag{3.1.40}$$

There will then be $n + 1$ unknowns $v_0, \ldots, v_n$, and the coefficient matrix for the problem (3.1.4) can be obtained, using (3.1.39) and (3.1.40), in the form

$$A = \begin{bmatrix} 2 + c_0h^2 & -1 & & -1 \\ -1 & 2 + c_1h^2 & \ddots & \\ & \ddots & \ddots & -1 \\ -1 & & -1 & 2 + c_nh^2 \end{bmatrix}. \tag{3.1.41}$$

The tridiagonal structure has now been lost because of the outlying -1's. As with (3.1.37), it follows from Theorem 3.1.1 that the matrix A of (3.1.41) is non-singular if all c_i are positive, but singular if all c_i are zero. (As before, $A\mathbf{e} = 0$.) This singularity again reflects the differential equation, since if $c(x) \equiv 0$ in (3.1.4) and v is a solution satisfying (3.1.38), then $v + \gamma$ is also a solution for any constant.

Whatever difference approximation or boundary conditions are involved, the basic computational problem is to solve the resulting system of linear equations. We will address this question in the next section for the particular systems arising from two-point boundary value problems. The next chapter will consider linear systems in more generality.

Supplementary Discussion and References: 3.1

We have discussed in the text only the simplest differencing procedures for rather simple problems. As we saw, the central difference approximation (3.1.5) gives rise to discretization error proportional to h^2. It is frequently desirable to use more accurate methods. One approach is to use higher-order approximations; for example,

$$v''(x_i) \doteq \frac{1}{12h^2}(-v_{i-2} + 16v_{i-1} - 30v_i + 16v_{i-1} - v_{i+2}) \tag{3.1.42}$$

is a fourth-order approximation (the error is proportional to h^4), provided that v is sufficiently differentiable. One difficulty with applying approximations of this type to two-point boundary-value problems occurs near the boundary. For example, if we apply the approximation at the first interior grid point x_1, it requires values of v not only at x_0 but also at x_{-1}, which is outside the interval. However, second-order approximations can be applied at x_1 and x_{n-1} and still retain fourth-order accuracy of the solution. This was proved by Shoosmith [1973] for various second order approximations, depending on (3.1.3), using techniques developed by Bramble and Hubbard [1964] for partial differential equations.

Another approach to obtaining higher-order approximations to the solution, while using only second-order approximations to the derivatives is Richardson extrapolation. This was discussed in Section 2.2 for initial value problems and applies also to boundary value problems. Let x^* be some fixed point in the interval $[a, b]$ and let $v(x^*; h)$ denote the approximate solution at x^* as a function of h. Assume that

$$v(x^*; h) = v(x^*) + c_2h^2 + c_3h^3 \cdots + c_ph^p + 0(h^{p+1}) \tag{3.1.43}$$

where $v(x^*)$ is the solution of the differential equation at x^*, and the c_i are functions of x^* and the solution. Note that the expansion (3.1.43) starts with h^2 since we are assuming that the method is second order accurate. If $v(x^*; \frac{h}{2})$ is the approximate solution obtained with step length $\frac{h}{2}$, then

$$\bar{v}(x^*; h) \equiv \frac{1}{3}[4v(x^*; \frac{h}{2}) - v(x^*; h)] = d_3h^3 + \cdots + 0(h^{p+1}),$$

so that this new approximation is third-order accurate. In certain cases, such as when the first derivative is absent in (3.1.3), the error expansion (3.1.43) will contain only even powers of h, and one application of the extrapolation principle will then give fourth-order accuracy and is particularly effective.

Still another approach to higher-order accuracy is *deferred correction.* In this procedure one computes an approximation to the local discretization error by means of the current approximate solution, and then uses this to obtain a new approximate solution with $0(h^4)$ accuracy. The process can then be repeated to obtain still higher order approximations. For further discussion of this method, see, for example, Ascher et al. [1988].

We next give more details on "symmetric differencing" for the "self-adjoint" equation (3.1.31). As in the derivation of (3.1.5), we use the auxiliary grid points $x_i \pm \frac{h}{2}$ and approximate the outermost derivative of $(av')'$ at x_i by

$$[a(x)v'(x)]'_i \doteq \frac{1}{h}(a_{i+\frac{1}{2}}v'_{i+\frac{1}{2}} - a_{i-\frac{1}{2}}v'_{i-\frac{1}{2}}), \tag{3.1.44}$$

where the subscripts indicate the grid point at which the evaluation is done. We approximate the first derivatives by the centered differences

$$v'_{i+\frac{1}{2}} \doteq \frac{1}{h}(v_{i+1} - v_i) \qquad v'_{i-\frac{1}{2}} \doteq \frac{1}{h}(v_i - v_{i-1})$$

and use these in (3.1.44) to obtain

$$\begin{aligned} [a(x)v'(x)]'_i &\doteq \frac{1}{h^2}[a_{i+\frac{1}{2}}(v_{i+1} - v_i) - a_{i-\frac{1}{2}}(v_i - v_{i-1})] \\ &= \frac{1}{h^2}[a_{i-\frac{1}{2}}v_{i-1} - (a_{i+\frac{1}{2}} + a_{i-\frac{1}{2}})v_i + a_{i+\frac{1}{2}}v_{i+1}]. \end{aligned} \tag{3.1.45}$$

Therefore the system of difference equations corresponding to (3.1.31) is of the form $A\mathbf{v} = \mathbf{q}$ where

$$A = \begin{bmatrix} a_{3/2} + a_{1/2} & -a_{3/2} & & & \\ -a_{3/2} & a_{5/2} + a_{3/2} & -a_{5/2} & & \\ & \ddots & \ddots & \ddots & \\ & & & & -a_{n-1/2} \\ & & & -a_{n-1/2} & a_{n-1/2} + a_{n+1/2} \end{bmatrix}. \tag{3.1.46}$$

Clearly, this matrix is both symmetric and diagonally dominant. Note also that if $a(x) \equiv 1$, (3.1.46) reduces to the $(2, -1)$ matrix (3.1.10). If a term $c(x)v$ is also present in (3.1.31), this will just add $c_i h^2$ to the ith diagonal element of (3.1.46).

A matrix A is *reducible* if there is a permutation matrix P (see Section 4.3) so that PAP^T has the form

$$PAP^T = \begin{bmatrix} A_1 & A_2 \\ 0 & A_3 \end{bmatrix},$$

where A_1 and A_3 are square submatrices. If no such permutation matrix can be found, then A is *irreducible*. It can be shown that any tridiagonal matrix whose off-diagonal elements are non-zero is irreducible; in particular, the matrix (3.1.10) is irreducible. A matrix that is irreducible and row diagonally dominant is *irreducibly diagonally dominant* if strict inequality holds in (3.1.24) for at least one i. The proof of Theorem 3.1.1 can be extended to show that irreducibly diagonally dominant matrices are non-singular; in particular, the matrix (3.1.10) is non-singular.

For further reading on two-point boundary value problems, see Ortega [1990] and the excellent books by Ascher et al. [1988] and Keller [1968].

EXERCISES 3.1

3.1.1. Consider the boundary value problem $u''(z) = g(z, u(z), u'(z))$ with $u(a) = \alpha$, $u(b) = \beta$. By the change of variable $t = (b-a)x + a$, show that this problem is equivalent to (3.1.1) and (3.1.2) with $v(x) = u((b-a)x + a)$ and $f(x, v(x), v'(x)) = (b-a)^2 g((b-a)x + a, v(x), (b-a)^{-1}v'(x))$. Specialize this result to the linear problem (3.1.3).

3.1.2. Assume that the function $v(x)$ is suitably differentiable. By expanding $v(x+h)$ and $v(x-h)$ in Taylor series, verify that (3.1.14) holds.

3.1.3 Assume that the function $v(x)$ is twice differentiable. Show that the error in the centered difference approximation (3.1.21) is proportional to h^2.

3.1.4. Consider the two-point boundary-value problem

$$v'' + 2xv' - x^2 v = x^2, v(0) = 1, v(1) = 0.$$

a. Let $h = \frac{1}{4}$ and explicitly write out the difference equations (3.1.23).

b. Repeat part **a** using the one-sided approximations (3.1.28) for v'.

c. Repeat parts **a** and **b** for the boundary conditions $v'(0) = 1$, $v(1) = 0$, and then $v'(0) + v(0) = 1$, $v'(1) + \frac{1}{2}v(1) = 0$.

3.1.5. Assume that the function a of (3.1.31) is twice differentiable. By the change of variable $w(x) = \sqrt{a(x)}v(x)$, show that (3.1.31) can be replaced by an equation of the form $w''(x) = c(x)w(x) + f(x)$. Give an expression for $c(x)$ and show how it can be approximated numerically.

3.1.6. Suppose the equation (3.1.4) is defined on the interval $[\delta, \gamma]$. Show that the equations (3.1.8) are still correct provided that $h = (\gamma - \delta/(n+1))$ and v_i is interpreted as the approximate solution at $x_i = \delta + ih$.

3.1.7. Assuming that v is sufficiently differentiable, show that (3.1.35) is, in general, only a first-order approximation to $v''(0)$. However, if $\alpha = 0$, argue that it is second-order by assuming that v has been extended outside the interval $[0, 1]$ so that $v(-h) = v(h)$.

3.1.8. Let A and B be the tridiagonal matrices

$$A = \begin{bmatrix} a_1 & b_1 & & \\ c_1 & \ddots & \ddots & \\ & \ddots & & b_{n-1} \\ & & c_{n-1} & a_n \end{bmatrix}, \qquad B = \begin{bmatrix} a_1 & \gamma_1 & & \\ \gamma_1 & \ddots & \ddots & \\ & \ddots & & \gamma_{n-1} \\ & & \gamma_{n-1} & a_n \end{bmatrix},$$

where $b_j c_j > 0$ and $\gamma_i = \sqrt{b_i c_i}$, $i = 1, \ldots, n-1$. Show that if

$$D = \mathrm{diag}\left(1, \frac{b_1}{c_1}, \frac{b_1 b_2}{c_1 c_2}, \ldots, \frac{b_1 \cdots b_{n-1}}{c_1 \cdots c_{n-1}}\right)$$

then $B = D^{1/2} A D^{-1/2}$, where $D^{1/2} = \mathrm{diag}(d_1^{1/2}, d_2^{1/2}, \ldots, d_n^{1/2})$.

3.1.9. Derive the approximations (3.1.5) and (3.1.21) by means of interpolation polynomials: let l be the polynomial of degree 1 that satisfies $l(x \pm h) = v(x \pm h)$, and show that $l'(x)$ is the approximation (3.1.21). Then let q be the quadratic polynomial that satisfies $q(x) = v(x)$ and $q(x \pm h) = v(x \pm h)$, and show that $q''(x)$ gives the approximation (3.1.5).

3.1.10. Assume that the b_i in (3.1.22) are all equal to a constant b. Give a condition that ensures that the matrix of (3.1.23) can be symmetrized as in Exercise 3.1.8. Also, if the $c_i = 0$ in (3.1.22), derive an expansion for the quantities d_n of Exercise 3.1.8 as $n \to \infty$.

3.1.11. Show that the matrix (3.1.46) can be written as $A = E^T D E$, where E is the $(n+1) \times n$ matrix

$$E = \begin{bmatrix} 1 & & & \\ -1 & 1 & & \\ & & \ddots & \\ & & \ddots & 1 \\ & & & -1 \end{bmatrix}.$$

3.1.12. Let $A = I - B$, where $\|B\|_\infty < 1$ (See Appendix 2). Modify the proof of Theorem 3.1.1 to show that A is nonsingular.

3.2 Solution of the Discretized Problem

In the previous section we saw that the use of finite difference discretization of the two-point boundary value problem (3.1.3) led to a system of linear equations. The exact form of this system depends on the boundary conditions, but in all the cases we considered, except periodic boundary conditions, the system was of the tridiagonal form

$$\begin{bmatrix} a_{11} & a_{12} & & & \\ a_{21} & a_{22} & a_{23} & & \\ & a_{32} & \ddots & \ddots & \\ & & \ddots & & a_{n-1,n} \\ & & & a_{n,n-1} & a_{nn} \end{bmatrix} \begin{bmatrix} v_1 \\ \vdots \\ v_n \end{bmatrix} = \begin{bmatrix} d_1 \\ \vdots \\ d_n \end{bmatrix}. \tag{3.2.1}$$

Gaussian Elimination

We will solve the system (3.2.1) by the Gaussian elimination method. This method, along with several variants, will be considered in detail for general linear systems in the next chapter. For the tridiagonal system (3.2.1) it turns out to be particularly simple. We consider first an example for $n = 3$ in which the equations are

$$\begin{aligned} 2v_1 - v_2 \quad &= 1 \\ -v_1 + 2v_2 - v_3 &= 2 \\ -v_2 + 2v_3 &= 3\,. \end{aligned} \tag{3.2.2}$$

Multiply the first equation by $\frac{1}{2}$ and add to the second equation. This eliminates the coefficient of v_1 in the second equation, which becomes

$$\frac{3}{2}v_2 - v_3 = \frac{5}{2}. \tag{3.2.3}$$

Now multiply this equation by $\frac{2}{3}$ and add to the original third equation. This eliminates the coefficient of v_2 in the third equation and gives the modified equation

$$\frac{4}{3}v_3 = \frac{14}{3},$$

in which v_3 is the only unknown. Thus $v_3 = \frac{7}{2}$. Then (3.2.3) gives $v_2 = 4$, and the first equation of (3.2.2) gives $v_1 = \frac{5}{2}$.

The above example illustrates the main ideas of Gaussian elimination for the general tridiagonal system (3.2.1), which we write out in equation form as

$$\begin{aligned} a_{11}v_1 + a_{12}v_2 \qquad\qquad &= d_1 \\ a_{21}v_1 + a_{22}v_2 + a_{23}v_3 \qquad &= d_2 \\ a_{32}v_2 + a_{33}v_3 + a_{34}v_4 &= d_3 \\ \ddots \qquad\qquad &\ \ \vdots \\ a_{nn-1}v_{n-1} + a_{nn}v_n &= d_n. \end{aligned} \tag{3.2.4}$$

We assume that $a_{11} \neq 0$, multiply the first equation by a_{21}/a_{11}, and subtract it from the second. This eliminates the coefficient of v_1 in the second equation so that the last $n-1$ equations are

$$\begin{aligned}
a'_{22}v_2 + a_{23}v_3 \qquad\qquad\qquad\qquad &= d'_2 \\
a_{32}v_2 + a_{33}v_3 \quad + a_{34}v_4 \qquad\qquad &= d_3 \\
\ddots \qquad\qquad\qquad\qquad &\ \ \vdots \\
a_{nn-1}v_{n-1} + a_{nn}v_n &= d_n,
\end{aligned}$$

where the primes indicate that the elements have been changed by the first row operations. This system is again tridiagonal, and we repeat the process: multiply the first row by a_{32}/a'_{22} and subtract from the next row to eliminate the coefficient of v_2. Note that in this step, as in all subsequent ones, we must assume that the divisor element is non-zero. We will consider this assumption in much more detail in the next chapter.

At the end of the above process, the system has been transformed to

$$\begin{aligned}
a_{11}v_1 + a_{12}v_2 \qquad\qquad\qquad\qquad\qquad &= d_1 \\
a'_{22}v_2 + a_{23}v_3 \qquad\qquad\qquad &= d'_2 \\
\ddots \qquad\qquad\qquad &\ \ \vdots \\
a'_{n-1,n-1}v_{n-1} + a_{n-1,n}v_n &= d'_{n-1} \\
a'_{n,n}v_n &= d'_n.
\end{aligned} \tag{3.2.5}$$

Since the last equation contains only a single unknown, it can be solved. Then with v_n computed, the next to last equation can be solved for v_{n-1}, and so on, working our way back up the equations. This process is known as the *back substitution*, at the end of which we have computed the solution of the system (3.2.4).

As we shall see in the next chapter, Gaussian elimination for a general linear system of n equations requires approximately $\frac{2}{3}n^3$ arithmetic operations. Because of the simple form of tridiagonal systems, far fewer operations are needed. To reduce the original system (3.2.4) to (3.2.5) requires only $2(n-1)$ additions, $2(n-1)$ multiplications, and $n-1$ divisions, while the back substitution requires $n-1$ additions, $n-1$ multiplications, and n divisions. Hence the total operation count to solve a tridiagonal system is

$$3(n-1) \text{ additions } + \ 3(n-1) \text{ multiplications } + \ (2n-1) \text{ divisions.}$$

Because of this relatively small operation count, very large systems can be solved very rapidly. Suppose, for example, that multiplication and addition each require $1\mu s$ ($=10^{-6}$ seconds) and division requires $5\mu s$. Then the arithmetic operation time for $n = 10,000$ is approximately

$$(60,000 + 20,000 \times 5)\mu s = 0.16 \text{ seconds.}$$

A Numerical Example

To illustrate the methods of this chapter we solve a particular two-point boundary value problem numerically. The equation is

$$[(1+x^2)v']' = 2 + 6x^2 + 2x\cos x - (1+x^2)\sin x, \tag{3.2.6}$$

with the boundary conditions

$$v(0) = 1, \qquad v(1) = 2 + \sin 1. \tag{3.2.7}$$

To evaluate the accuracy of the methods we have chosen the right hand side of (3.2.6) so that the exact solution is known:

$$v(x) = x^2 + \sin x + 1. \tag{3.2.8}$$

The equation (3.2.6) is of the form (3.1.31), but we can carry out the differentiation to write it as

$$(1+x^2)v'' + 2xv' = d(x), \tag{3.2.9}$$

where

$$d(x) = 2 + 6x^2 + 2x\cos x - (1+x^2)\sin x. \tag{3.2.10}$$

We could divide (3.2.9) through by $1+x^2$ to obtain the form (3.1.3). Alternatively, we can modify the difference equations by multiplying the approximation of v'' by $1+x^2$; that is,

$$(1+x^2)v''\Big|_{x_i} \doteq \frac{(1+x_i^2)}{h^2}(v_{i+1} - 2v_i + v_{i-1}). \tag{3.2.11}$$

Using (3.2.11) and central differences to approximate v', the difference equations have the matrix form (3.1.23), where if $a_i = 1 + x_i^2$

$$p_i = 2a_i, \quad q_i = -a_i - x_i h, \quad r_i = -a_i + x_i h. \tag{3.2.12}$$

Since $x_i = ih$, and $a_i = 1 + (ih)^2$, the coefficient matrix is then

$$\begin{bmatrix} 2+2h^2 & -1-2h^2 & & \\ -1-2h^2 & 2+8h^2 & -1-6h^2 & \\ \ddots & \ddots & \ddots & \\ & & & -1-n(n-1)h^2 \\ & & -1-n(n-1)h^2 & 2+2n^2h^2 \end{bmatrix}, \tag{3.2.13}$$

and the right hand side is

$$-h^2\left\{d_1 - \frac{1}{h^2}, d_2, \ldots, d_{n-1}, d_n - \left[n(n+1) + \frac{1}{h^2}\right](2+\sin 1)\right\}^T. \tag{3.2.14}$$

On the other hand, if we use the one-sided differences (3.1.28) to approximate v', the coefficient matrix and the right hand side are

$$\begin{bmatrix} 2+4h^2 & -1-3h^2 & & & \\ -1-4h^2 & 2+12h^2 & -1-8h^2 & & \\ & \ddots & \ddots & \ddots & \\ & & & & -1-(n-1)^2h^2 \\ & & & -1-n^2h^2 & 2+2n(n+1)h^2 \end{bmatrix} \tag{3.2.15}$$

$$-h^2\left[d_1-\left(1+\frac{1}{h^2}\right), d_2, \ldots, d_n-\left(n^2+2n+\frac{1}{h^2}\right)(2+\sin 1)\right]^T. \tag{3.2.16}$$

It is left to Exercise 3.2.2 to verify (3.2.13) - (3.2.16).

We could also use the discretization of (3.2.6) discussed in the Supplementary Discussion of Section 3.1. In this case the matrix (3.1.46) becomes

$$\begin{bmatrix} 2+\frac{5}{2}h^2 & -1-\frac{9}{4}h^2 & & & \\ -1-\frac{9}{4}h^2 & 2+\frac{17}{2}h^2 & -1-\frac{25}{4}h^2 & & \\ \ddots & \ddots & \ddots & & \\ & & & & -1-(n-\frac{1}{2})^2h^2 \\ & & & -1-(n-\frac{1}{2})^2h^2 & 2+(2n^2+\frac{1}{2})h^2 \end{bmatrix}, \tag{3.2.17}$$

and the right-hand side is

$$-h^2\left\{d_1-\left(\frac{1}{4}+\frac{1}{h^2}\right), d_2, \ldots, d_n-\left[(n+\tfrac{1}{2})^2+\frac{1}{h^2}\right](2+\sin 1)\right\}^T. \tag{3.2.18}$$

All three systems were solved for various values of h. Figure 3.2 shows the approximate solution using (3.2.13) and (3.2.14) with a step size $h = 1/32$. This agrees very closely with the exact solution (3.2.8). The other methods generated visibly indistinguishable approximate solutions.

Periodic Boundary Conditions

We conclude this chapter by considering other boundary conditions. As discussed in Section 3.1, Neumann boundary conditions also lead to tridiagonal systems of equations, but periodic boundary conditions introduce matrix elements outside of the tridiagonal part, as shown in (3.1.41). We could solve linear systems with such coefficient matrices by Gaussian elimination, but we wish to consider an alternative.

For any non-zero column vectors $\mathbf{u}$ and $\mathbf{v}$ of length n, the product $\mathbf{u}\mathbf{v}^T$ is called a *rank-one* matrix and is an $n \times n$ matrix whose i, j entry is $u_i v_j$. Now

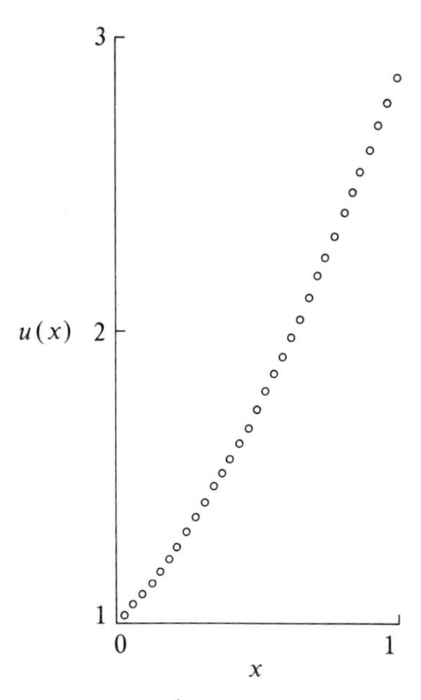

Figure 3.2: *The Computed Solution to* (3.2.9), (3.2.7), *using* $h = 1/32$ *for the Difference Method Corresponding to* (3.2.13), (3.2.14)

let A be the matrix of (3.1.41) and B the tridiagonal part of the matrix. Then A may be written as (Exercise 3.2.7)

$$A = B - \mathbf{e}_1\mathbf{e}_n^T - \mathbf{e}_n\mathbf{e}_1^T, \tag{3.2.19}$$

where $\mathbf{e}_i$ is the vector with 1 in the ith position and zero elsewhere. Thus A may be written as the sum of its tridiagonal part plus two rank-one matrices that bring in the outlying -1's. We may also write A as

$$A = T - \mathbf{w}\mathbf{w}^T, \tag{3.2.20}$$

where $T = B + \mathrm{diag}(1, 0, \cdots, 0, 1)$ and $\mathbf{w} = \mathbf{e}_1 + \mathbf{e}_n$. This changes the original tridiagonal matrix but has the advantage that A is now a tridiagonal matrix plus a single rank-one matrix.

Next let C be any nonsingular matrix and $\mathbf{u}\mathbf{v}^T$ a rank-one matrix. Then the *Sherman-Morrison* formula is

$$(C + \mathbf{u}\mathbf{v}^T)^{-1} = C^{-1} - \alpha^{-1}C^{-1}\mathbf{u}\mathbf{v}^TC^{-1}, \quad \alpha = 1 + \mathbf{v}^TC^{-1}\mathbf{u}, \tag{3.2.21}$$

which is easily verified (Exercise 3.2.7). The condition for nonsingularity of $C + \mathbf{u}\mathbf{v}^T$ is that $\alpha \neq 0$. Note that the matrix added to C^{-1} on the right side of (3.2.21) is also a rank-one matrix. To solve a linear system of the form

$$(C + \mathbf{u}\mathbf{v}^T)\mathbf{x} = \mathbf{b}, \tag{3.2.22}$$

we do not wish to form the inverse as in (3.2.21), but only solve linear systems. Thus we use (3.2.21) to write the solution of (3.2.22) in the form

$$\begin{aligned}\mathbf{x} &= (C+\mathbf{u}\mathbf{v}^T)^{-1}\mathbf{b} = C^{-1}\mathbf{b} - \alpha^{-1}C^{-1}\mathbf{u}\mathbf{v}^TC^{-1}\mathbf{b} \\ &= \mathbf{y} - \alpha^{-1}(\mathbf{v}^T\mathbf{y})\mathbf{z}, \qquad \alpha = 1+\mathbf{v}^T\mathbf{z},\end{aligned} \tag{3.2.23}$$

where $\mathbf{y}$ is the solution of $C\mathbf{y}=\mathbf{b}$ and $\mathbf{z}$ the solution of $C\mathbf{z}=\mathbf{u}$. In particular, if C is a tridiagonal matrix, we can solve the system (3.2.22) by solving two tridiagonal systems and then combining those solutions as shown in (3.2.23). Thus, for the matrix of (3.2.20), the solution of $A\mathbf{x}=\mathbf{b}$ would consist of the steps:

$$\text{Solve } T\mathbf{y}=\mathbf{b},\ T\mathbf{z}=\mathbf{w}, \tag{3.2.24}$$

$$\text{Form } \alpha = 1+\mathbf{w}^T\mathbf{z},\ \mathbf{x}=\mathbf{y}-\alpha^{-1}(\mathbf{w}^T\mathbf{y})\mathbf{z}. \tag{3.2.25}$$

In the next chapter we will see (Exercises 4.2.20,21) that this approach has a slightly lower operation count than applying Gaussian elimination to A itself. More importantly, it requires only a code for solving tridiagonal systems (plus the additional operations of (3.2.25)) whereas a Gaussian elimination code that takes advantage of the zeros in A is somewhat more complicated.

Although we were able to convert the matrix A of (3.2.19) to the form (3.2.20), which involved only a single rank-one matrix, in many situations we wish to deal with a matrix of the form $C+R$, where R is a matrix of rank m. A rank m matrix may be written in the form $R=UV^T$, where U and V are $n\times m$ matrices. Then (3.2.21) extends to the *Sherman-Morrison-Woodbury* formula

$$(C+UV^T)^{-1} = C^{-1} - C^{-1}U(I+V^TC^{-1}U)^{-1}V^TC^{-1}. \tag{3.2.26}$$

The matrix $I+V^TC^{-1}U$ is $m\times m$, and the Sherman-Morrison formula (3.2.21) is the special case $m=1$ of (3.2.26). We could apply (3.2.26) to solve the system $A\mathbf{x}=\mathbf{b}$, where A is given by (3.2.19), although it is slightly more efficient to use (3.2.20) and (3.2.21). To use the formula (3.2.26) we take $C=B$, $U=(\mathbf{e}_1,\mathbf{e}_n)$;and $V=-(\mathbf{e}_n,\mathbf{e}_1)$. The details of the computation are left to Exercise 3.2.8.

Supplementary Discussion and References: 3.2

For a nice review of the history and many applications of the formulas (3.2.21) and (3.2.26), see Hager [1989]. In particular, (3.2.21) was first given by J. Sherman and W. Morrison in 1949 for the special case of changing the elements in one column of C; in this case $\mathbf{v}=\mathbf{e}_i$ if the ith column is changed. The general formula (3.2.21) was given by M. Bartlett in 1951. Simultaneously,

M. Woodbury gave the still more general formula (3.2.26) in a 1950 report, although it had appeared in earlier work in the mid 1940's. For a discussion of rounding error analysis of these formulas, see Yip [1986].

EXERCISES 3.2

3.2.1. Solve the linear system

$$\begin{bmatrix} 4 & 3 & 0 \\ 3 & 5 & 4 \\ 0 & 4 & 6 \end{bmatrix} \begin{bmatrix} u_1 \\ u_2 \\ u_3 \end{bmatrix} = \begin{bmatrix} 1 \\ 2 \\ 3 \end{bmatrix}$$

by Gaussian elimination.

3.2.2. Verify that the coefficient matrix and right hand side of the difference approximation of (3.2.9) using (3.2.11) and central difference approximations for v' are given by (3.2.13) and (3.2.14). Verify (3.2.15) and (3.2.16) if one-sided approximations to v' are used.

3.2.3. Write a computer program to implement Gaussian elimination for tridiagonal systems. Use one-dimensional arrays to store the matrix.

3.2.4. Multiply two $n \times n$ tridiagonal matrices. How many arithmetic operations does this require? Is the product matrix tridiagonal?

3.2.5. Solve numerically the system with coefficient matrix and right-hand side (3.2.13), (3.2.14) for $n = 10$ and $n = 20$. Discuss the accuracy of your approximate solutions.

3.2.6. Repeat Exercise 3.2.5 for (3.2.15), (3.2.16) and (3.2.17), (3.2.18).

3.2.7. Verify formulas (3.2.19), (3.2.21), and (3.2.26).

3.2.8. Verify that the formula (3.2.26) can be used to solve the system $A\mathbf{x} = \mathbf{b}$, where A is given by (3.2.19), by carrying out the following steps:

1. Solve $B\mathbf{y} = \mathbf{b}$, $B\mathbf{w}_1 = \mathbf{e}_1$, $B\mathbf{w}_n = \mathbf{e}_n$.

2. Form the 2×2 matrix

$$U^T W = \begin{pmatrix} \mathbf{e}_1^T \\ \mathbf{e}_n^T \end{pmatrix} (\mathbf{w}_1, \mathbf{w}_n) = \begin{pmatrix} \mathbf{e}_1^T \mathbf{w}_1 & \mathbf{e}_1^T \mathbf{w}_n \\ \mathbf{e}_n^T \mathbf{w}_1 & \mathbf{e}_n^T \mathbf{w}_n \end{pmatrix}.$$

3. Form the 2-vector

$$\mathbf{q} = (I + U^T W)^{-1} U^T \mathbf{y}.$$

4. Form the solution

$$\mathbf{x} = \mathbf{y} - W\mathbf{q}.$$

3.2.9. Consider the problem on the infinite interval $(0, \infty)$:

$$y''(x) = (2x+2)/(x+2)y(x), \ y(0) = 2, \ y(\infty) = 0.$$

a. Approximate a solution to this problem by replacing the boundary condition at ∞ by $y(5) = 0$ and using finite differences for $h = 1/10, 1/20$ and $1/32$.

b. Now note that $(2x+2)/(x+2) \to 2$ as $x \to \infty$. Hence consider the problem

$$z''(x) = 2z(x), \ z(0) = 2, \ z(\infty) = 0.$$

Solve this problem exactly and find an expression for $z'(x)$ in terms of $z(x)$. Using this expression for the value of $y'(x)$ as $x \to \infty$, solve the truncated tridiagonal system and compare the solution to the truncated system used in part a.

Chapter 4

More on Linear Systems of Equations

4.1 Introduction and Least Squares Problems

In the previous chapter we saw that two-point boundary value problems led to solving systems of linear equations. These linear equations were of a very special form: the coefficient matrix had only three non-zero diagonals. Such matrices are a special case of *banded* matrices, to be discussed in Section 4.2. Banded matrices in turn are special cases of *full* or *dense* matrices in which all elements of the matrix are non-zero. The primary purpose of this chapter is to discuss solution techniques for banded or full matrices.

Most approaches to the solution of ordinary or partial differential equations give rise to linear systems in which the coefficient matrix is banded or very *sparse*; that is, it has few non-zero elements. How such matrices arise for partial differential equations will be discussed in Chapter 9. In the remainder of this section we will consider another important class of problems that generally lead to full coefficient matrices. These *least square problems* are not usually related to differential equations, although they may be. For example, given data on the number of predators and prey at different times, can we estimate the coefficients α, β, γ, and δ in the predator-prey equations (2.1.3)? Least squares techniques would be one approach to making such estimates.

Least Squares Polynomials

Recall from Section 2.3 that if $x_0, x_1, \ldots, x_n$ are $n+1$ distinct points and f is a given function, then there is a unique polynomial p of degree n such that

$$p(x_i) = f(x_i), \qquad i = 0, \ldots, n.$$

Now suppose that f itself is a polynomial of degree n and that we wish to determine its coefficients. Then, by the preceding interpolation result, it suffices to know f at $n+1$ distinct points, provided that the determination of the values $f(x_i)$ can be made exactly. In many situations, however, the values of f can be found only by measurements and may be in error. In this case it is common to take many more than $n+1$ measurements in the hope that these measurement errors will "average out." The way in which these errors average out is determined by the method used to combine the measurements to obtain the coefficients of f. For both computational and statistical reasons, the method of choice is often that of least squares, and this method also enjoys an elegant mathematical simplicity.

We now assume that we have m points $x_1, \ldots, x_m$ where $m \geq n+1$ and at least $n+1$ of the points are distinct. Let $f_1, \ldots, f_m$ be approximate values of a function f, not necessarily a polynomial, at the points $x_1, \ldots x_m$. Then we wish to find a polynomial $p(x) = a_0 + a_1 x + \cdots + a_n x^n$ such that

$$\sum_{i=1}^{m} w_i [f_i - p(x_i)]^2 \tag{4.1.1}$$

is a minimum over all polynomials of degree n. That is, we wish to find $a_0, a_1, \ldots, a_n$ so that the weighted sum of the squares of the "errors" $f_i - p(x_i)$ is minimized. In (4.1.1) the w_i are given positive constants, called *weights*, that may be used to assign greater or lesser emphasis to the terms of (4.1.1). For example, if the f_i are measurements and we have great confidence in, say, $f_1, \ldots, f_{10}$, but rather little confidence in the rest, we might set $w_1 = w_2 = \cdots = w_{10} = 5$ and $w_{11} = \cdots = w_m = 1$.

The simplest case of a least squares problem is when $n = 0$, so that p is just a constant. Suppose, for example, that we have m measurements $l_1, \ldots, l_m$ of the length of some object, where the measurements are obtained from m different rulers. Here the points $x_1, \ldots, x_m$ are all identical and, indeed, do not enter explicitly. If we invoke the least-squares principle, then we wish to minimize

$$g(l) = \sum_{i=1}^{m} w_i (l_i - l)^2.$$

From the calculus we know that g takes on a (relative) minimum at a point $\hat{l}$ that satisfies $g'(\hat{l}) = 0$ and $g''(\hat{l}) \geq 0$. Since

$$g'(l) = -2 \sum_{i=1}^{m} w_i (l_i - l), \qquad g''(l) = 2 \sum_{i=1}^{m} w_i,$$

it follows that

$$\hat{l} = \frac{1}{s}\sum_{i=1}^{m} w_i l_i, \qquad s = \sum_{i=1}^{m} w_i,$$

and because this is the only solution of $g'(l) = 0$ it must be the unique point that minimizes g. Thus if the weights w_i are all 1, so that $s = m$, the least-squares approximation to l is just the average of the measurements $l_1, \ldots, l_m$.

The next-simplest situation is if we use a linear polynomial $p(x) = a_0 + a_1 x$. Problems of this type arise very frequently under the assumption that the data are obeying some linear relationship. In this case the function (4.1.1) is

$$g(a_0, a_1) = \sum_{i=1}^{m} w_i (f_i - a_0 - a_1 x_i)^2, \tag{4.1.2}$$

which we wish to minimize over the coefficients a_0 and a_1. Again from the calculus, we know that a necessary condition for g to be minimized is that the partial derivatives of g at the minimizer must vanish:

$$\begin{aligned} \frac{\partial g}{\partial a_0} &= -2\sum_{i=1}^{m} w_i (f_i - a_0 - a_1 x_i) = 0, \\ \frac{\partial g}{\partial a_1} &= -2\sum_{i=1}^{m} w_i x_i (f_i - a_0 - a_1 x_i) = 0. \end{aligned}$$

Collecting coefficients of a_0 and a_1 gives the system of two linear equations

$$\begin{aligned} \left(\sum_{i=1}^{m} w_i\right) a_0 + \left(\sum_{i=1}^{m} w_i x_i\right) a_1 &= \sum_{i=1}^{m} w_i f_i \\ \left(\sum_{i=1}^{m} w_i x_i\right) a_0 + \left(\sum_{i=1}^{m} w_i x_i^2\right) a_1 &= \sum_{i=1}^{m} w_i x_i f_i \end{aligned} \tag{4.1.3}$$

for the unknowns a_0 and a_1.

For polynomials of degree n, the function (4.1.1) that we wish to minimize is

$$g(a_0, a_1, \ldots, a_n) = \sum_{i=1}^{m} w_i (a_0 + a_1 x_i + \cdots + a_n x_i^n - f_i)^2. \tag{4.1.4}$$

Proceeding as in the $n = 2$ case, we know from the calculus that a necessary condition for g to be minimized is that

$$\frac{\partial g}{\partial a_j}(a_0, a_1, \ldots, a_n) = 0, \qquad j = 0, 1 \ldots, n.$$

Writing these partial derivatives out explicitly gives the conditions

$$\sum_{i=1}^{m} w_i x_i^j (a_0 + a_1 x_i + \cdots + a_n x_i^n - f_i) = 0, \qquad j = 0, 1, \ldots, n,$$

which is a system of $n+1$ linear equations in the $n+1$ unknowns $a_0, a_1, \ldots, a_n$; these equations are known as the *normal equations.* Collecting the coefficients of the a_j and rewriting the equations in matrix-vector form gives the system

$$\begin{bmatrix} s_0 & s_1 & s_2 & \cdots & s_n \\ s_1 & s_2 & & & \\ s_2 & & \ddots & & \vdots \\ \vdots & & & & \\ s_n & & \cdots & & s_{2n} \end{bmatrix} \begin{bmatrix} a_0 \\ a_1 \\ \vdots \\ \\ a_n \end{bmatrix} = \begin{bmatrix} c_0 \\ c_1 \\ \vdots \\ \\ c_n \end{bmatrix}, \tag{4.1.5}$$

where

$$s_j = \sum_{i=1}^{m} w_i x_i^j, \qquad c_j = \sum_{i=1}^{m} w_i x_i^j f_i. \tag{4.1.6}$$

Equation (4.1.3) is the special case of (4.1.5) for $n = 1$. Note that the matrix in (4.1.5) is determined by only $2n + 1$ quantities $s_0, \ldots, s_{2n}$, and the "cross-diagonals" of the matrix are constant. Such a matrix is called a *Hankel matrix* and has many interesting properties.

The system (4.1.5) can also be written in the form

$$E^T W E \mathbf{a} = E^T W \mathbf{f}, \tag{4.1.7}$$

where W is a diagonal matrix containing the weights and

$$E = \begin{bmatrix} 1 & x_1 & \cdots & x_1^n \\ 1 & x_2 & \cdots & x_2^n \\ \vdots & & & \vdots \\ 1 & x_m & \cdots & x_m^n \end{bmatrix}, \quad \mathbf{a} = \begin{bmatrix} a_0 \\ a_1 \\ \vdots \\ a_n \end{bmatrix}, \quad \mathbf{f} = \begin{bmatrix} f_1 \\ f_2 \\ \vdots \\ f_m \end{bmatrix}. \tag{4.1.8}$$

The matrix E is $m \times (n+1)$ and is of Vandermonde type; in particular, if $m = n + 1$, it is precisely the Vandermonde matrix of (2.3.11), which we showed was non-singular. We now extend that argument to show that the matrix of (4.1.7) is non-singular provided that at least $n + 1$ of the points x_i are distinct.

A symmetric matrix A is *positive definite* if

$$\mathbf{x}^T A \mathbf{x} > 0 \quad \text{for all} \quad \mathbf{x} \neq 0. \tag{4.1.9}$$

A positive definite matrix is nonsingular, since otherwise there would be an $\mathbf{x} \neq 0$ so that $A\mathbf{x} = 0$ and thus (4.1.9) would be violated.

We now show that the matrix $E^T W E$ of (4.1.7) is symmetric positive definite. Clearly it is symmetric since it is the matrix of (4.1.5). Let $\mathbf{y}$ be any m-vector and consider

$$\mathbf{y}^T W \mathbf{y} = \sum_{i=1}^{m} w_i y_i^2. \tag{4.1.10}$$

Since the w_i are assumed positive, $\mathbf{y}^T W \mathbf{y} \geq 0$ and is equal to zero if and only if $\mathbf{y} = 0$. Thus, if we set $\mathbf{y} = E\mathbf{a}$, we conclude that

$$\mathbf{a}^T E^T W E \mathbf{a} > 0 \quad \text{for all} \quad \mathbf{a} \neq 0$$

provided that $E\mathbf{a} \neq 0$. But if $E\mathbf{a} = 0$, this implies that

$$a_0 + a_1 x_i + \cdots + a_n x_i^n = 0, \qquad i = 1, \ldots, m.$$

Recalling that we have assumed that at least $n+1$ of the x_i are distinct, the nth degree polynomial $a_0 + a_1 x + \cdots + a_n x^n$ would have at least $n+1$ distinct roots. This contradiction proves that $E^T W E$ is positive definite. Thus the system (4.1.7) has a unique solution, and the resulting polynomial with coefficients a_i is the unique polynomial of degree n which minimizes (4.1.1).

General Least Squares Problems

We next consider more general least-squares problems in which the approximating function is not necessarily a polynomial but is a linear combination

$$\phi(x) = \sum_{i=0}^{n} a_i \phi_i(x) \tag{4.1.11}$$

of given functions $\phi_0, \phi_1, \ldots, \phi_n$. If $\phi_j(x) = x^j$, $j = 0, \ldots, n$, then ϕ is the polynomial considered previously. Other common choices for the "basis functions" ϕ_j are

$$\phi_j(x) = \sin j\pi x, \qquad j = 0, 1 \ldots, n,$$

and

$$\phi_j(x) = e^{\alpha_j x}, \qquad j = 0, 1, \ldots, n,$$

where the α_j are given real numbers.

The *general linear least squares problem* is to find $a_0, \ldots, a_n$ such that

$$g(a_0, a_1, \cdots, a_n) = \sum_{i=1}^{m} w_i [\phi(x_i) - f_i]^2 \tag{4.1.12}$$

is minimized, where ϕ is given by (4.1.11). We can proceed exactly as before in obtaining the normal equations for (4.1.12). The first partial derivatives of g are

$$\frac{\partial g}{\partial a_j} = 2\sum_{i=1}^{m} w_i \phi_j(x_i)[a_0\phi_0(x_i) + a_1\phi_1(x_1) + \cdots + a_n\phi_n(x_i) - f_i].$$

Setting these equal to zero, collecting coefficients of the a_i, and writing the resulting linear system in matrix-vector form gives the system

$$\begin{bmatrix} s_{00} & s_{10} & \cdots & s_{n0} \\ s_{01} & s_{11} & & \\ \vdots & & \ddots & \vdots \\ s_{0n} & \cdots & & s_{nn} \end{bmatrix} \begin{bmatrix} a_0 \\ a_1 \\ \vdots \\ a_n \end{bmatrix} = \begin{bmatrix} c_0 \\ c_1 \\ \vdots \\ c_n \end{bmatrix}, \tag{4.1.13}$$

where

$$s_{ij} = \sum_{k=1}^{m} w_k \phi_i(x_k)\phi_j(x_k), \qquad c_j = \sum_{k=1}^{m} w_k \phi_j(x_k) f_k.$$

Note that in this case the coefficient matrix of (4.1.13) is not necessarily a Hankel matrix. As before, we can write (4.1.13) in the form (4.1.7) where now

$$E = \begin{bmatrix} \phi_0(x_1) & \phi_1(x_1) & \cdots & \phi_n(x_1) \\ \phi_0(x_2) & \phi_1(x_2) & \cdots & \phi_n(x_2) \\ \vdots & & & \vdots \\ \phi_0(x_m) & \cdots & & \phi_n(x_m) \end{bmatrix}, \mathbf{f} = \begin{bmatrix} f_1 \\ f_2 \\ \vdots \\ f_m \end{bmatrix}.$$

Clearly, E^TWE is again symmetric, but in order to conclude that it is positive definite, suitable conditions must be imposed on the functions $\phi_0, \ldots, \phi_n$ as well as $x_1, \ldots, x_m$.

Orthogonal Polynomials

The normal equations are very useful for theoretical purposes or for computation purposes when n is small. But they have a tendency to become very ill-conditioned (see Section 4.4) for n at all large, even $n \geq 5$. We now describe an alternative approach to computing the least squares polynomial by means of orthogonal polynomials. In what follows we assume that $w_i = 1$, although the inclusion of weights presents no problem (Exercise 4.1.5).

Let $q_0, q_1, \ldots, q_n$ be polynomials of degree $0, 1, \ldots, n$, respectively. Then we will say that the q_i are mutually *orthogonal* with respect to the points $x_1, \ldots, x_m$ if

$$\sum_{i=1}^{m} q_k(x_i) q_j(x_i) = 0, \qquad k, j = 0, 1, \ldots, n, \quad k \neq j. \tag{4.1.14}$$

We shall return shortly to the question of how one obtains such a set of orthogonal polynomials. For the moment assume that we have them and take $\phi_i = q_i$, $i = 0, 1, \ldots, n$, in the normal equations (4.1.13), in which the weights w_i are all equal to 1. Then, because of (4.1.14), all elements of the coefficient matrix of (4.1.13) off the main diagonal are zero, and the system of equations reduces to

$$\sum_{i=1}^{m}[q_k(x_i)]^2 a_k = \sum_{i=1}^{m} q_k(x_i) f_i, \qquad k = 0, 1, \ldots, n.$$

Thus

$$a_k = \frac{1}{\gamma_k}\sum_{i=1}^{m} q_k(x_i) f_i, \qquad k = 0, 1, \ldots, n, \tag{4.1.15}$$

where

$$\gamma_k = \sum_{i=1}^{m}[q_k(x_i)]^2. \tag{4.1.16}$$

Therefore the least squares polynomial is

$$q(x) = \sum_{k=0}^{n} a_k q_k(x). \tag{4.1.17}$$

An obvious question is whether the polynomial q of (4.1.17) is the same as the polynomial obtained from the normal equations (4.1.5) (with the $w_i = 1$). The answer is yes, under our standard assumption that at least $n+1$ of the points x_i are distinct. This follows from the fact that – as shown earlier – there is a unique polynomial of degree n or less that minimizes (4.1.1). Therefore to show that the polynomial q of (4.1.17) is this same minimizing polynomial, it suffices to show that any polynomial of degree n can be written as a linear combination of the q_i; that is, given a polynomial

$$\hat{p}(x) = b_0 + b_1 x + \cdots + b_n x^n, \tag{4.1.18}$$

we can find coefficients $c_0, c_1, \ldots, c_n$ so that

$$\hat{p}(x) = c_0 q_0(x) + c_1 q_1(x) + \cdots + c_n q_n(x). \tag{4.1.19}$$

This can be done as follows. Let

$$q_i(x) = d_{i,0} + d_{i,1}x + \cdots + d_{i,i}x^i, \qquad i = 0, 1, \ldots, n,$$

where $d_{i,i} \neq 0$. Equating the right-hand sides of (4.1.18) and (4.1.19) gives

$$\begin{aligned} &b_0 + b_1 x + \cdots + b_n x^n \\ &\quad = c_0 d_{0,0} + c_1(d_{1,0} + d_{1,1}x) + \cdots + c_n(d_{n,0} + \cdots + d_{n,n}x^n), \end{aligned}$$

and equating coefficients of powers of x then gives

$$\begin{aligned} b_n &= c_n d_{n,n} \\ b_{n-1} &= c_n d_{n,n-1} + c_{n-1} d_{n-1,n-1} \\ &\vdots \\ b_0 &= c_n d_{n,0} + c_{n-1} d_{n-1,0} + \cdots + c_0 d_{0,0}. \end{aligned} \tag{4.1.20}$$

These are necessary and sufficient conditions that the polynomials of (4.1.18) and (4.1.19) be identical. Given $b_0, b_1, \ldots, b_n$, (4.1.20) is a triangular linear system of equations for the c_i and is solvable since $d_{i,i} \neq 0$, $i = 0, 1, \ldots, n$. Hence the polynomial of (4.1.17) is just another representation of the unique least-squares polynomial obtained by solving the normal equations (4.1.5).

The use of orthogonal polynomials reduces the normal equations to a diagonal system of equations which is trivial to solve. However, the burden is now shifted to the computation of the q_i. There are several possible ways to construct orthogonal polynomials; we will describe one which is particularly suitable for computation.

Let

$$q_0(x) \equiv 1, \qquad q_1(x) \equiv x - \alpha_1, \tag{4.1.21}$$

where α_1 is to be determined so that q_0 and q_1 are orthogonal with respect to the x_i. Thus we must have

$$0 = \sum_{i=1}^{m} q_0(x_i) q_1(x_i) = \sum_{i=1}^{m} (x_i - \alpha_1) = \sum_{i=1}^{m} x_i - m\alpha_1,$$

so that

$$\alpha_1 = \frac{1}{m} \sum_{i=1}^{m} x_i. \tag{4.1.22}$$

Now let

$$q_2(x) = x q_1(x) - \alpha_2 q_1(x) - \beta_1,$$

where α_2 and β_1 are to be determined so that q_2 is orthogonal to both q_0 and q_1; that is,

$$\begin{aligned} \sum_{i=1}^{m} [x_i q_1(x_i) - \alpha_2 q_1(x_i) - \beta_1] &= 0 \\ \sum_{i=1}^{m} [x_i q_1(x_i) - \alpha_2 q_1(x_i) - \beta_1] q_1(x_i) &= 0. \end{aligned}$$

Noting that $\sum q_1(x_i) = 0$, these relations reduce to

$$\sum_{i=1}^{m} x_i q_1(x_i) - m\beta_1 = 0, \qquad \sum_{i=1}^{m} x_i [q_1(x_i)]^2 - \alpha_2 \gamma_1 = 0,$$

where γ_1 is given by (4.1.16). Thus

$$\beta_1 = \frac{1}{m}\sum_{i=1}^{m} x_i q_1(x_i), \qquad \alpha_2 = \frac{1}{\gamma_1}\sum_{i=1}^{m} x_i [q_1(x_i)]^2.$$

The computation for the remaining q_i proceeds in an analogous fashion. Assume that we have determined $q_0, q_1, \ldots, q_j$, and define q_{j+1} by the *three-term recurrence relation*

$$q_{j+1}(x) = xq_j(x) - \alpha_{j+1}q_j(x) - \beta_j q_{j-1}(x), \tag{4.1.23}$$

where α_{j+1} and β_j are to be determined from the orthogonality requirements

$$\sum_{i=1}^{m} q_{j+1}(x_i)q_j(x_i) = 0, \qquad \sum_{i=1}^{m} q_{j+1}(x_i)q_{j-1}(x_i) = 0. \tag{4.1.24}$$

If these two relations are satisfied, then q_{j+1} must also be orthogonal to all the previous q_k, $k < j-1$, since by (4.1.23)

$$\begin{aligned}\sum_{i=1}^{m} q_{j+1}(x_i)q_k(x_i) &= \sum_{i=1}^{m} x_i q_j(x_i)q_k(x_i) - \alpha_{j+1}\sum_{i=1}^{m} q_j(x_i)q_k(x_i) \qquad (4.1.25)\\ &\quad -\beta_j \sum_{j=1}^{m} q_{j-1}(x_i)q_k(x_i).\end{aligned}$$

The last two terms in (4.1.25) are zero by assumption, whereas $xq_k(x)$ is a polynomial of degree $k+1$ and can be expressed as a linear combination of $q_0, q_1, \ldots, q_{k+1}$. Hence the first term on the right-hand side of (4.1.25) is also zero.

Returning to the conditions (4.1.24) and inserting q_{j+1} from (4.1.23) leads to the expressions

$$\alpha_{j+1} = \frac{1}{\gamma_j}\sum_{i=1}^{m} x_i [q_j(x_i)]^2, \tag{4.1.26}$$

$$\beta_j = \frac{1}{\gamma_{j-1}}\sum_{i=1}^{m} x_i q_j(x_i)q_{j-1}(x_i),$$

for α_{j+1} and β_j, where the γ's are given by (4.1.16). The β_i, however, can be computed in a better way if we substitute for $x_i q_{j-1}(x_i)$ using (4.1.23), and then note that

$$\sum_{i-1}^{m} q_j(x_i)[q_j(x_i) + \alpha_j q_{j-1}(x_i) + \beta_{j-1}q_{j-2}(x_i)] = \sum_{i=1}^{m} [q_j(x_i)]^2 = \gamma_j$$

by the orthogonality of the q's. Thus

$$\beta_j = \frac{\gamma_j}{\gamma_{j-1}}. \tag{4.1.27}$$

Note that the denominator in (4.1.27) can vanish only if $q_{j-1}(x_i) = 0$, $i = 1, \ldots, m$. But since at least $n+1$ of the x_i are assumed to be distinct, this would imply that q_{j-1} is identically zero, which contradicts the definition of q_{j-1}. Hence $\gamma_{j-1} \neq 0$.

We can summarize the orthogonal polynomial algorithm as follows:

1. Set $q_0(x) \equiv 1$, $q_1(x) = x - \frac{1}{m}\sum_{i=1}^m x_i$.

2. For $j = 1, \ldots, n-1$, define q_{j+1} by (4.1.23), where α_{j+1} is given by (4.1.26), and β_j by (4.1.27).

3. Compute the coefficients $a_0, a_1, \ldots, a_n$ of the least squares polynomial $a_0q_0(x) + a_1q_1(x) + \cdots + a_nq_n(x)$ by (4.1.15).

As mentioned previously, this approach is preferred numerically because it avoids the necessity of solving the (possibly) ill-conditioned system (4.1.5). Another advantage is that we are able to build up the least-squares polynomial degree by degree. For example, if we do not know what degree polynomial we wish to use, we might start with a first-degree polynomial, then a second-degree, and so on, until we obtain a fit that we believe is suitable. With the orthogonal polynomial algorithm the coefficients a_i are independent of n, and as soon as we compute q_j we can compute a_j, and hence the least squares polynomial of degree j.

A Numerical Example

We now give a simple example, using the data

$$\begin{array}{ccccc} x_1 = 0 & x_2 = \frac{1}{4} & x_3 = \frac{1}{2} & x_4 = \frac{3}{4} & x_5 = 1 \\ f_1 = 1 & f_2 = 2 & f_3 = 1 & f_4 = 0 & f_5 = 1. \end{array}$$

Since there are five points x_i, these data will uniquely determine a fourth-degree interpolating polynomial. We now compute the linear and quadratic least-squares polynomials by both the normal equations and the orthogonal polynomial approaches.

For the normal equations for the linear polynomial, we will need the following quantities:

$$\sum_{i=1}^5 x_i = \tfrac{5}{2}, \qquad \sum_{i=1}^5 x_i^2 = \tfrac{15}{8}, \qquad \sum_{i=1}^5 f_i = 5, \qquad \sum_{i=1}^5 x_i f_i = 2. \tag{4.1.28}$$

Then the coefficients a_0 and a_1 are the solutions of the system (4.1.3) (with the $w_i = 1$):

$$a_0 = \frac{7}{5}, \qquad a_1 = \frac{-4}{5}.$$

Thus the linear least squares polynomial is

$$p_1(x) = \tfrac{7}{5} - \tfrac{4}{5}x. \tag{4.1.29}$$

If we compute the same polynomial by orthogonal polynomials, the polynomial is given in the form

$$a_0q_0(x) + a_1q_1(x) = a_0 + a_1(x - \alpha_1), \tag{4.1.30}$$

where a_0 and a_1 are given by (4.1.15), and α_1 by (4.1.22):

$$a_0 = 1, \qquad a_1 = \frac{-4}{5}, \qquad \alpha_1 = \tfrac{1}{2}.$$

Therefore the polynomial (4.1.30) is $1 - \frac{4}{5}(x - \frac{1}{2})$, which is, as expected, the same as (4.1.29).

To compute the least squares quadratic polynomial by the normal equations, we need to solve the system (4.1.5) for $n = 2$, which for our data is

$$\begin{bmatrix} 640 & 320 & 240 \\ 320 & 240 & 200 \\ 240 & 200 & 177 \end{bmatrix} \begin{bmatrix} a_0 \\ a_1 \\ a_2 \end{bmatrix} = \begin{bmatrix} 640 \\ 256 \\ 176 \end{bmatrix}.$$

The solution of this system is

$$a_0 = \tfrac{7}{5}, \qquad a_1 = \tfrac{-4}{5}, \qquad a_2 = 0. \tag{4.1.31}$$

Thus the best least squares polynomial approximation turns out to be the linear least-squares polynomial; that is, no improvement of the linear approximation can be made by adding a quadratic term. That (4.1.31) is correct is verified by computing the least squares quadratic by orthogonal polynomials. The orthogonal polynomial representation will be

$$a_0q_0(x) + a_1q_1(x) + a_2q_2(x) = \tfrac{7}{5} - \tfrac{4}{5}x + a_2[x(x - \tfrac{1}{2}) - \alpha_2(x - \tfrac{1}{2}) - \beta_1],$$

where α_2 and β_1 are computed from (4.1.26) and (4.1.27) as

$$\alpha_2 = \frac{\sum x_i(x_i - \frac{1}{2})^2}{\sum(x_i - \frac{1}{2})^2} = \tfrac{1}{2}, \qquad \beta_1 = \tfrac{1}{5}\sum x_i(x_i - \tfrac{1}{2}) = \tfrac{1}{8}.$$

Thus $q_2(x) = x^2 - x - \frac{1}{8}$, and from (4.1.15) we find that $a_2 = 0$.

The orthogonal polynomial approach can be extended to other basis functions that are orthogonal, such as trigonometric functions. However, for most basis functions the system (4.1.13) must be solved and, in general, all elements of this coefficient matrix will be non-zero. Hence we must solve a "full" linear system, which is the topic of the next two sections.

Supplementary Discussion and References: 4.1

An alternative approach to solving the least squares polynomial problem is to deal directly with the system of linear equations $E\mathbf{a} = \mathbf{f}$, where E and $\mathbf{f}$ are given in (4.1.8). This is an $m \times (n+1)$ system, where m is usually greater that $n+1$; hence the matrix E is not square. Techniques for dealing with this type of system are given in Section 4.5. For further discussions of least squares problems see Golub and Van Loan [1989] and Lawson and Hanson [1974].

EXERCISES 4.1

4.1.1. Assume that f is a given function for which the following values are known: $f(1) = 2$, $f(2) = 3$, $f(3) = 5$, $f(4) = 3$. Find the constant, linear, and quadratic least-squares polynomials by both the normal-equation and orthogonal-polynomial approaches.

4.1.2. Write a computer program to obtain the least-squares polynomial of degree n using $m \geq n+1$ data points by the orthogonal-polynomial approach. Test your program on the polynomials of Exercise 4.1.1.

4.1.3. Let $y_1 \dots y_m$ be a set of data that we wish to approximate by a constant c. Determine c so that

a. $\sum_{i=1}^{m} |c - y_i| = \text{minimum}.$

b. $\max_{1 \leq i \leq m} |c - y_i| = \text{minimum}.$

4.1.4. If $y_1, \dots, y_m$ are m observations, define the mean and variance by

$$\bar{y} = \frac{1}{m} \sum_{i=1}^{m} y_i, \qquad v = \sum_{i=1}^{m} (y_i - \bar{y})^2.$$

Let $y_{m+1}, \dots, y_{m+n}$ be n additional observations with mean and variance y_a and v_a. Show how to combine v and v_a by $\alpha_1 v + \alpha_2 v_a$ to obtain the variance of the combined set of observations. Specialize this to $n = 1$ so as to obtain an updating formula for the variance each time a new observation is added.

4.1.5. Modify the orthogonal polynomial approach to handle the problem (4.1.1) with given weights w_i. Replace the relation (4.1.14) by $\sum w_i q_k(x_i) q_j(x_i)$ and then modify (4.1.15), (4.1.26), and (4.1.27) accordingly.

4.1.6. Let f be a twice differentiable function on $[0,8]$, and let $x_i = (i-9)$, $i = 1,\ldots,17$, and $f_i = f(x_i)$.
a. Using orthogonal polynomials, find the polynomial of degree 5 which minimizes (4.1.1) with the weights given by $w_i = \gamma|i-9|$ where $0 < \gamma \le 1$.
b. Show how $f'(0)$ and $f''(0)$ can be approximated by using the recursion relations for the orthogonal polynomials. In particular, show how to obtain coefficients a_i so that $f'(0)$ is approximated by $\sum_{i=1}^{17} a_i f_i$.

4.2 Gaussian Elimination

In the previous chapter we discussed how to solve a system of tridiagonal linear equations by the Gaussian elimination method. We now apply Gaussian elimination to a general linear system

$$A\mathbf{x} = \mathbf{b}, \tag{4.2.1}$$

where A is a given $n \times n$ matrix assumed to be nonsingular, $\mathbf{b}$ is a given column n vector, and $\mathbf{x}$ is the solution vector to be determined.

We write the system (4.2.1) as

$$\begin{aligned} a_{11}x_1 \quad + \cdots + \quad a_{1n}x_n &= b_1 \\ a_{21}x_1 \quad + \cdots + \quad a_{2n}x_n &= b_2 \\ \vdots \qquad\qquad\qquad & \vdots \\ a_{n1}x_1 \quad + \cdots + \quad a_{nn}x_n &= b_n. \end{aligned} \tag{4.2.2}$$

As with tridiagonal systems we first subtract a_{21}/a_{11} times the first equation from the second equation to eliminate the coefficient of x_1 in the second equation. For tridiagonal systems this completed the first stage but in the general case we wish to eliminate the coefficient of x_1 in all the remaining equations. Thus we subtract a_{31}/a_{11} times the first equation from the third equation, a_{41}/a_{11} times the first equation from the fourth equation, and so on, until the coefficients of x_1 in the last $n-1$ equations have all been eliminated. This gives the reduced system of equations

$$\begin{aligned} a_{11}x_1 + a_{12}x_2 + \cdots + a_{1n}x_n &= b_1 \\ a_{22}^{(1)}x_2 + \cdots + a_{2n}^{(1)}x_n &= b_2^{(1)} \\ \vdots \qquad\qquad & \\ a_{n2}^{(1)}x_2 + \cdots + a_{nn}^{(1)}x_n &= b_n^{(1)}, \end{aligned} \tag{4.2.3}$$

where

$$a_{ij}^{(1)} = a_{ij} - a_{1j}\frac{a_{i1}}{a_{11}}, \qquad b_i^{(1)} = b_i - b_1\frac{a_{i1}}{a_{11}}, \qquad i,j = 2,\ldots,n. \tag{4.2.4}$$

Precisely the same process is now applied to the last $n-1$ equations of the system (4.2.3) to eliminate the coefficients of x_2 in the last $n-2$ equations, and so on, until the entire system has been reduced to the *triangular form*

$$\begin{bmatrix} a_{11} & a_{12} & \cdots & a_{1n} \\ & a_{22}^{(1)} & \cdots & a_{2n}^{(1)} \\ & & \ddots & \vdots \\ & & & a_{nn}^{(n-1)} \end{bmatrix} \begin{bmatrix} x_1 \\ x_2 \\ \vdots \\ x_n \end{bmatrix} = \begin{bmatrix} b_1 \\ b_2^{(1)} \\ \vdots \\ b_n^{(n-1)} \end{bmatrix}. \qquad (4.2.5)$$

The superscripts indicate the number of times the elements have, in general, been changed. This completes the *forward reduction* (or *forward elimination* or *triangular reduction*) phase of the Gaussian elimination algorithm. Note that we have tacitly assumed that a_{11} and the $a_{ii}^{(i-1)}$ are all non-zero since we divide by these elements. In the following section we will consider the important question of how to handle zero or small divisors.

The Gaussian elimination method is based on the fact (usually established in an introductory linear algebra course) that replacing any equation of the original system (4.2.2) by a linear combination of itself and another equation does not change the solution of (4.2.2). Thus the triangular system (4.2.5) has the same solution as the original system. The purpose of the forward reduction is to reduce the original system to one which is easy to solve; this is a common theme in much of scientific computing. The second part of the Gaussian elimination method then consists of the solution of (4.2.5) by *back substitution*, in which the equations are solved in reverse order:

$$\begin{aligned} x_n &= \frac{b_n^{(n-1)}}{a_{nn}^{(n-1)}} \\ x_{n-1} &= \frac{b_{n-1}^{(n-2)} - a_{n-1,n}^{(n-2)} x_n}{a_{n-1,n-1}^{(n-2)}} \\ &\vdots \\ x_1 &= \frac{b_1 - a_{12}x_2 - \cdots - a_{1n}x_n}{a_{11}}. \end{aligned} \qquad (4.2.6)$$

The Gaussian elimination algorithm can be written in algorithmic form as follows:

Forward Reduction

$$
\begin{aligned}
&\text{For } k = 1, \ldots, n-1 \\
&\quad \text{For } i = k+1, \ldots, n \\
&\qquad l_{ik} = \frac{a_{ik}}{a_{kk}} \\
&\qquad \text{For } j = k+1, \ldots, n \\
&\qquad\qquad a_{ij} = a_{ij} - l_{ik} a_{kj} \\
&\qquad b_i = b_i - l_{ik} b_k
\end{aligned}
\tag{4.2.7}
$$

Back Substitution

$$
\begin{aligned}
&\text{For } k = n, n-1, \ldots, 1 \\
&\quad x_k = \frac{b_k - \sum_{j=k+1}^{n} a_{kj} x_j}{a_{kk}}.
\end{aligned}
\tag{4.2.8}
$$

To translate the preceding algorithm into a computer code one should note that the $a_{ij}^{(k)}$ can be overwritten on the same storage spaces occupied by the original elements a_{ij}. If this is done, the original matrix will, of course, be destroyed during the process. Similarly, the new $b_i^{(k)}$ may be overwritten on the original storage spaces of the b_i. The multiplier l_{ik} can be written into the corresponding storage space for a_{ik}, which is no longer needed after l_{ik} is computed.

LU Factorization

Gaussian elimination is related to a factorization

$$A = LU \tag{4.2.9}$$

of the matrix A. Here U is the upper triangular matrix of (4.2.5) obtained in the forward reduction, and L is a *unit* lower triangular matrix (all main diagonal elements are 1) in which the subdiagonal element l_{ij} is the multiplier used for eliminating the jth variable from the ith equation. For example, if the original system is

$$
\begin{bmatrix} 4 & -9 & 2 \\ 2 & -4 & 4 \\ -1 & 2 & 2 \end{bmatrix}
\begin{bmatrix} x_1 \\ x_2 \\ x_3 \end{bmatrix}
=
\begin{bmatrix} 2 \\ 3 \\ 1 \end{bmatrix},
\tag{4.2.10}
$$

then the reduced system (4.2.5) is

$$\begin{bmatrix} 4 & -9 & 2 \\ 0 & 0.5 & 3 \\ 0 & 0 & 4 \end{bmatrix} \begin{bmatrix} x_1 \\ x_2 \\ x_3 \end{bmatrix} = \begin{bmatrix} 2 \\ 2 \\ 2.5 \end{bmatrix}. \tag{4.2.11}$$

The multipliers used to obtain (4.2.11) from (4.2.10) are 0.5, −0.25, and −0.5. Thus

$$L = \begin{bmatrix} 1 & 0 & 0 \\ 0.5 & 1 & 0 \\ -0.25 & -0.5 & 1 \end{bmatrix}, \tag{4.2.12}$$

and A is the product of (4.2.12) and the matrix U of (4.2.11). The calculations needed for the above assertions are left to Exercise 4.2.1.

More generally, it is easy to verify (Exercise 4.2.2) that the elimination step that produces (4.2.3) from (4.2.2) is equivalent to multiplying (4.2.2) by the matrix

$$L_1 = \begin{bmatrix} 1 & & & \\ -l_{21} & 1 & & \\ \vdots & & \ddots & \\ -l_{n1} & & & 1 \end{bmatrix}. \tag{4.2.13}$$

Continuing in this fashion, the reduced system (4.2.5) may be written as

$$\hat{L}A\mathbf{x} = \hat{L}\mathbf{b}, \qquad \hat{L} = L_{n-1} \cdots L_2 L_1, \tag{4.2.14}$$

where

$$L_i = \begin{bmatrix} 1 & & & & \\ & \ddots & & & \\ & & 1 & & \\ & & -l_{i+1,i} & & \\ & & \vdots & \ddots & \\ & & -l_{n,i} & & 1 \end{bmatrix}. \tag{4.2.15}$$

Each of the matrices L_i has determinant equal to 1 and so is non-singular. Therefore the product $\hat{L}$ is non-singular. Moreover, $\hat{L}$ is unit lower triangular and, therefore, so is $\hat{L}^{-1}$. By construction, the coefficient matrix of (4.2.14) is U, and if we set $L = \hat{L}^{-1}$ we have

$$U = \hat{L}A = L^{-1}A,$$

which is equivalent to (4.2.9). The verification of the above statements is left to Exercise 4.2.3.

The factorization (4.2.8) is known as the *LU decomposition* (or *LU factorization*) of A. The right-hand side of (4.2.13) is $L^{-1}\mathbf{b}$, which is the solution of $L\mathbf{y} = \mathbf{b}$. Thus the Gaussian elimination algorithm for solving $A\mathbf{x} = \mathbf{b}$ is mathematically equivalent to the three-step process:

$$\begin{aligned} &1. \quad \text{Factor} \quad A = LU. \\ &2. \quad \text{Solve} \quad L\mathbf{y} = \mathbf{b}. \\ &3. \quad \text{Solve} \quad U\mathbf{x} = \mathbf{y}. \end{aligned} \tag{4.2.16}$$

This matrix-theoretic view of Gaussian elimination is very useful for theoretical purposes and also forms the basis for some computational variants of the elimination process. In particular, the classical Crout and Doolittle forms of LU decomposition are based on formulas for the elements of L and U obtained from equating LU and A. Another use of (4.2.16) is when there are many right hand sides; in this case, step 1. is done once and the factors saved so that steps 2. and 3. may be performed repeatedly. This will be discussed in more detail later.

Operation Counts

An important question is the efficiency of the Gaussian elimination algorithm, and we next estimate the number of arithmetic operations needed to compute the solution vector $\mathbf{x}$. The major part of the work is in updating elements of A and we first count the number of operations to carry out the arithmetic statement $a_{ij} = a_{ij} - l_{ik}a_{kj}$ in (4.2.7). There is one addition and one multiplication in this statement and we count the number of additions. In the j loop of (4.2.7) there are $n-k$ additions. The i loop repeats these $n-k$ times so that the total over the k loop is

$$\sum_{k=1}^{n-1} (n-k)^2 = \sum_{k=1}^{n-1} k^2.$$

Thus using the summation formula in Exercise 4.2.4 we obtain

$$\text{Number of additions} = \sum_{k=1}^{n-1} k^2 = \frac{(n-1)(n)(2n-1)}{6} \doteq \frac{n^3}{3}, \tag{4.2.17}$$

and the same count holds for the number of multiplications. We also need the number of divisions to compute the l_{ik}, and the number of operations to modify the right hand side **b** and to do the back substitution. All of these involve no more than order n^2 operations (see Exercise 4.2.5). Hence for sufficiently large n the work involved in the triangular factorization makes the dominant contribution to the computing time and is proportional to n^3. Note that this implies that if n is doubled, the amount of work increases by approximately a factor of 8.

To obtain an understanding of the amount of time that Gaussian elimination might require on a moderate-size problem, suppose that $n = 100$ and that addition and multiplication times are $1\mu s$ and $2\mu s$, respectively (μs= microsecond = 10^{-6} second). Then the time for the additions and multiplications in the factorization is approximately

$$\frac{100^3}{3}(3\mu s) = 10^6\mu s = 1 \ second.$$

There will also be time required for the other arithmetical operations of the complete elimination process, but this will be much less than 1 second. More importantly, there will be various "overhead" costs such as moving data back and forth from storage, and indexing. This could easily double or triple the total computing time, but on a computer of this speed only a few seconds would be required to solve a 100×100 system.

Banded Matrices

The previous discussion has assumed that the matrix of the system is "full," that is, it has few zero elements. For many matrices that arise in practice, and particularly in the solution of differential equations, the elements of the matrix are primarily zero. Perhaps the simplest nontrivial examples of this are the tridiagonal matrices discussed in Section 3.1; here there are no more than three non-zero elements in each row regardless of the size of n.

Tridiagonal matrices are special cases of *banded* matrices in which the non-zero elements are all contained in diagonals about the main diagonal, as illustrated in Figure 4.1. A matrix $A = (a_{ij})$ is a banded matrix of *bandwidth* $p+q+1$ if $a_{ij} = 0$ for all i, j such that $i-j > p$ or $j-i > q$. Such a matrix has all of its non-zero elements on the main diagonal, the closest p subdiagonals, and the closest q superdiagonals. If $p = q = 1$, there are only three non-zero diagonals, and the matrix is tridiagonal. In Section 3.1 tridiagonal matrices arose in the finite difference solution of two-point boundary-value problems. When derivatives are approximated by higher-order difference approximations, the matrices will have larger bandwidths. For example, the fourth-order approximation given by (3.1.42) leads to a matrix with bandwidth 5 ($p = q = 2$).

Matrices with larger bandwidths will be discussed in Chapter 9 in connection with the numerical solution of partial differential equations. In most cases of interest $p = q$, and p is called the *semi-bandwidth.* We will restrict ourselves to such matrices in the sequel, although the case $p \neq q$ presents no difficulties.

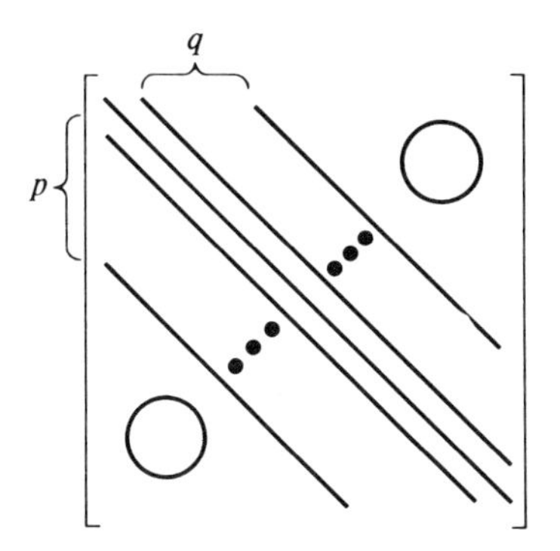

Figure 4.1: *Matrix with Bandwidth* $p + q + 1$

Just as for tridiagonal matrices, the Gaussian elimination algorithm for banded matrices benefits by not having to deal with the zero elements outside of the band. If A has semi-bandwidth p, there will be p coefficients to be eliminated in the first column, and this elimination will alter only the elements in the second through $(p+1)$st rows and columns of A. The number of operations required will be p^2 multiplications and additions, and p divisions (not counting operations on the right-hand side of the system). The first reduced matrix will be a banded matrix with the same bandwidth, and hence the same count applies. After $n - p - 1$ of these reductions, there will remain a full $(p+1) \times (p+1)$ matrix to be reduced. Hence the number of additions (or multiplications) required in the triangular reduction is

$$(n-p-1)p^2 + (\tfrac{1}{6}p)(p+1)(2p+1) \doteq np^2 - \tfrac{2}{3}p^3.$$

If n is large with respect to p (for example, $n = 1000$, $p = 7$), the dominant term in this operation count is np^2. As with full matrices, the number of operations to modify the right-hand side and perform the back substitution is of lower order, namely $0(np)$; see Exercise 4.2.7. However, as the bandwidth decreases, the number of operations for the right-hand side, the back substitution, and the divisions in the forward reduction constitute a larger fraction of the total operation count. In particular, for tridiagonal matrices, as discussed in Section 3.2, the forward reduction requires only $n-1$ addition/multiplication pairs and $n-1$ divisions, whereas the right-hand side and back substitution require $(2n-1)$ additions/multiplications and n divisions.

The storage of a banded matrix does not require reserving a full $n \times n$ two-dimensional array, which would be very inefficient. All that is needed is $2p+1$ one-dimensional arrays, each holding one of the non-zero diagonals. In particular, a tridiagonal matrix may be stored in three one-dimensional arrays. However, if p is at all large it is probably better to store the diagonals of A as columns in a $(2p+1) \times n$ two-dimensional array, as indicated in Figure 4.2 for $p = 2$ and $n = 6$. In any case, the total storage required, including the right-hand side and the solution vector, is no more than $(2p+3)n$; in particular, for tridiagonal systems it is no more than $5n$.

$$\begin{bmatrix} 0 & 0 & a_{11} & a_{12} & a_{13} \\ 0 & a_{21} & a_{22} & a_{23} & a_{24} \\ a_{31} & a_{32} & a_{33} & a_{34} & a_{35} \\ a_{42} & a_{43} & a_{44} & a_{45} & a_{46} \\ a_{53} & a_{54} & a_{55} & a_{56} & 0 \\ a_{64} & a_{65} & a_{66} & 0 & 0 \end{bmatrix}$$

Figure 4.2: *Storage for a Banded Matrix*

For a banded matrix, the factors L and U of the LU decomposition retain the same bandwidth (Exercise 4.2.8). In particular, for tridiagonal matrices

$$L = \begin{bmatrix} 1 & & & \\ l_2 & 1 & & \\ & \ddots & \ddots & \\ & & & \\ & & l_n & 1 \end{bmatrix}, U = \begin{bmatrix} u_1 & a_{12} & & \\ & u_2 & a_{23} & \\ & & \ddots & \ddots \\ & & & a_{n-1,n} \\ & & & u_n \end{bmatrix}. \tag{4.2.18}$$

Note that the superdiagonal elements in U are the original elements of A. Matrices of the form (4.2.18) are called *bidiagonal*.

Determinants and Inverses

We return now to general matrices (not necessarily banded). We note first that the determinant of the coefficient matrix A, denoted by $\det A$, is an easy by-product of the elimination process. By the LU decomposition of A, and using the facts that the determinant of a product of two matrices is the product of the determinants and that the determinant of a triangular matrix is the product of its diagonal elements, we have

$$\det A = \det LU = \det L \det U = u_{11}u_{22}\cdots u_{nn}\ . \tag{4.2.19}$$

Thus, the determinant is just the product of the diagonal elements of the reduced triangular matrix and is computed by an additional $n-1$ multiplications. Even if only the determinant of the matrix is desired – and not the solution of a linear system – the Gaussian elimination reduction to triangular form is still the best general method for its computation.

The Gaussian elimination process is also the best way, in general, to compute the inverse of A, if that is needed. Let $\mathbf{e}_i$ be the vector with 1 in the ith position and zeros elsewhere. Then $\mathbf{e}_i$ is the ith column of the identity matrix I, and from the basic relation $AA^{-1} = I$ it follows that the ith column of A^{-1} is the solution of the linear system of equations $A\mathbf{x} = \mathbf{e}_i$. Hence we can obtain A^{-1} by solving the n systems of equations

$$A\mathbf{x}_i = \mathbf{e}_i, \qquad i = 1, \ldots, n, \tag{4.2.20}$$

where the solution vectors $\mathbf{x}_1, \ldots, \mathbf{x}_n$ will be the columns of A^{-1}.

We note that one does *not* wish to solve a system $A\mathbf{x} = \mathbf{b}$ by computing A^{-1} and then forming $\mathbf{x} = A^{-1}\mathbf{b}$. This would generally require considerably more work than just solving the system.

Several Right-Hand Sides

The above procedure for computing A^{-1} extends to the more general problem of solving several systems with the same coefficient matrix:

$$A\mathbf{x}_i = \mathbf{b}_i, \qquad i = 1, \ldots, m. \tag{4.2.21}$$

Recall that this was the case is carrying out the Sherman-Morrison or Sherman-Morrison-Woodbury formulas in Section 3.2 (See Exercise 3.2.8 and (3.2.24)). In terms of the LU decomposition of A, (4.2.21) can be carried out efficiently by the following modification of (4.2.16):

$$\begin{aligned} &1.\quad \text{Factor } A = LU. \\ &2.\quad \text{Solve } L\mathbf{y}_i = \mathbf{b}_i, \qquad i = 1, \ldots, m. \\ &3.\quad \text{Solve } U\mathbf{x}_i = \mathbf{y}_i, \qquad i = 1, \ldots, m. \end{aligned} \tag{4.2.22}$$

Note that the matrix A is factored only once, regardless of the number of right-hand sides. Hence the operation count is $0(n^3) + 0(mn^2)$, the latter term representing parts 2 and 3 in (4.2.22). Only when m becomes nearly as large as n does the amount of work in parts 2 and 3 approach that of the factorization, at least for full matrices. In the case of computing A^{-1}, $m = n$, but the total operation count is still $0(n^3)$.

To carry out the intent of (4.2.21) in terms of the elimination process, we can either do parts 1 and 2 simultaneously, modifying the right-hand sides as the elimination proceeds, or we can first complete the factorization and save the multipliers l_{ij} to do 2.

The various algorithms in this section have all been predicated on the assumption that a_{11} and the subsequent diagonal elements of the reduced matrix do not vanish. In practice, it is not sufficient that these divisors be non-zero; they must also be large enough in some sense or severe rounding error problems may occur. In Section 4.3 we will consider these problems and the modifications that are necessary for the elimination process to be a viable procedure.

Supplementary Discussion and References: 4.2

There are many books on numerical linear algebra and these, as well as more elementary books on numerical methods, all discuss the problem of solving linear systems of equations. For an advanced treatment and more references, see Golub and van Loan [1989].

Although the Gaussian elimination method is very efficient, there are algorithms that have a lower operation count as a function of n. In particular, the number of multiplications required to solve a linear system of size n by a method due to Strassen [1969] is $0(n^{2.8\cdots})$. But the constant multiplying the high-order term is larger than for Gaussian elimination, and the method is more complicated; consequently it has not been a serious competitor to Gaussian elimination for practical computation. Recent research, however (Higham [1990]), indicates that this approach can be superior to Gaussian elimination, at least in certain circumstances.

EXERCISES 4.2

4.2.1. Verify the forward reduction of (4.2.10) to obtain (4.2.11). Then verify that $A = LU$, where L is given by (4.2.12) and U is the upper triangular matrix in (4.2.11)

4.2.2. Verify that multiplication of (4.2.2) by (4.2.13) gives (4.2.3). Then verify that (4.2.14) holds.

4.2.3. Verify the following statements:

a. The determinants of the L_i of (4.2.15) are all 1.

b. Products of lower (upper) triangular matrices are lower (upper) triangular.

c. Inverses of lower (upper) triangular matrices are lower (upper) triangular.

d. The inverse of L_i of (4.2.15) is the same matrix with the signs of the off-diagonal elements changed.

4.2.4. Verify by induction the following summation formulas:

$$\sum_{i=1}^{n} i = \tfrac{1}{2}n(n+1), \qquad \sum_{i=1}^{n} i^2 = \tfrac{1}{6}n(n+1)(2n+1).$$

4.2.5. Show that the following operation counts are correct for Gaussian elimination:

a. Number of additions (multiplications) to compute new right hand side $= n(n-1)/2$.

b. Number of divisions to compute the multipliers $l_{ik} = n(n-1)/2$.

c. Number of divisions in back substitution $= n$.

d. Number of additions (multiplications) in back substitution $= n(n-1)/2$.

4.2.6. Using a Gaussian elimination code (either a package or one that you write), measure the time to solve full linear systems of sizes $n = 50$ and $n = 100$. Discuss why the larger time is not exactly a factor of 8 larger than the smaller, as suggested by the $0(n^3)$ operation count.

4.2.7. If A is a banded matrix with semi-bandwidth p, show that $0(pn)$ operations are required to modify the right-hand side and do the back substitution.

4.2.8. Let A be a matrix with bandwidth $p+q+1$, as illustrated in Figure 4.1. If $A = LU$ is the LU decomposition, show that L has bandwidth $p+1$ and U has bandwidth $q+1$. In particular, for tridiagonal matrices show that L and U have the form (4.2.18).

4.2.9. Verify that the product of L and U in (4.2.18) is tridiagonal.

4.2.10. Write a computer program to implement Gaussian elimination for a banded matrix with p subdiagonals and q superdiagonals. Use the storage pattern of Figure 4.2.

4.2.11. If $A = LU$, show that $A^{-1} = U^{-1}L^{-1}$. Using this, describe an algorithm for computing A^{-1}, taking advantage of the triangular structure of U^{-1} and L^{-1}. How does your algorithm compare with solving the systems (4.2.20)?

4.2.12. We have shown that Gaussian elimination on a full matrix requires $0(n^3)$ operations and on a tridiagonal matrix it requires $0(n)$ operations. For what size semi-bandwidth will it require $0(n^2)$ operations?

4.2.13. Consider the $n \times n$ complex system $(A + iB)(\mathbf{u} + i\mathbf{v}) = \mathbf{b} + i\mathbf{c}$ where A, B, $\mathbf{u}$, $\mathbf{v}$, $\mathbf{b}$, and $\mathbf{c}$ are real. Show how to write this as a real $2n \times 2n$ system. Compare the operation counts of Gaussian elimination applied to the real system and the original complex system, assuming that complex multiplication takes four real multiplications and two additions. (What does complex division require?)

4.2.14. (Uniqueness of LU Decomposition). Let A be non-singular and suppose that $A = LU = \hat{L}\hat{U}$ where L and $\hat{L}$ are unit lower triangular and U and $\hat{U}$ are upper triangular. Show that $L = \hat{L}$ and $U = \hat{U}$.

4.2.15. Let $A = LU$. Use Exercise 4.2.17 to give the LU decomposition of AT, where T is upper triangular. Specialize this to $T = D$, a diagonal matrix. What does this imply about the multipliers in Gaussian elimination when the columns of A are scaled?

4.2.16. (Jordan elimination). Show that it is possible to eliminate the elements above the main diagonal as well as below, so that the reduced system is $D\mathbf{x} = \hat{\mathbf{b}}$ where D is diagonal. How many operations does this method require to solve a linear system?

4.2.17. Show that if Gaussian elimination is done on the matrix (3.1.41) the factor L of the LU decomposition has the same non-zero structure as the lower triangular part of A except that the last row has, in general, all non-zero elements. Assuming that no operations are done on elements known to be zero, show that the operation count for this algorithm is $8(n-1)$ multiplications plus $6(n-1)$ additions plus $3n-2$ divisions.

4.2.18. Again for the matrix (3.1.41) show that solving $A\mathbf{x} = \mathbf{b}$ by the Sherman-Morrison formula (3.2.21) requires the following operations, assuming that the LU decomposition of T is done only once: $6n-1$ multiplications, $6n-1$ additions, and $3n-2$ divisions. Compare this operation count with that of Exercise 4.2.20. Show also that the use of the Sherman-Morrison-Woodbury formula (3.2.26) to solve this system, as described in Exercise 3.2.8, requires $8n-6$ multiplications, $8n-5$ additions, and $4n-3$ divisions.

4.2.19. Let T be a symmetric nonsingular tridiagonal matrix. Give an algorithm to find numbers $p_1, \cdots, p_n$ and $q_1, \cdots, q_n$ so that the i,j element of T^{-1} is $p_i q_j$ if $i \geq j$ and $p_j q_i$ if $i < j$. (Hint: Look at the first column of T^{-1}.)

4.3 Interchanges

In our discussion of the Gaussian elimination process in the previous section we assumed that a_{11} and all subsequent divisors were non-zero. However, we do not need to make such an assumption provided that we revise the algorithm so as to interchange equations if necessary, as we shall now describe.

We assume, as usual, that the coefficient matrix A is nonsingular. Suppose that $a_{11} = 0$. Then some other element in the first column of A must be non-zero or else A is singular (see Exercise 4.3.1). If, say, $a_{k1} \neq 0$, then we interchange the first equation in the system with the kth; clearly, this does not change the solution. In the new system the $(1,1)$ coefficient is now non-zero, and the elimination process can proceed. Similarly, an interchange can be done if any computed diagonal element that is to become a divisor in the next stage should vanish. Suppose, for example, that the elimination has progressed to the point

$$\begin{matrix} a_{11} & & & & \\ & \ddots & & & \\ & & a_{ii}^{(i-1)} & \cdots & a_{in}^{(i-1)} \\ & & \vdots & & \vdots \\ & & a_{ni}^{(i-1)} & \cdots & a_{nn}^{(i-1)} \end{matrix},$$

and that $a_{ii}^{(i-1)} = 0$. If all of the remaining elements below $a_{ii}^{(i-1)}$ in the ith column are also zero, this matrix is singular (see Exercise 4.3.2). Since the operations of adding a multiple of one row to another row, which produced this reduced matrix, do not affect the determinant, and an interchange of rows only changes the sign of the determinant, it follows that the original matrix is also singular, contrary to assumption. Hence at least one of the elements $a_{ki}^{(i-1)}$, $k = i+1, \ldots, n$, is non-zero, and we can interchange a row that contains a non-zero element with the ith row, thus ensuring that the new i, i element is non-zero. Again, this interchange of rows does not change the solution of the system. However, an interchange of rows does change the sign of the determinant of the coefficient matrix, so that if the determinant is to be computed a record must be kept of whether the number of interchanges is even or odd. In any case, in exact arithmetic we can ensure that Gaussian elimination can be carried out by interchanging rows so that no divisions are zero.

Rounding Error and Instability

Since in exact arithmetic the Gaussian elimination process produces the solution of the linear system in a finite number of steps, and there is no discretization error associated with the process, the only thing that can affect the accuracy of our computed solution is rounding error. There are two possibilities. The first is an accumulation of rounding errors during a large number of arithmetic operations. For example, if $n = 1,000$, the operation count of the previous section shows that on the order of $n^3 = 10^9$ operations will be required; even though the error in each individual operation may be small, the total buildup could be large. We shall see later that this potential accumulation of rounding error is not as serious as one might expect.

An Example of Instability

The second possibility involves catastrophic rounding errors. If an algorithm has this unfortunate characteristic, it is called *numerically unstable* and is not suitable as a general method. Although the interchange process described previously ensures that Gaussian elimination can be carried out mathematically for any nonsingular matrix, the algorithm can give rise to catastrophic rounding errors and is numerically unstable. We shall analyze a simple 2×2 example in order to see how this can occur.

Consider the system

$$\begin{bmatrix} -10^{-5} & 1 \\ 2 & 1 \end{bmatrix} \begin{bmatrix} x_1 \\ x_2 \end{bmatrix} = \begin{bmatrix} 1 \\ 0 \end{bmatrix}, \tag{4.3.1}$$

whose exact solution is

$$x_1 = -0.4999975\cdots, \qquad x_2 = 0.999995.$$

Now suppose that we have a decimal computer with a word length of four digits; that is, numbers are represented in the form $0.**** \times 10^p$. Let us carry out Gaussian elimination on this hypothetical computer. First, we note that $a_{11} \neq 0$, and no interchange is needed. The multiplier is

$$l_{21} = \frac{-0.2 \times 10^1}{0.1 \times 10^{-4}} = -0.2 \times 10^6,$$

which is exact, and the calculation for the new a_{22} is

$$\begin{aligned} a_{22}^{(1)} &= 0.1 \times 10^1 - (-0.2 \times 10^6)(0.1 \times 10^1) \\ &= 0.1 \times 10^1 + 0.2 \times 10^6 = 0.2 \times 10^6. \end{aligned} \tag{4.3.2}$$

The exact sum in (4.3.2) is, of course, 0.200001×10^6, but since the computer has a word length of only four digits this must be represented as 0.2000×10^6; this is the first error in the calculation.

The new b_2 is

$$b_2^{(1)} = -(-0.2 \times 10^6)(0.1 \times 10^1) = 0.2 \times 10^6. \tag{4.3.3}$$

No rounding errors occurred in this computation, nor do any occur in the back substitution:

$$\begin{aligned} x_2 &= \frac{b_2^{(1)}}{a_{22}^{(1)}} = \frac{0.2 \times 10^6}{0.2 \times 10^6} = 0.1 \times 10^1, \\ x_1 &= \frac{0.1 \times 10^1 - 0.1 \times 10^1}{-0.1 \times 10^{-4}} = 0. \end{aligned}$$

The computed x_2 agrees excellently with the exact x_2, but the computed x_1 has no digits of accuracy. Note that the only error made in the calculation is in $a_{22}^{(1)}$, which has an error in the sixth decimal place. Every other operation was exact. How, then, can this one "small" error cause the computed x_1 to deviate so drastically from its exact value?

Backward Error Analysis

The answer lies in the principle of *backward error analysis*, one of the most important concepts in scientific computing. The basic idea of backward error analysis is to "ask not what the error is, but what problem have we really solved." We shall invoke this principle here in the following form. Note that the quantity 0.000001×10^6 that was dropped from the computed $a_{22}^{(1)}$ in (4.3.2) is the original element a_{22}. Since this is the only place that a_{22} enters the calculation, the computed solution would have been the same if a_{22} were zero. Therefore the calculation on our four-digit computer has computed the exact solution of the system

$$\begin{bmatrix} -10^{-5} & 1 \\ 2 & 0 \end{bmatrix} \begin{bmatrix} x_1 \\ x_2 \end{bmatrix} = \begin{bmatrix} 1 \\ 0 \end{bmatrix}. \tag{4.3.4}$$

Intuitively, we would expect the two systems (4.3.1) and (4.3.4) to have rather different solutions, and this is indeed the case.

But why did this occur? The culprit is the large multiplier l_{21}, which made it impossible for a_{22} to be included in the sum in (4.3.2) because of the word length of the machine. The large multiplier was due to the smallness of a_{11} relative to a_{21}, and the remedy is, again, an interchange of the equations. Indeed, if we solve the system

$$\begin{bmatrix} 2 & 1 \\ -10^{-5} & 1 \end{bmatrix} \begin{bmatrix} x_1 \\ x_2 \end{bmatrix} = \begin{bmatrix} 0 \\ 1 \end{bmatrix} \tag{4.3.5}$$

on our hypothetical four-digit computer, we obtain

$$\begin{aligned}
l_{21} &= \frac{0.1 \times 10^{-4}}{0.2 \times 10^{1}} = -0.5 \times 10^{-5} \\
a_{22}^{(1)} &= 0.1 \times 10^{1} - (-0.5 \times 10^{-5})(1) = 0.1 \times 10^{1} \\
b_2^{(1)} &= 0.1 \times 10^{1} - (-0.5 \times 10^{-5})(0) = 0.1 \times 10^{1} \\
x_2 &= \frac{0.1 \times 10^{1}}{0.1 \times 10^{1}} = 1.0 \\
x_1 &= \frac{-(0.1 \times 10^{1})(1)}{0.2 \times 10^{1}} = -0.5\,.
\end{aligned}$$

The computed solution now agrees excellently with the exact solution.

Partial Pivoting

By a relatively simple strategy we can always arrange to keep the multipliers in the elimination process less than or equal to 1 in absolute value. This is known as *partial pivoting*: at the kth stage of the elimination process an interchange of rows is made, if necessary, to place in the main diagonal position the element of largest absolute value from the kth column below or on the main diagonal. If we include this interchange strategy in the forward reduction algorithm (4.2.7), we have:

Forward Reduction with Partial Pivoting

For $k = 1, \ldots, n-1$

  Find $m \geq k$ such that $|a_{mk}| = \max\{|a_{ik}| : i \geq k\}$.

  If $a_{mk} = 0$, then A is singular, and stop.

    else interchange a_{kj} and a_{mj}, $j = k, k+1, \ldots, n$.

    interchange b_k and b_m. (4.3.6)

  For $i = k+1, k+2, \ldots, n$

   $l_{ik} = a_{ik}/a_{kk}$

   For $j = k+1, k+2, \ldots, n$

    $a_{ij} = a_{ij} - l_{ik}a_{kj}$

   $b_i = b_i - l_{ik}b_k$.

Gaussian elimination with partial pivoting has proved to be an extremely reliable algorithm in practice. However, there are two major precautions that should be kept in mind. First, the matrix must be properly scaled before the algorithm is used. To illustrate this point, consider the system

$$\begin{bmatrix} 10 & -10^6 \\ 2 & 1 \end{bmatrix} \begin{bmatrix} x_1 \\ x_2 \end{bmatrix} = \begin{bmatrix} -10^6 \\ 0 \end{bmatrix}. \tag{4.3.7}$$

No interchange is called for by the partial pivoting strategy since the $(1, 1)$ element is already the largest in the first column. However, if we carry out the elimination on our hypothetical four-digit computer (see Exercise 4.3.5), we will encounter exactly the same problem that we did with the system (4.3.1). Indeed, (4.3.7) is just (4.3.1) with the first equation multiplied by -10^6.

The use of the partial pivoting strategy is predicated on the coefficient matrix being properly scaled so that the maximum element in each row and column is the same order of magnitude. This scaling is called *equilibration* or *balancing* of the matrix. Unfortunately, there is no known foolproof general procedure for such scaling, but usually it will be clear that some rows or columns of the matrix need scaling, and this can be done before the elimination starts. For example, if we were given the system (4.3.7), we should scale the first row so that its maximum element is approximately 1. Then a_{11} will be small, and the partial pivoting strategy will cause an interchange of the first and second rows.

The second precaution regarding the partial pivoting strategy is that even with an equilibrated matrix, it can be numerically unstable. Examples in which this can happen have been given, but the occurrence of such matrices in practical computations seems to be sufficiently rare that the danger can be safely ignored. (For additional remarks, see the Supplementary Discussion.)

LU with Interchanges

If row interchanges are made, the Gaussian elimination process is not equivalent to a factorization of the matrix A into the product of lower- and upper-triangular matrices; the lower-triangular matrix must be modified in the following way. A *permutation matrix*, P, is an $n \times n$ matrix that has exactly one element equal to 1 in each row and column and zeros elsewhere. A 4×4 example is

$$P = \begin{bmatrix} 1 & 0 & 0 & 0 \\ 0 & 0 & 0 & 1 \\ 0 & 0 & 1 & 0 \\ 0 & 1 & 0 & 0 \end{bmatrix}. \tag{4.3.8}$$

Interchange of rows of a matrix can be effected by multiplication on the left by a permutation matrix. For example, multiplication of a 4×4 matrix by the permutation matrix (4.3.8) will leave the first and third rows the same and interchange the second and fourth rows (see Exercise 4.3.6). Thus the row interchanges of the coefficient matrix A that are required by the partial pivoting strategy can be represented by multiplication of A on the left by suitable permutation matrices. If P_i denotes the permutation matrix corresponding to the interchange required at the ith stage, then conceptually we are generating the triangular factorization of the matrix

$$P_{n-1}P_{n-2}\cdots P_2P_1A = PA = LU \tag{4.3.9}$$

rather than A itself. Thus the factorization is $A = (P^{-1}L)U$. Since the product of permutation matrices is again a permutation matrix and the inverse of a permutation matrix is a permutation matrix (Exercise 4.3.7), the first factor is a permutation of a lower-triangular matrix, whereas the second is again upper-triangular. Note that if no interchange is required at the ith stage, the permutation matrix P_i is simply the identity matrix.

Banded Systems

Row interchanges require additional time and, in the case of banded matrices, also complicate the storage. Consider first a tridiagonal system. If an interchange is made at the first stage, the elements in the first two rows will be

$$\begin{array}{l} * \ * \ * \ 0 \cdots \\ * \ * \ 0 \cdots \end{array}$$

The elimination process will then reenter a (generally) non-zero element into the $(2,3)$ position, and the reduced $(n-1) \times (n-1)$ matrix will again be tridiagonal. Thus, the effect of the interchanges will be to introduce possible non-zero elements into the second superdiagonal of the reduced triangular matrix. Then the factor U in the decomposition of A will no longer be bidiagonal but will have, in general, three non-zero diagonals. Perhaps the simplest way of handling the storage is to add an additional one-dimensional array to hold these elements in the second superdiagonal.

For a banded matrix with semibandwidth p the same kind of problem occurs. If an interchange is made at the first stage between the first and $(p+1)$st rows, an additional p elements will be introduced into the first row, and these, in turn, will be propagated into rows 2 through $p+1$ during the elimination process. Thus we need to provide storage for a possible additional p superdiagonals. The simplest way to handle this is to allow for an additional $n \times p$ array of storage at the outset. Of course this requires an additional np storage locations. An alternative method is based on the observation that the amount of additional storage needed is no more than the amount of storage required for the non-zero subdiagonals. As the subdiagonals are eliminated, we no longer will need that storage, and the new superdiagonals elements can be stored in those positions. However, it is this subdiagonal space that is normally used to store the multipliers if their retention is desired; in this case we have no alternative but to set aside additional storage.

Diagonally Dominant and Positive Definite Matrices

Although for general nonsingular matrices it is necessary to use the partial pivoting strategy, there are some types of matrices for which it is known that no interchanges are necessary. The most important of these are diagonally dominant matrices [recall (3.1.24) and (3.1.25)] and symmetric positive definite matrices [recall (4.1.9)]. In both cases, it is safe to use Gaussian elimination with no interchanges at all (see the Supplementary Discussion), although for positive definite matrices accuracy is often improved slightly by using interchanges. Not needing to interchange is especially advantageous for banded matrices, and it is a fortunate fact that most banded matrices arising from differential equations are either diagonally dominant or symmetric and positive definite. In particular, no interchanges are needed for those tridiagonal matrices that were shown to be diagonally dominant in Chapter 3.

In this section, we have discussed various questions concerning the accuracy of computed solutions of systems of linear equations. In the following section we will consider additional questions, including the important topic of "ill-conditioning."

Supplementary Discussion and References: 4.3

The interchange of rows required by partial pivoting need not be done explicitly. Instead, the interchanges may be carried out implicitly by using a permutation vector that keeps track of which rows are interchanged. Whether one should use explicit interchanges depends on the computer's "interchange" time, time required for indexing, program clarity, and other considerations.

In those cases in which partial pivoting is not sufficient to guarantee accuracy, we can use another strategy called *complete pivoting,* in which both rows and columns are interchanged so as to bring into the diagonal divisor position the largest element in absolute value in the remaining submatrix to be processed. This adds a significant amount of time to the Gaussian elimination process and is rarely incorporated into a standard program. See Wilkinson [1961] for further discussion.

In a very important paper, Wilkinson [1961] showed that the effect of rounding errors in Gaussian elimination is such that the computed solution is the exact solution of a perturbed system $(A+E)\mathbf{x} = \mathbf{b}$ (see also, for example, Golub and van Loan [1989] and Ortega [1990] for textbook discussions). A bound on the matrix E is of the form

$$||E||_\infty \leq p(n)g(n)\varepsilon||A||_\infty,$$

where $p(n)$ is a cubic polynomial in the size of the matrix, ε is the basic rounding error of the computer (for example, 2^{-27}), and $g(n)$ is the *growth factor* defined by

$$g = \frac{\max_{i,j,k} |a_{ij}^{(k)}|}{a}, \qquad a = \max_{i,j} |a_{ij}|,$$

where the $a_{ij}^{(k)}$ are the elements of the successive reduced matrices formed in the elimination process. The growth factor depends crucially on the interchange strategy used. With no interchanges, g may be arbitrarily large. With partial pivoting (and in exact arithmetic), $g(n)$ is bounded by 2^{n-1}, which for large n completely dominates $p(n)$. Wilkinson has exhibited matrices for which $g(n) = 2^{n-1}$, but such matrices seem to be very rare in practice; indeed, the actual size of g has been monitored extensively by Wilkinson and others for a large number of practical problems and has seldom exceeded 10, regardless of the size of the matrix. For the complete pivoting strategy a complicated but much better bound for g has been given by Wilkinson, and a long-standing conjecture was that $g(n) \leq n$. This conjecture has recently been shown by Gould [1991] to be false if rounding error in Gaussian elimination is allowed. Subsequently Edelman and Ohlrich [1991] used Mathematica to show that the conjecture is also false in exact arithmetic. The form of the best bound for g using complete pivoting remains an open question.

For matrices that are (column) diagonally dominant, the growth factor g is bounded by 2, without any interchanges. For symmetric positive definite matrices, g is equal to 1. This explains why for these two important classes of matrices, no interchange strategy is necessary.

The partial pivoting strategy ensures that the multipliers are all less than or equal to one in magnitude. However, for problems where the matrix has few non-zero elements it is sometimes desirable to require only that $|l_{ij}| \leq \alpha$ for some "threshold" parameter $\alpha > 1$. For further discussion, see Section 9.2.

EXERCISES 4.3

4.3.1. Suppose that the ith column of the matrix A consists of zero elements. Show that A is singular by the following different arguments:

a. The determinant of A is zero.

b. $A\mathbf{e}_i = 0$, where $\mathbf{e}_i$ is the vector with 1 in the ith position and zeros elsewhere.

c. A has, at most, $n - 1$ linearly independent columns.

4.3.2. Let A be a matrix of the form

$$\begin{bmatrix} a_{11} & \cdots & & & & a_{1n} \\ & \ddots & & & & \vdots \\ & & a_{i-1,i-1} & & & \\ & & & a_{ii} & & \\ & & & \vdots & \ddots & \\ & & & a_{i+1,i} & & \\ & & & a_{ni} & \cdots & a_{nn} \end{bmatrix}.$$

If $a_{ii} = a_{i+1,i} = \cdots = a_{ni} = 0$, show that A is singular.

4.3.3. Solve the following 3×3 system by Gaussian elimination by making row interchanges where needed to avoid division by zero:

$$\begin{aligned} 2x_1 + 2x_2 + 3x_3 &= 1 \\ x_1 + x_2 + 2x_3 &= 2 \\ 2x_1 + x_2 + 2x_3 &= 3 \end{aligned}$$

4.3.4. Translate the algorithm (4.3.6) into a computer program for Gaussian elimination using partial pivoting. Include back substitution.

4.3.5. Apply Gaussian elimination to the system (4.3.7) using the four-digit decimal computer of the text. Repeat the calculation after interchanging the equations.

4.3.6. **a.** Show that multiplication of a 4×4 matrix on the left by the permutation matrix (4.3.8) interchanges the second and fourth rows and leaves the first and third rows the same.

b. Show that multiplication on the right by the permutation matrix interchanges the second and fourth columns.

c. Give the 4×4 permutation matrix that interchanges the first and third rows and leaves the second and fourth rows the same.

4.3.7. Show that the product of two $n \times n$ permutation matrices is a permutation matrix. Show that the inverse of a permutation matrix is a permutation matrix.

4.3.8. Let $A = LL^T$ be a factorization of a symmetric positive definite matrix A, where L is lower-triangular and has positive main diagonal elements. Show that if $\hat{L}$ is obtained from L by changing the sign of every element of the ith row, then $A = \hat{L}\hat{L}^T$. (This shows that the LL^T factorization is not unique, although there is a unique L with positive main diagonal elements.)

4.3.9. A matrix H is called *Hessenberg* (see Chapter 7) if $h_{ij} = 0$ when $i > j + 1$. How many operations are required to solve $H\mathbf{x} = \mathbf{b}$ by Gaussian elimination? If H is normalized so that $|h_{ij}| \leq 1$ for all i, j, and partial pivoting is used in Gaussian elimination, show that the elements of U are less than n in magnitude.

4.3.10. Consider a matrix of the form

$$\begin{bmatrix} * & * & \cdots & * \\ * & * & & \\ \vdots & & \ddots & \\ * & & & * \end{bmatrix},$$

in which all elements are zero except in the first row and column and the main diagonal. How many operations will Gaussian elimination require if no pivoting is used? Does pivoting change this? Can you find a reordering of the equations and unknowns so that only $0(n)$ operations are required?

4.3.11. Assume that the positive definite condition $\mathbf{x}^T A\mathbf{x} > 0$ if $\mathbf{x} \neq 0$ holds for all real $\mathbf{x}$ even though A is not symmetric (such matrices are sometimes called *positive real*). Show that an LU decomposition of A exists.

4.4 Ill-Conditioning and Error Analysis

The Gaussian elimination algorithm with partial pivoting has proved to be an efficient and reliable method in practice. Nevertheless, it may fail to compute accurate solutions of systems of equations that are "ill-conditioned". A linear system of equations is said to be *ill-conditioned* if small changes in the elements of the coefficient matrix and/or right-hand side cause large changes in the solution. In this case no numerical method can be expected to produce an accurate solution, nor, in many cases, should a solution even be attempted.

An Example of an Ill-conditioned System

We begin with a 2×2 example. Consider the system

$$\begin{aligned} 0.832x_1 + 0.448x_2 &= 1.00 \\ 0.784x_1 + 0.421x_2 &= 0, \end{aligned} \tag{4.4.1}$$

and assume that we use a three-digit decimal computer to carry out Gaussian elimination. Since a_{11} is the largest element of the matrix no interchange is required, and the computation of the new elements $a_{22}^{(1)}$ and $b_1^{(1)}$ is

$$\begin{aligned} l_{21} &= \frac{0.784}{0.832} = 0.942 \mid 308 \cdots = 0.942 \\ a_{22}^{(1)} &= 0.421 - 0.942 \times 0.448 = 0.421 - 0.422 \mid 016 = -0.001 \\ b_2^{(1)} &= 0 - 1.00 \times 0.942 = -0.942, \end{aligned} \tag{4.4.2}$$

where we have indicated by the vertical bars those digits lost in the computation. Hence the computed triangular system is

$$\begin{aligned} 0.832x_1 + 0.448x_2 &= 1.00 \\ -0.001x_2 &= -0.942, \end{aligned}$$

and the back substitution produces the approximate solution

$$x_1 = -506, \qquad x_2 = 942. \tag{4.4.3}$$

But the exact solution of (4.4.1), correct to three figures, is

$$x_1 = -439, \qquad x_2 = 817, \tag{4.4.4}$$

so the computed solution is incorrect by about 15%. Why has this occurred?

The first easy answer is that we have lost significance in the calculation of $a_{22}^{(1)}$. Indeed, it is clear that the computed value of $a_{22}^{(1)}$ has only one significant figure, so our final computed solution will have no more than one significant figure. But this is only the manifestation of the real problem. We invoke again the principle of backward error analysis. By carrying out a more detailed computation we can show that the computed solution (4.4.3) is the exact solution of the system

$$\begin{aligned} 0.832x_1 + 0.447974\cdots x_2 &= 1.00 \\ 0.783744\cdots x_1 + 0.420992\cdots x_2 &= 0. \end{aligned} \tag{4.4.5}$$

The maximum percentage change between the elements of this system and the original system (4.4.1) is only 0.03%; therefore errors in the data are magnified by a factor of about 500.

The root cause of this ill-conditioning is that the coefficient matrix of (4.4.1) is "almost singular." Geometrically, this means that the lines defined by the two equations (4.4.1) are almost parallel, as indicated in Figure 4.3. Consider now the system of equations

$$\begin{aligned} 0.832x_1 + 0.448x_2 &= 1.00 \\ 0.784x_1 + (0.421 + \varepsilon)x_2 &= 0. \end{aligned} \tag{4.4.6}$$

The second equation defines a family of lines depending on the parameter ε. As ε increases from zero to approximately 0.0012, the line rotates counterclockwise and its intersection with the line defined by the first equation recedes to infinity until the two lines become exactly parallel and no solution of the linear system exists.

For only one value of ε, say ε_0, is the coefficient matrix of (4.4.6) singular, but for infinitely many values of ε near ε_0 the matrix is almost singular. In general, the probability of a matrix being exactly singular is very small unless it was constructed in such a way that singularity is ensured. For example, we saw in Section 3.2 that periodic boundary conditions can give rise to singular coefficient matrices. In many situations, however, it may not be obvious in the formulation of the problem that the resulting matrix will be singular or almost singular. This must be detected during the course of the solution and a warning issued. We will consider such detection mechanisms later, but we point out here that, in general, it is extremely difficult to ascertain computationally if a given matrix is exactly singular. For example, if LU is the computed factorization of A and $u_{nn} = 0$, then U is singular. But u_{nn} may be contaminated by rounding error so we cannot claim that A itself is singular. Conversely, if the computed u_{nn} is not zero, this does not guarantee that A is non-singular. The fundamental problem is the detection of zero in the presence of rounding error.

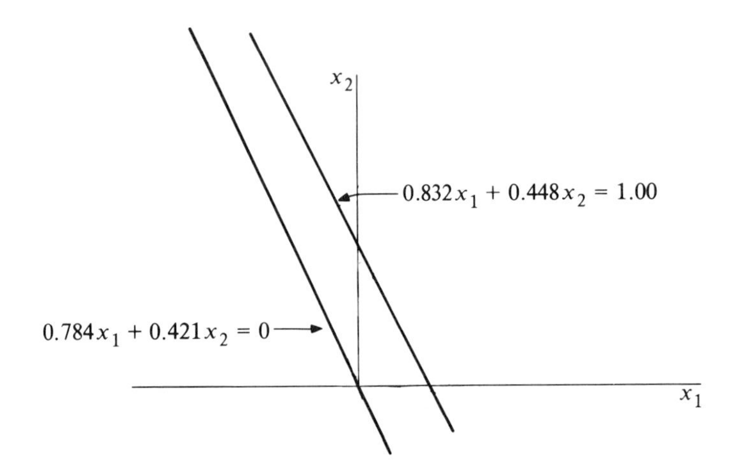

Figure 4.3: *Almost-Parallel Lines Defined by* (4.4.1). *The Intersection of the Lines is at* (-439,817)

However, near-singularity is easier to detect, and if that is the case, it is likely that the problem should be reformulated. For example, we may have chosen variables that are close to being dependent, and we should remove some of them or choose another set of variables.

Determinants and Ill-conditioning

Since a matrix is singular if its determinant is zero, it is sometimes suggested that the smallness of the determinant is a measure of the ill-conditioning of the system. This is not however generally true as the following example shows:

$$\det\begin{bmatrix} 10^{-10} & 0 \\ 0 & 10^{-10} \end{bmatrix} = 10^{-20}, \qquad \det\begin{bmatrix} 10^{10} & 0 \\ 0 & 10^{10} \end{bmatrix} = 10^{20}. \tag{4.4.7}$$

The values of the two determinants are very different, but the lines defined by the two corresponding sets of equations

$$\begin{aligned} 10^{-10}x_1 &= 0 \qquad & 10^{10}x_1 &= 0 \\ 10^{-10}x_2 &= 0 \qquad & 10^{10}x_2 &= 0 \end{aligned} \tag{4.4.8}$$

are the same and are the coordinate axes. As we shall see more clearly in a moment, if the lines defined by the equations of a system are perpendicular, that system is "perfectly conditioned." Thus the magnitude of the determinant of the coefficient matrix is not a good measure of the near-singularity of the matrix. It can, however, become the basis of such a measure if the matrix is suitably scaled, as we shall now see.

For two equations it is clear that a good measure of the "almost parallelness" of the corresponding two lines is the angle between them. An essentially equivalent measure is the area of the parallelogram shown in Figure 4.4, in which the sides of the parallelogram are of length 1 and the height is denoted by h. The area of the parallelogram is then equal to h since the base is 1, and the angle θ between the lines defined by the two equations is related to h by $h = \sin\theta$. The area, h, varies between zero, when the lines coalesce, and 1, when they are perpendicular.

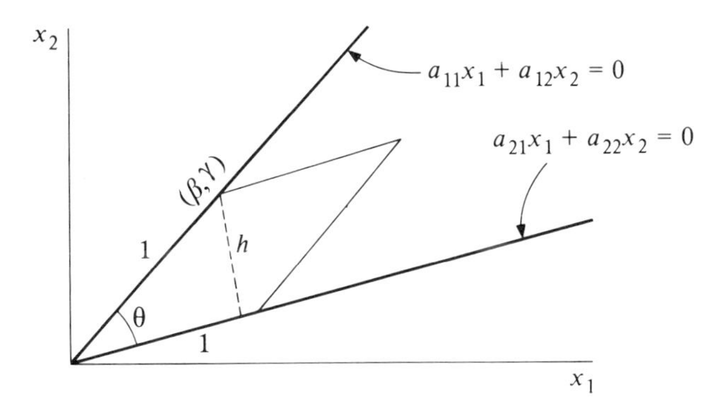

Figure 4.4: *The Unit Parallelogram*

From analytic geometry, the distance from the point (β, γ) to the line $a_{21}x_1 + a_{22}x_2 = 0$ is

$$h = \frac{|a_{21}\beta + a_{22}\gamma|}{\alpha_2}, \qquad \alpha_2 = (a_{21}^2 + a_{22}^2)^{1/2}.$$

If we assume that $a_{11} \geq 0$, the coordinates (β, γ) are given by

$$\beta = \frac{-a_{12}}{\alpha_1}, \qquad \gamma = \frac{a_{11}}{\alpha_1}, \qquad \alpha_1 = (a_{11}^2 + a_{12}^2)^{1/2},$$

so that

$$h = \frac{|a_{11}a_{22} - a_{21}a_{12}|}{\alpha_1\alpha_2} = \frac{|\det A|}{\alpha_1\alpha_2}. \tag{4.4.9}$$

Hence we see that the area, h, is just the determinant divided by the product $\alpha_1\alpha_2$.

This measure easily extends to n equations. Let $A = (a_{ij})$ be the coefficient matrix, and set

$$V = \frac{|\det A|}{\alpha_1\alpha_2\cdots\alpha_n} = \left|\det \begin{bmatrix} a_{11}/\alpha_1 & \cdots & a_{1n}/\alpha_1 \\ \vdots & & \vdots \\ a_{n1}/\alpha_n & \cdots & a_{nn}/\alpha_n \end{bmatrix}\right|, \tag{4.4.10}$$

where

$$\alpha_i = (a_{i1}^2 + a_{i2}^2 + \cdots + a_{in}^2)^{1/2}.$$

We have called the quantity in (4.4.10) V instead of h because it is the volume of the n-dimensional unit parallelepiped circumscribed by the lines defined by the rows of matrix A; that is,

$$\frac{1}{\alpha_i}(a_{i1}, a_{i2}, \ldots, a_{in}), \qquad i = 1, \ldots, n,$$

are the coordinates of n points in n-dimensional space located Euclidean distance 1 from the origin, and these n points define a parallelepiped whose sides are of length 1. It is intuitively clear, and can be proved rigorously, that the volume of this parallelepiped is between zero, when two or more of the edges coincide, and 1, when the edges are all mutually perpendicular. If $V = 0$, then $\det A = 0$, and the matrix is singular. If $V = 1$, then the edges are as far from being parallel as possible, and in this case the matrix is called *perfectly conditioned.*

Ramifications of Ill-Conditioning

There are various ramifications of ill-conditioning of a matrix besides the difficulty in computing an accurate solution of the corresponding linear system. Consider again the system (4.4.1) and suppose that (4.4.5) is the "real" system that we wish to solve but that the coefficients of this system must be measured by some physical apparatus accurate to only the third decimal place. Thus (4.4.1) is not the system that we really wish to solve but is the best approximation to it that we can make. Suppose that we can also claim that the coefficients of (4.4.1) are accurate to at least 0.05%, as indeed they are, compared with (4.4.5). Then, it is an often-heard argument that we should be able to compute the solution of the system to about the same accuracy. But we have seen that this is not true; the ill-conditioning of the coefficient matrix magnifies small errors in the coefficients by a factor of about 500 in the case of (4.4.1). Hence no matter how accurately the system (4.4.1) is solved, we will still have the error that has come from the measurement error in the coefficients. If, for example, we need the solution of the "real" system

(4.4.5) accurate to less than 1%, we need to measure the coefficients much more accurately than three decimal places.

In some cases, however, the coefficient matrix may be exact. A famous example of a class of ill-conditioned matrices is the *Hilbert matrices* (or *Hilbert segments*), in which the elements of the matrix are exact rational numbers:

$$H_n = \begin{bmatrix} 1 & 1/2 & \cdots & 1/n \\ 1/2 & & & \\ \vdots & & & \vdots \\ 1/n & & \cdots & 1/(2n-1) \end{bmatrix}. \tag{4.4.11}$$

These matrices are increasingly ill-conditioned as n increases. If for $n = 8$ the coefficients are entered in the computer as binary fractions exact to the extent possible with 27 binary digits (equivalent to about 8 decimal digits), the exact inverse of the matrix in the computer differs from the exact inverse of H_8 in the first figure!

The following is another manifestation of ill-conditioning. Suppose that $\bar{\mathbf{x}}$ is a computed solution of the system $A\bar{\mathbf{x}} = \mathbf{b}$. One way to try to ascertain the accuracy of $\bar{\mathbf{x}}$ is to form the *residual vector*,

$$\mathbf{r} = \mathbf{b} - A\bar{\mathbf{x}}. \tag{4.4.12}$$

If $\bar{\mathbf{x}}$ were the exact solution, then $\mathbf{r}$ would be zero. Thus we would expect $\mathbf{r}$ to be "small" if $\bar{\mathbf{x}}$ were a good approximation to the exact solution, and, conversely, that if $\mathbf{r}$ were small, then $\bar{\mathbf{x}}$ would be a good approximation. This is true in some cases, but if A is ill-conditioned, the magnitude of $\mathbf{r}$ can be very misleading. As an example, consider the system

$$\begin{aligned} 0.780x_1 + 0.563x_2 &= 0.217 \\ 0.913x_1 + 0.659x_2 &= 0.254, \end{aligned} \tag{4.4.13}$$

and the approximate solution

$$\bar{\mathbf{x}} = \begin{bmatrix} 0.341 \\ -0.087 \end{bmatrix}. \tag{4.4.14}$$

Then, the residual vector is

$$\mathbf{r} = \begin{bmatrix} 10^{-6} \\ 0 \end{bmatrix}. \tag{4.4.15}$$

Now consider another very different approximate solution

$$\bar{\mathbf{x}} = \begin{bmatrix} 0.999 \\ -1.001 \end{bmatrix}, \tag{4.4.16}$$

and the corresponding residual vector

$$\mathbf{r} = \begin{bmatrix} 0.0013\cdots \\ -0.0015\cdots \end{bmatrix}. \tag{4.4.17}$$

By comparing the residuals (4.4.15) and (4.4.17) we could easily conclude that (4.4.14) is the better approximate solution. However, the exact solution of (4.4.13) is $(1, -1)$, so the residuals give completely misleading information.

Condition Numbers Based on Norms

We turn now to another way of measuring the ill-conditioning of a matrix by means of norms (see Appendix 2 for a review of vector and matrix norms). Suppose first that $\hat{\mathbf{x}}$ is the solution of $A\mathbf{x} = \mathbf{b}$ and that $\hat{\mathbf{x}} + \Delta\mathbf{x}$ is the solution of the system with the right-hand side $\mathbf{b} + \Delta\mathbf{b}$:

$$A(\hat{\mathbf{x}} + \Delta\mathbf{x}) = \mathbf{b} + \Delta\mathbf{b}. \tag{4.4.18}$$

Since $A\hat{\mathbf{x}} = \mathbf{b}$, it follows that $A(\Delta\mathbf{x}) = \Delta\mathbf{b}$ and $\Delta\mathbf{x} = A^{-1}(\Delta\mathbf{b})$, assuming, as usual, that A is nonsingular. Thus

$$||\Delta\mathbf{x}|| \le ||A^{-1}||\ ||\Delta\mathbf{b}||, \tag{4.4.19}$$

which shows that the change in the solution due to a change in the right-hand side is bounded by $||A^{-1}||$. Thus a small change in $\mathbf{b}$ may cause a large change in $\hat{\mathbf{x}}$ if $||A^{-1}||$ is large. The notion of "large" is always relative, however, and it is more useful to deal with the relative change $||\Delta\mathbf{x}||/||\hat{\mathbf{x}}||$. From $A\hat{\mathbf{x}} = \mathbf{b}$, it follows that

$$||\mathbf{b}|| \le ||A||\ ||\hat{\mathbf{x}}||,$$

and combining this with (4.4.19) yields

$$||\Delta\mathbf{x}||\ ||\mathbf{b}|| \le ||A||\ ||A^{-1}||\ ||\Delta\mathbf{b}||\ ||\hat{\mathbf{x}}||,$$

or, equivalently (if $\mathbf{b} \neq 0$),

$$\frac{||\Delta\mathbf{x}||}{||\hat{\mathbf{x}}||} \le ||A||\ ||A^{-1}||\frac{||\Delta\mathbf{b}||}{||\mathbf{b}||}. \tag{4.4.20}$$

This inequality shows that the relative change in $\hat{\mathbf{x}}$ due to a change in $\mathbf{b}$ is bounded by the relative change in $\mathbf{b}$, $||\Delta\mathbf{b}||/||\mathbf{b}||$, multiplied by $||A||\ ||A^{-1}||$. The latter quantity is of great importance and is called the *condition number* of A (with respect to the norm being used); it will be denoted by cond(A). This is the condition number for the problem of solving $A\mathbf{x} = \mathbf{b}$ or computing A^{-1}. Other problems will have different condition numbers. For example, in Chapter 7 we discuss the computation of eigenvalues, and the condition number of an eigenvalue of A is different than cond(A).

Consider next the case in which the elements of A are changed so that the perturbed equations are

$$(A + \delta A)(\hat{\mathbf{x}} + \delta\mathbf{x}) = \mathbf{b}. \tag{4.4.21}$$

Thus, since $A\hat{\mathbf{x}} = \mathbf{b}$,

$$A\delta\mathbf{x} = \mathbf{b} - A\hat{\mathbf{x}} - \delta A(\hat{\mathbf{x}} + \delta\mathbf{x}) = -\delta A(\hat{\mathbf{x}} + \delta\mathbf{x}),$$

or

$$-\delta\mathbf{x} = A^{-1}\delta A(\hat{\mathbf{x}} + \delta\mathbf{x}).$$

Therefore

$$||\delta\mathbf{x}|| \le ||A^{-1}||\ ||\delta A||\ ||\hat{\mathbf{x}} + \delta\mathbf{x}|| = \text{cond}(A)\frac{||\delta A||}{||A||}||\hat{\mathbf{x}} + \delta\mathbf{x}||,$$

so that

$$\frac{||\delta\mathbf{x}||}{||\hat{\mathbf{x}} + \delta\mathbf{x}||} \le \text{cond}(A)\frac{||\delta A||}{||A||}. \tag{4.4.22}$$

Once again, the condition number plays a major role in the bound. Note that (4.4.22) expresses the change in $\hat{\mathbf{x}}$ relative to the perturbed solution, $\hat{\mathbf{x}} + \delta\mathbf{x}$, rather than $\hat{\mathbf{x}}$ itself, as in (4.4.20), although it is possible to obtain a bound relative to $\hat{\mathbf{x}}$.

The inequalities (4.4.20) and (4.4.22) need to be interpreted correctly. Note first that cond(A) ≥ 1 (see Exercise 4.4.2). If cond(A) is close to 1, then small relative changes in the data can lead to only small relative changes in the solution. In this case we say that the problem is *well-conditioned.* This also guarantees that the residual vector provides a valid estimate of the accuracy of an approximate solution $\bar{\mathbf{x}}$. From (4.4.12)

$$\mathbf{r} = A(A^{-1}\mathbf{b} - \bar{\mathbf{x}}), \tag{4.4.23}$$

so that, if $\mathbf{e} = A^{-1}\mathbf{b} - \bar{\mathbf{x}}$ is the error in the approximate solution,

$$\mathbf{e} = A^{-1}\mathbf{r}. \tag{4.4.24}$$

This is the fundamental relation between the residual and the error. Then

$$||\mathbf{e}|| \le ||A^{-1}||\ ||\mathbf{r}|| = \text{cond}(A)\frac{||\mathbf{r}||}{||A||}, \tag{4.4.25}$$

so that the error is bounded by cond(A) times a normalized residual vector. Note that it is necessary to normalize the residual vector somehow since we can multiply the equation $A\mathbf{x} = \mathbf{b}$ by any constant without changing the solution, and such a multiplication would change the residual by the same amount.

On the other hand, if the condition number is large, then small changes in the data may cause large changes in the solution, but not necessarily, depending on the particular perturbation. The practical effect of a large condition number depends on the accuracy of the data and the word length of the computer being used. If, for example, $\text{cond}(A) = 10^6$, then the equivalent of 6 decimal digits could possibly be lost. On a computer with a word length equivalent to 8 decimal digits, this could be disastrous; on the other hand if the word length were the equivalent of 16 decimal digits, it might not cause much of a problem. If the data are measured quantities, however, the computed solution may not have any meaning even if computed accurately.

Computation of the Condition Number

In general, it is very difficult to compute the condition number $||A||\ ||A^{-1}||$ without knowing A^{-1}, although the packages LINPACK and LAPACK (see the Supplementary Discussion) are able to estimate $\text{cond}(A)$ in the course of solving a linear system. In some cases of interest, however, it is relatively easy to compute the condition number explicitly, and we give an example of this for the $(2, -1)$ tridiagonal matrix of (3.1.10).

As given in Appendix 2, the l_2 norm of a symmetric matrix is its spectral radius $\rho(A)$. Thus

$$\text{cond}_2(A) = ||A||_2||A^{-1}||_2 = \rho(A)\rho(A^{-1}). \tag{4.4.26}$$

For the matrix of (3.1.10), we can compute explicitly (Exercise 4.4.5) the eigenvalues as

$$\lambda_k = 2 - 2\cos\frac{k\pi}{n+1} = 2 - 2\cos kh, \tag{4.4.27}$$

where we have set $h = \pi/(n+1)$. Thus, the largest eigenvalue of A is

$$\rho(A) = \lambda_n = 2 - 2\cos nh,$$

and the smallest is

$$\lambda_1 = 2 - 2\cos h > 0.$$

(This shows, incidentally, that A is positive definite; see Appendix 2). Since the eigenvalues of A^{-1} are $\lambda_1^{-1}, \ldots, \lambda_n^{-1}$ (see Exercise 4.4.6), the spectral radius of A^{-1} is λ_1^{-1}. Thus

$$\text{cond}_2(A) = \lambda_n\lambda_1^{-1} = \frac{1-\cos nh}{1-\cos h} = \frac{1+\cos h}{1-\cos h}.$$

For small h we can approximate the cosine by the first-order Taylor expansion

$$\cos h \doteq 1 - \frac{h^2}{2}.$$

Thus

$$\text{cond}_2(A) \doteq \frac{4 - h^2}{h^2} = 0(h^{-2}) = 0(n^2).$$

This shows that A is moderately ill-conditioned and that the condition number grows approximately as the square of the dimension of the matrix, or as h^{-2}. This is typical for matrices arising from boundary value problems.

Supplementary Discussion and References: 4.4

The state of the art in solving linear equations has now reached a very high level, especially for full and banded systems that can be stored in fast memory. Probably the best current set of codes is LINPACK, a package of FORTRAN subroutines. Recently LINPACK has evolved to a new package, LAPACK, specifically designed for use on vector and parallel computers. See Dongarra et al. [1979] for a discussion of LINPACK and Dongarra and Anderson et al. [1990] for LAPACK.

One way to attempt to obtain an accurate solution of ill-conditioned systems – and also to detect the ill-conditioning – is *iterative refinement*, which we now describe. Let $\mathbf{x}_1$ be the computed solution of the system $A\mathbf{x} = \mathbf{b}$ and $\mathbf{r}_1 = A\mathbf{x}_1 - \mathbf{b}$. If $\mathbf{x}_1$ is not the exact solution, then $\mathbf{r}_1 \neq 0$. Now solve the system $A\mathbf{z}_1 = -\mathbf{r}_1$. If $\mathbf{z}_1$ were the exact solution of this system, then

$$\mathbf{A}(\mathbf{x}_1 + \mathbf{z}_1) = A\mathbf{x}_1 - \mathbf{r}_1 = \mathbf{b}$$

so that $\mathbf{x}_1 + \mathbf{z}_1$ is the exact solution of the original system. Of course, we will not be able to compute $\mathbf{z}_1$ exactly, but we hope that $\mathbf{x}_2 = \mathbf{x}_1 + \mathbf{z}_1$ will be a better approximation to the exact solution than $\mathbf{x}_1$. For this to be the case, it is usual to compute the residual in double precision, although single precision can be sufficient in certain cases (see Skeel [1980] and Arioli et al. [1989]). The process can be repeated: form $\mathbf{r}_2 = A\mathbf{x}_2 - \mathbf{b}$, solve $A\mathbf{z}_2 = -\mathbf{r}_2$, set $\mathbf{x}_3 = \mathbf{x}_2 + \mathbf{z}_2$, and so on. One or two iterations will usually suffice to obtain an accurate solution, unless the problem is very ill-conditioned. For further discussion of iterative refinement, see Golub and Van Loan [1989].

For additional perturbation results such as (4.4.20) see Stewart and Sun [1990].

EXERCISES 4.4

4.4.1. Compute the determinant and the normalized determinant (4.4.10) for the matrix of (4.4.1) and for the matrix

$$A = \begin{bmatrix} 1 & 2 & 3 \\ 2 & 3 & 4 \\ 3 & 4 & 4 \end{bmatrix}.$$

4.4.2. Using properties of matrix norms, prove that $\text{cond}(A) \geq 1$.

4.4.3. Compute $\text{cond}(A)$ for the matrices in Exercise 4.4.1 using both the l_1 and l_∞ norms (see Appendix 2 for definitions of these norms).

4.4.4. Solve the system (4.4.1) for different right-hand sides. Compare the differences in these solutions to the bound (4.4.20), using the l_∞ norm.

4.4.5. Use the trigonometric identity $\sin(\alpha \pm \beta) = \sin\alpha\cos\beta \pm \cos\alpha\sin\beta$ to verify that the eigenvalues of the matrix (3.1.10) are given by (4.4.27) with corresponding eigenvectors

$$\mathbf{x}_k = (\sin kh, \sin 2kh, \ldots, \sin nkh)^T.$$

That is, verify that $A\mathbf{x}_k = \lambda_k \mathbf{x}_k$, $k = 1, \ldots, n$.

4.4.6. If A is nonsingular and $A\mathbf{x} = \lambda\mathbf{x}$, show that $A^{-1}\mathbf{x} = \lambda^{-1}\mathbf{x}$.

4.4.7. If A has eigenvalues $\lambda_1, \ldots, \lambda_n$ and corresponding eigenvectors $\mathbf{v}_n \ldots, \mathbf{v}_n$, show that $(A + cI)\mathbf{v}_k = (\lambda_k + c)\mathbf{v}_k$, $k = 1, \ldots, n$, so that $A + cI$ has eigenvalues $\lambda_1 + c, \ldots, \lambda_n + c$.

4.4.8. Let A be the matrix of (3.1.9) in which $c_1 = c_2 = \cdots = c_n = c$. Use Exercise 4.4.7 to show that A has eigenvalues $2 + c - 2coskh$, $k = 1, \ldots, n$, and use them to find $\text{cond}_2(A)$. Discuss how $\text{cond}_2(A)$ varies with c.

4.4.9 Let

$$A = \begin{pmatrix} 1.6384 & 0.8065 \\ 0.8321 & 0.4096 \end{pmatrix}, \mathbf{b} = \begin{pmatrix} 0.8319 \\ 0.4225 \end{pmatrix}.$$

Verify that $(1, -1)^T$ is the exact solution of $A\mathbf{x} = \mathbf{b}$. If $\mathbf{r} = A\bar{\mathbf{x}} - \mathbf{b}$, construct an $\bar{\mathbf{x}}$ for which $\mathbf{r} = (-10^{-8}, 10^{-8})$ exactly. Find $\text{cond}_\infty(A)$. If $\mathbf{b}$ is exact, how small should the relative error in A be so that the solution can be guaranteed to have a relative error which is $\leq 10^{-8}$?

4.4.10 Let $\mathbf{r} = \mathbf{A}\bar{\mathbf{x}} - \mathbf{b}, \bar{\mathbf{x}} = \mathbf{x} + \delta\mathbf{x}, \mathbf{A}\mathbf{x} = \mathbf{b}$ and $R = AC - I$, where C is an approximate inverse of A and $\|R\| < 1$. Prove that

$$\|\delta\mathbf{x}\| \leq \frac{\|\mathbf{r}\|\|\mathbf{C}\|}{\mathbf{1} - \|\mathbf{R}\|}.$$

4.4.11 Let A be a non-singular diagonal matrix. Show that the quantity V of (4.4.10) is always equal to 1, but that $\|A\|\|A^{-1}\|$ may be arbitrarily large.

4.5 Other Factorizations

So far in this chapter we have considered only Gaussian elimination and the corresponding LU factorization. But there are other factorizations of the matrix A which are sometimes very useful.

Cholesky Factorization

In the case of a symmetric positive definite matrix there is an important variant of Gaussian elimination, *Cholesky's method*, which is based on a factorization (or decomposition) of the form

$$A = LL^T. \tag{4.5.1}$$

Here L is a lower-triangular matrix but does not necessarily have 1's on the main diagonal as in the LU factorization. The factorization (4.5.1) is unique, provided that L is required to have positive diagonal elements (see Exercise 4.5.1).

The product in (4.5.1) is

$$\begin{bmatrix} l_{11} & & & & \\ \vdots & \ddots & & & \\ l_{il} & & l_{ii} & & \\ \vdots & & & \ddots & \\ l_{n1} & & \cdots & & l_{nn} \end{bmatrix} \begin{bmatrix} l_{11} & \cdots & l_{i1} & \cdots & l_{n1} \\ & \ddots & & & \\ & & l_{ii} & & \vdots \\ & & & \ddots & \\ & & & & l_{nn} \end{bmatrix}. \tag{4.5.2}$$

By equating elements of the first column of (4.5.2) with corresponding elements of A, we see that $a_{i1} = l_{i1}l_{11}$, so the first column of L is determined by

$$l_{11} = (a_{11})^{1/2}, \qquad l_{i1} = \frac{a_{i1}}{l_{11}}, \qquad i = 2, \ldots, n. \tag{4.5.3}$$

In general,

$$a_{ii} = \sum_{k=1}^{i} l_{ik}^2, \qquad a_{ij} = \sum_{k=1}^{j} l_{ik}l_{jk}, \qquad j < i, \tag{4.5.4}$$

which forms the basis for determining the columns of L in sequence . Once L is computed, the solution of the linear system can proceed just as in the LU decomposition (4.2.16): solve $L\mathbf{y} = \mathbf{b}$ and then solve $L^T\mathbf{x} = \mathbf{y}$. The algorithm for the factorization is given in Figure 4.5. For the Cholesky decomposition to be carried out, it is necessary that the quantities $a_{jj} - \sum l_{jk}^2$ all be positive so that the square roots may be taken. If the coefficient matrix A is positive definite, these quantities are indeed positive; moreover, the algorithm is numerically stable.

For $j = 1, \ldots, n$

$$l_{jj} = \left(a_{jj} - \sum_{k=1}^{j-1} l_{jk}^2 \right)^{1/2}$$

For $i = j+1, \ldots, n$

$$l_{ij} = \frac{a_{ij} - \sum_{k=1}^{j-1} l_{ik} l_{jk}}{l_{jj}}$$

Figure 4.5: *Cholesky Factorization*

The Cholesky factorization enjoys two advantages over LU factorization. First, there are approximately half as many arithmetic operations (Exercise 4.5.2). Square roots are also required in the Cholesky factorization, although there is a variant of the algorithm which avoids these (see Exercise 4.5.3). The second advantage is that by utilizing symmetry only the lower triangular part of A needs to be stored. As with the LU factorization, the l_{ij} can be overwritten onto the corresponding portions of A as they are computed. Finally, the Cholesky factorization extends readily to banded matrices and preserves the bandwidth, just as LU factorization without interchanges (see Exercise 4.5.4).

The QR Factorization

The Cholesky factorization applies only to symmetric positive definite matrices. We next consider a factorization that applies to any matrix. Indeed, later in this section we will use this factorization for rectangular matrices, but for the moment we will assume that A is $n \times n$ and real. The *QR factorization* (or *decomposition* or *reduction*) is then

$$A = QR, \tag{4.5.5}$$

where Q is an orthogonal matrix (See Appendix 2) and R is upper triangular. (We could denote R by U, but historically R has always been used in this context.)

We will obtain the matrix Q as a product of simpler orthogonal matrices based upon the rotation matrix

$$\begin{bmatrix} \cos\theta & \sin\theta \\ -\sin\theta & \cos\theta \end{bmatrix}. \tag{4.5.6}$$

We generalize such rotation matrices to $n \times n$ matrices of the form

$$P_{ij} = \begin{bmatrix} 1 & & & & & & & & & \\ & \ddots & & & & & & & & \\ & & 1 & & & & & & & \\ & & & c_{ij} & & & & s_{ij} & & \\ & & & & 1 & & & & & \\ & & & & & \ddots & & & & \\ & & & & & & 1 & & & \\ & & & -s_{ij} & & & & c_{ij} & & \\ & & & & & & & & 1 & \\ & & & & & & & & & \ddots \\ & & & & & & & & & & 1 \end{bmatrix}, \tag{4.5.7}$$

where $c_{ij} = \cos\theta_{ij}$ and $s_{ij} = \sin\theta_{ij}$ are located in the ith and jth rows and columns, as indicated. Such matrices are called *plane rotation matrices* or *Givens transformations.* Just as (4.5.6) defines a rotation of the plane, a matrix of the form (4.5.7) gives a rotation in the (i, j) plane in n-space. It is easy to show (Exercise 4.5.5) that the matrices (4.5.6) and (4.5.7) are orthogonal.

We now use the matrices P_{ij} in the following way to achieve the QR factorization (4.5.5). Let $\mathbf{a}_i$ denote the ith row of A. Then multiplication of A by P_{12} gives the matrix

$$A_1 = P_{12}A = \begin{bmatrix} c_{12}\mathbf{a}_1 + s_{12}\mathbf{a}_2 \\ -s_{12}\mathbf{a}_1 + c_{12}\mathbf{a}_2 \\ \mathbf{a}_3 \\ \vdots \\ \mathbf{a}_n \end{bmatrix}. \tag{4.5.8}$$

Note that only the first two rows of A are changed by this multiplication. If we choose s_{12} and c_{12} so that

$$-a_{11}s_{12} + a_{21}c_{12} = 0, \tag{4.5.9}$$

then A_1 has a zero in the $(2, 1)$ position, and the other elements in the first two rows of A_1 differ, in general, from those of A. We do not actually compute the angle θ_{12} to achieve (4.5.9) since we can obtain the desired sine and cosine directly by

$$c_{12} = a_{11}(a_{11}^2 + a_{21}^2)^{-1/2}, \qquad s_{12} = a_{21}(a_{11}^2 + a_{21}^2)^{-1/2}. \tag{4.5.10}$$

The denominator in (4.5.10) is non-zero unless both a_{11} and a_{21} are zero; but if $a_{21} = 0$, this step can be bypassed since a zero is already in the desired position.

The transformation (4.5.8) is analogous to the first step of an LU factorization in which a multiple of the first row of A is subtracted from the second row to achieve a zero in the $(2,1)$ position. We next proceed, as in LU, to obtain zeros in the remaining positions of the first column. We form $A_2 = P_{13}A$, which modifies the first and third rows of A_1 while leaving all other rows the same; in particular, the zero produced in the first stage in the $(2,1)$ position remains unchanged. The elements c_{13} and s_{13} of P_{13} are chosen analogously to (4.5.10) so that the $(3,1)$ element of A_2 is zero. We continue in this fashion, zeroing the remaining elements in the first column one after another and then zeroing elements in the second column in the order $(3,2), (4,2), \ldots, (n,2)$, and so on. In all, we will use $(n-1)+(n-2)+\cdots+1$ plane rotation matrices P_{ij}, and the result is that

$$PA \equiv P_{n-1,n} \cdots P_{12}A = R \tag{4.5.11}$$

is upper triangular.

We now need two basic facts about orthogonal matrices. First, if U and V are orthogonal, then

$$(UV)^T UV = V^T U^T UV = V^T V = I,$$

so that the product of orthogonal matrices is orthogonal. Each of the matrices P_{ij} in (4.5.11), and thus their product, P, is orthogonal. The second fact is that the inverse of an orthogonal matrix is orthogonal. This follows from the definition $U^T U = I$ since this implies that $I = (U^T U)^{-1} = U^{-1}U^{-T}$. Thus, if we set $Q = P^{-1}$, then Q is orthogonal, and multiplying (4.5.11) by Q gives (4.5.5).

We next count the operations to carry out the above QR factorization. The majority of the work is in modifying the elements of the two rows that are changed at each rotation. From (4.5.8) it is clear that modification of the first two rows requires $4n$ multiplications and $2n$ additions. (For simplicity we have also counted the operations used to produce the zero in the $(2,1)$ position even though they do not need to be performed.) The same count is true for producing zeros in the remaining $n-2$ elements in the first column. Hence the first stage requires $4n(n-1)$ multiplications and $2n(n-1)$ additions. In each of the subsequent $n-2$ stages, n decreases by 1 in this count so that the total is

$$4\sum_{k=2}^{n} k(k-1) \text{ mult } + 2\sum_{k=2}^{n} k(k-1) \text{ add } \doteq \frac{4}{3}n^3 \text{ mult } + \frac{2}{3}n^3 \text{ add}, \tag{4.5.12}$$

where we have used the summation formulas of Exercise 4.2.4. Computation of the c_{ij} and s_{ij} is also needed, but this requires only $0(n^2)$ operations. Thus we see by (4.5.12) that this QR factorization requires approximately four times the multiplications and twice the additions of LU factorization (see Section 4.2). We will discuss the relative merits of QR and LU factorization shortly, but first we show that by using other orthogonal transformations the QR factorization can be computed more economically.

Householder Transformations

A *Householder tranfsormation* is a matrix of the form $I - 2\mathbf{w}\mathbf{w}^T$, where $\mathbf{w}^T\mathbf{w} = 1$. It is easy to see (Exercise 4.5.6) that such matrices are symmetric and orthogonal. They are also called *elementary reflection* matrices (see Exercise 4.5.6). We now show how Householder transformations can be used to obtain the QR factorization of A. Let $\mathbf{a}_1$ be the first column of A and define (see Exercise 4.5.11)

$$\mathbf{w}_1 = \mu_1\mathbf{u}_1, \quad \mathbf{u}^T = (a_{11} - s_1, a_{21}, \ldots, a_{n1}), \tag{4.5.13}$$

where

$$s_1 = \pm(\mathbf{a}_1^T\mathbf{a}_1)^{1/2}, \qquad \mu_1 = (2s_1^2 - 2a_{11}s_1)^{-1/2}. \tag{4.5.14}$$

The sign of s_1 must be chosen to be opposite that of a_{11} so that there is no cancellation in the computation of μ_1; otherwise, the algorithm would be numerically unstable. The vector $\mathbf{w}_1$ satisfies

$$\mathbf{w}_1^T\mathbf{w}_1 = \mu_1^2[(a_{11} - s_1)^2 + \sum_{j=2}^n a_{j1}^2] = \mu_1^2(a_{11}^2 - 2a_{11}s_1 + 2s_1^2 - a_{11}^2) = 1,$$

so that $P_1 = I - 2\mathbf{w}_1\mathbf{w}_1^T$ is a Householder transformation. Moreover,

$$\mathbf{w}_1^T\mathbf{a}_1 = \mu[(a_{11} - s_1)a_{21} + \sum_{j=2}^n a_{j1}^2] = \mu_1(s_1^2 - a_{11}s_1) = \frac{1}{2\mu_1},$$

so that

$$a_{11} - 2w_1\mathbf{w}_1^T\mathbf{a}_1 = a_{11} - \frac{2(a_{11} - s_1)\mu_1}{2\mu_1} = s_1,$$

and

$$a_{i1} - 2w_i\mathbf{w}_1^T\mathbf{a}_1 = a_{i1} - \frac{2a_{i1}\mu_1}{2\mu_1} = 0, \qquad i = 2, 3, \ldots, n.$$

This shows that the first column of P_1A is

$$P_1\mathbf{a}_1 = \mathbf{a}_1 - 2\mathbf{w}_1^T\mathbf{a}_1\mathbf{w}_1 = (s_1, 0, \ldots, 0)^T.$$

Thus with this one orthogonal transformation, zeros have been introduced into the subdiagonal positions of the first column, as was done by $n-1$ Givens transformations.

The second stage is analogous. A Householder transformation $P_2 = I - 2\mathbf{w}_2\mathbf{w}_2^T$ is defined by a vector $\mathbf{w}_2$ whose first component is zero and whose remaining components are defined as in (4.5.13, 4.5.14), using now the second column of P_1A from the main diagonal element down. The matrix P_2P_1A then has zeros below the main diagonal in each of its first two columns. We continue in this way to zero elements below the main diagonal by Householder matrices $P_i = I - 2\mathbf{w}_i\mathbf{w}_i^T$, where $\mathbf{w}_i$ has zeros in its first $i-1$ components. Thus

$$P_{n-1}\cdots P_1A = R,$$

where R is upper triangular. The matrices P_i are all orthogonal so that $P = P_{n-1}\cdots P_1$ and P^{-1} are also orthogonal. Therefore, $Q = P^{-1}$ is the orthogonal matrix of (4.5.5).

The above discussion has considered only the formation of the vectors $\mathbf{w}_i$ that define the Householder transformations, and we next consider the remainder of the computation. If $\mathbf{a}_1, \mathbf{a}_2, \ldots, \mathbf{a}_n$ are the columns of A, then

$$P_1A = A - 2\mathbf{w}_1\mathbf{w}_1^TA = A - 2\mathbf{w}_1(\mathbf{w}_1^T\mathbf{a}_1, \mathbf{w}_1^T\mathbf{a}_2, \ldots, \mathbf{w}_1^T\mathbf{a}_n). \tag{4.5.15}$$

Thus, the ith column of P_1A is

$$\mathbf{a}_i - 2\mathbf{w}_1^T\mathbf{a}_i\mathbf{w}_1 = \mathbf{a}_i - \gamma_1\mathbf{u}_1^T\mathbf{a}_i\mathbf{u}_1, \tag{4.5.16}$$

where

$$\gamma_1 = 2\mu_1^2 = (s_1^2 - s_1a_{11})^{-1}.$$

It is more efficient computationally not to form the vector $\mathbf{w}_1$ explicitly but to work with γ_1 and $\mathbf{u}_1$ as shown in (4.5.16). Analogous computations are performed to obtain the remaining reduced matrices $P_2P_1A, \ldots$; the complete algorithm is summarized in Figure 4.6.

We next count the operations in the Householder reduction. The bulk of the work is in the formation of the new columns of the reduced matrices. Referring to the inner loop in Figure 4.6, at the kth stage the inner product $\mathbf{u}_k^T\mathbf{a}_j$ requires $n-k+1$ multiplications and $n-k$ additions and the operation $\mathbf{a}_j - \alpha_j\mathbf{u}_k$ requires $n-k+1$ additions and multiplications. Since there are $n-k$ columns to update at the kth stage, this gives approximately $2(n-k)$ additions and multiplications. Summing this over all $n-1$ stages, we obtain

$$2\sum_{k=1}^{n-1}(n-k)^2 = \frac{n(n-1)(2n-1)}{3} \doteq \frac{2}{3}n^3 \tag{4.5.17}$$

additions and multiplications. The number of other operations is no more than order n^2. Comparing this count with (4.5.12) for the QR factorization using

For $k = 1$ to $n-1$

$$s_k = -\mathrm{sgn}(a_{kk}) \left(\sum_{l=k}^{n} a_{lk}^2 \right)^{1/2}$$

$$\mathbf{u}_k^T = (0, \ldots, 0, a_{kk} - s_k, a_{k+1,k}, \ldots, a_{nk})$$

$$\gamma_k = (s_k^2 - s_k a_{kk})^{-1}$$

$$a_{kk} = s_k$$

For $j = k+1$ to n

$$\alpha_j = \gamma_k \mathbf{u}_k^T \mathbf{a}_j$$

$$\mathbf{a}_j = \mathbf{a}_j - \alpha_j \mathbf{u}_k$$

Figure 4.6: *Householder Reduction*

Givens matrices, we see that while the additions are the same, the number of multiplications has been roughly halved. The number of square roots is also reduced considerably. Hence, the Householder reduction is clearly more efficient, although there are some situations in which it may be preferable to use Givens matrices.

Both the Givens and Householder factorizations are numerically stable, and they can be considered as alternatives to Gaussian elimination with partial pivoting. However, the operation count of even the Householder reduction is approximately twice that of Gaussian elimination, and the added expense of partial pivoting does not close this gap. Moreover, both the Givens and Householder transformations expand the bandwidth of a banded matrix (Exercise 4.5.13), just as partial pivoting does. Hence the QR factorization is rarely used for the solution of nonsingular systems of equations, although it is a key part of some of the best algorithms for computing eigenvalues, as we will see in Chapter 7. It is also useful for dealing with rectangular matrices which can be used in the formulation of least squares problems. We end this chapter by indicating this application.

Application to Least Squares Problems

Recall from Section 4.1 that in the least squares problem (4.1.11), (4.1.12), we wish to minimize over $a_1, \ldots, a_n$ the function

$$\sum_{i=1}^{m} \left[\sum_{j=0}^{n} a_j \phi_{ij} - f_i \right]^2, \tag{4.5.18}$$

where $\phi_{ij} = \phi_j(x_i)$ and where for simplicity we have taken the weights $w_i = 1$. If E is the matrix with elements ϕ_{ij}, then we can write (4.5.18) as (Exercise 4.5.12)

$$(E\mathbf{a} - \mathbf{f})^T(E\mathbf{a} - \mathbf{f}) = ||E\mathbf{a} - \mathbf{f}||_2^2. \tag{4.5.19}$$

Now suppose that we apply the QR factorization to the (generally rectangular) $m \times (n+1)$ matrix E, where $m \geq n+1$. The Householder factorization, for example, can be applied as previously described; just imagine that E is actually $m \times m$ but that we stop the process when we have completed zeroing elements in the first n columns. Thus we will have

$$E = Q\begin{bmatrix} R \\ 0 \end{bmatrix} \quad \text{or} \quad PE = \begin{bmatrix} R \\ 0 \end{bmatrix} = \hat{R}, \tag{4.5.20}$$

where R is an $(n+1) \times (n+1)$ upper triangular matrix. Since $||P\mathbf{x}||_2 = ||\mathbf{x}||_2$ for any vector $\mathbf{x}$ (Exercise 4.5.8), we have

$$||E\mathbf{a} - \mathbf{f}||_2 = ||P(E\mathbf{a} - \mathbf{f})||_2 = ||\hat{R}\mathbf{a} - \hat{\mathbf{f}}||_2, \tag{4.5.21}$$

where we have set $\hat{\mathbf{f}} = P\mathbf{f}$. If we partition $\hat{\mathbf{f}}$ commensurately with $\hat{R}$, we have

$$\hat{R}\mathbf{a} - \hat{\mathbf{f}} = \begin{bmatrix} R \\ 0 \end{bmatrix}\mathbf{a} - \begin{bmatrix} \hat{\mathbf{f}}_1 \\ \hat{\mathbf{f}}_2 \end{bmatrix} = \begin{bmatrix} R\mathbf{a} - \hat{\mathbf{f}}_1 \\ -\hat{\mathbf{f}}_2 \end{bmatrix}. \tag{4.5.22}$$

Thus

$$||E\mathbf{a} - \mathbf{f}||_2^2 = ||R\mathbf{a} - \hat{\mathbf{f}}_1||_2^2 + ||\hat{\mathbf{f}}_2||_2^2, \tag{4.5.23}$$

and it is clear that $||E\mathbf{a} - \mathbf{f}||_2$ is minimized when $||R\mathbf{a} - \hat{\mathbf{f}}_1||_2$ is minimized since $\hat{\mathbf{f}}_2$ is fixed.

Now recall from Section 4.1 that the condition for (4.5.18) to have a unique minimizer is that the matrix E^TE be nonsingular, which we assume. But from (4.5.20)

$$E^TE = \hat{R}^TQ^TQ\hat{R} = \hat{R}^T\hat{R} = R^TR. \tag{4.5.24}$$

Thus R^TR is also nonsingular, and since R is square R itself is nonsingular. (Note that if the signs of the diagonal elements of R are chosen to be positive, then R^TR is the Cholesky factorization of E^TE.) Therefore the system $R\mathbf{a} = \hat{\mathbf{f}}_1$ has a unique solution $\hat{\mathbf{a}}$. For this $\hat{\mathbf{a}}$ (4.5.23) is minimized and hence so is (4.5.18). We summarize the overall algorithm as

$$\text{find the } QR \text{ factorization (4.5.20)}, \tag{4.5.25}$$

$$\text{form the vector } P\mathbf{f} \text{ to obtain } \hat{\mathbf{f}}_1, \tag{4.5.26}$$

$$\text{solve the system } R\mathbf{a} = \hat{\mathbf{f}}_1. \tag{4.5.27}$$

This algorithm is much more numerically stable than forming E^TE explicitly and then solving the normal equations. The number of operations, however, is somewhat higher (Exercise 4.5.15).

Supplementary Discussion and References: 4.5

Further discussion of the use of the QR factorization in solving least squares problems may be found in Golub and Van Loan [1989], Lawson and Hanson [1974], and Stewart [1973].

Exercise 4.5.3 shows that the Cholesky factorization can be written in the form $A = LDL^T$. If A is positive definite, then so is D; conversely, if D is not positive definite then neither is A. Unfortunately this type of decomposition is not necessarily numerically stable if A is not positive definite. However, we can find a decomposition of the form $A = PLDL^TP^T$, where P is a permutation matrix and D is block diagonal with 1×1 or 2×2 blocks. This is the basis for the Bunch-Kaufman algorithm (and others also) for solving symmetric systems that are not positive definite. For further discussion, see Golub and Van Loan [1989].

EXERCISES 4.5

4.5.1. Let $A = LL^T$ be a factorization of a symmetric positive definite matrix A, where L is lower-triangular and has positive main diagonal elements. Show that if $\hat{L}$ is obtained from L by changing the sign of every element of the ith row, then $A = \hat{L}\hat{L}^T$. (This shows that the LL^T factorization is not unique although there is a unique L with positive main diagonal elements.)

4.5.2. Show that the number of additions and multiplications in Cholesky factorization is roughly half that of LU factorization.

4.5.3. Modify the Cholesky decomposition of Figure 4.5 to obtain the *root-free* Cholesky decomposition in which $A = LDL^T$, where L is unit lower triangular. What is the relation of this factorization to the LU factorization of A?

4.5.4. For a banded matrix A with semibandwidth p, show that the Cholesky factorization of Figure 4.5 can be written as

$$\begin{aligned}
&\text{For } j = 1, \ldots, n \\
&\quad q = \max(1, j - p) \\
&\quad l_{jj} = \left(a_{jj} - \sum_{k=q}^{j-1} l_{jk}^2 \right)^{1/2} \\
&\quad \text{For } i = j + 1, \ldots, \min(j + p, n) \\
&\qquad r = \max(1, i - p) \\
&\qquad l_{ij} = \frac{a_{ij} - \sum_{k=r}^{j-1} l_{ik} l_{jk}}{l_{jj}}.
\end{aligned}$$

Find the operation count for this algorithm as a function of p and n and compare with the LU factorization for banded matrices.

4.5.5. Show that the matrices (4.5.6) and (4.5.7) are orthogonal.

4.5.6. Show that the matrix $P = I - 2\mathbf{w}\mathbf{w}^T$ is always symmetric and is orthogonal if and only if $\mathbf{w}^T\mathbf{w} = 1$. If $\mathbf{w}^T\mathbf{w} = 1$, show also that $P\mathbf{w} = -\mathbf{w}$, and $P\mathbf{u} = \mathbf{0}$ if $\mathbf{u}^T\mathbf{w} = 0$. Interpret this geometrically in two dimensions as a reflection of the plane across the line orthogonal to $\mathbf{w}$.

4.5.7. What are the eigenvalues, eigenvectors, and determinant of a Householder transformation $I - 2\mathbf{w}\mathbf{w}^T$? More generally, what can you say about the eigenvalues and eigenvectors of $I + \mathbf{u}\mathbf{v}^T$ for given vectors $\mathbf{u}$ and $\mathbf{v}$?

4.5.8. If P is an orthogonal matrix, show that $\|P\mathbf{x}\|_2 = \|\mathbf{x}\|_2$, where $\| \ \|_2$ is the Euclidean norm.

4.5.9. If A is an orthogonal matrix and $A = QR$ is a QR factorization, what can you say about R?

4.5.10. If $A = QR$ is a QR factorization, what is the relationship of $\det R$ to $\det A$?

4.5.11. The vector $\mathbf{w}_1 = (w_1, \cdots, w_n)^T$ of (4.5.13) can be obtained as follows:

a. Since $P_1 = I - 2\mathbf{w}_1\mathbf{w}_1^T$ is orthogonal, use Exercise 4.5.8 to show that if $P_1\mathbf{a}_1 = s_1\mathbf{e}_1$, then $\|\mathbf{a}_1\|_2 = \|P_1\mathbf{a}_1\|_2 = |s_1|$ so that $s_1 = \pm\|\mathbf{a}_1\|_2$. (The vector $\mathbf{e}_1$ has 1 in its first component and zero elsewhere)

b. Since P_1 is symmetric, show that $\mathbf{a}_1 = s_1 P_1 \mathbf{e}_1 = s_1(\mathbf{e}_1 - 2\mathbf{w}_1\mathbf{w}_1)$ so that $a_{11} = s_1(1 - 2w_1^2), a_{1j} = -2s_1 w_1 w_j$, $j > 1$. Then show that (4.5.13) and (4.5.14) follow.

c. Extend the above argument to show how to find a Householder transformation P so that $P\mathbf{p} = \mathbf{q}$, where $\mathbf{p}$ and $\mathbf{q}$ are real vectors with $\|\mathbf{p}\|_2 = \|\mathbf{q}\|_2$. What can you do if $\mathbf{p}$ and $\mathbf{q}$ are complex?

4.5.12. Verify that (4.5.18) and (4.5.19) are the same.

4.5.13. Let A be a banded matrix with semibandwidth p. Show that if $A = QR$, and Q is obtained by either Givens or Householder transformations, then in general R will have $2p$ non-zero diagonals above the main diagonal.

4.5.14. Let

$$A = \begin{bmatrix} 1 & 1 & 1 & 1 \\ \varepsilon & 0 & 0 & 0 \\ 0 & \varepsilon & 0 & 0 \\ 0 & 0 & \varepsilon & 0 \\ 0 & 0 & 0 & \varepsilon \end{bmatrix}, A^T A = \begin{bmatrix} 1+\varepsilon^2 & 1 & 1 & 1 \\ 1 & 1+\varepsilon^2 & 1 & 1 \\ 1 & 1 & 1+\varepsilon^2 & 1 \\ 1 & 1 & 1 & 1+\varepsilon^2 \end{bmatrix}.$$

Show that $A^T A$ has eigenvalues $4 + \varepsilon^2, \varepsilon^2, \varepsilon^2, \varepsilon^2$ and thus is non-singular if $\varepsilon \neq 0$. Now let η be the largest positive number on the computer such that computed$(1 + \eta) = 1$. Show that if $|\varepsilon| < \sqrt{\eta}$,

$$\text{computed}(A^T A) = \begin{bmatrix} 1 & 1 & 1 & 1 \\ 1 & 1 & 1 & 1 \\ 1 & 1 & 1 & 1 \\ 1 & 1 & 1 & 1 \end{bmatrix},$$

a matrix of rank one, and consequently the numerical approximation to the normal equations will, for general $\mathbf{b}$, have no solution at all. (Note that ε need not be unrealistically small. For example, on machines with a 27-bit mantissa, $\eta \doteq 10^{-8}$, so that the above error would occur if $\varepsilon < 10^{-4}$.)

4.5.15. Modify (4.15.17) to show that the operation count of QR for an $m \times n$ matrix is approximately $2mn^2 - \frac{2}{3}n^3$. Show that this is also the high-order part of the operation count for the algorithm (4.5.25). Next show that the operation count for minimizing (4.5.19) by forming and solving the normal equations is $mn^2 + \frac{1}{3}n^3$. Thus, for $m >> n$, conclude that the QR approach is approximately twice as slow.

4.5.16. Consider the least squares problem of minimizing $\|E\mathbf{a} - \mathbf{f}\|_2$ subject to the constraint $F^T\mathbf{a} = \mathbf{g}$, where $\mathbf{g}$ is a given vector. Show how the QR decomposition of F can be used in solving this problem.

4.5.17. Consider the normal equations $E^T E\mathbf{a} = E^T\mathbf{f}$ and let $\mathbf{r} = \mathbf{f} - E\mathbf{a}$, $\mathbf{g} = \mathbf{f} + \alpha\mathbf{r}$, where α is a constant. Show that if $\mathbf{b}$ is the solution of $E^T E\mathbf{b} = E^T\mathbf{g}$, then $\mathbf{b} = \mathbf{a}$. Conclude from this that the vector $\mathbf{f}$ can be changed by arbitrarily large amounts without changing the solution to the normal equations.

Chapter 5

Life Is Really Nonlinear

5.1 Nonlinear Problems and Shooting

We consider in this chapter the solution of nonlinear problems. Suppose, for example, that the coefficients b, c and d in (3.1.1) are functions of v as well as x:

$$v''(x) = b(x, v(x))v'(x) + c(x, v(x))v(x) + d(x, v(x)). \tag{5.1.1}$$

Then (5.1.1) is a nonlinear equation for v. A simple example of this, which we will discuss later, is

$$v''(x) = 3v(x) + x^2 + 10[v(x)]^3, \qquad 0 \le x \le 1. \tag{5.1.2}$$

With either (5.1.1) or (5.1.2) we can have any of the boundary conditions discussed in Section 3.1, as well as others which are themselves nonlinear; for example,

$$[v(0)]^2 + v'(0) = 1, \qquad v(1) = [v'(1)]^2. \tag{5.1.3}$$

There are two basic approaches to such nonlinear problems. One is to discretize the differential equation as was done in Chapter 3 for linear equations; we will study this approach in Section 5.3. In the present section we will consider a method based on the solution of initial-value problems, and for this purpose we will first treat as an example the projectile problem of Chapter 2.

The Projectile Problem and Shooting

Recall that the projectile problem was given by equations (2.1.15), (2.1.17), and (2.1.18) with $\dot{m} = 0$ and $T = 0$:

$$\begin{aligned} \dot{x} &= v\cos\theta, \qquad \dot{y} = v\sin\theta, \\ \dot{v} &= \frac{-1}{2m}c\rho s v^2 - g\sin\theta, \qquad \dot{\theta} = -\frac{g}{v}\cos\theta. \end{aligned} \tag{5.1.4}$$

As before, we have the initial conditions

$$x(0) = y(0) = 0, \qquad v(0) = \bar{v}. \tag{5.1.5}$$

In Chapter 2 we also prescribed

$$\theta(0) = \bar{\theta}, \tag{5.1.6}$$

so that (5.1.4) - (5.1.6) was an initial-value problem. Suppose now, however, that in place of (5.1.6) we require that the projectile hit the ground at a given time t_f; that is,

$$y(t_f) = 0. \tag{5.1.7}$$

Since the equations (5.1.4) are nonlinear in the unknowns x, y, v, and θ, this is a nonlinear two-point boundary value problem. Note that it may not have a solution; for example, t_f may be too large for the given initial velocity $\bar{v}$.

We can base a numerical solution of this problem on the trial-and-error method that an artillery gunner might employ: choose a value of the launch angle, say $\bar{\theta}_1$, and "shoot," which, mathematically, means to solve the initial-value problem (5.1.4),(5.1.5) together with

$$\theta(0) = \bar{\theta}_1. \tag{5.1.8}$$

We follow the trajectory until $t = t_f$ and record the corresponding value of y at t_f, say y_1. If $y_1 \neq 0$ we choose another value of $\theta(0)$ and shoot again, continuing the process until a $\theta(0)$ has been found such that $y(t_f)$ is suitably close to zero. Shortly we will discuss systematic ways to choose new values of $\theta(0)$.

Other Boundary Value Problems

We can apply the above *shooting method* to other boundary value problems even though there may be no physical analogy to shooting. Consider, for example, the equation (5.1.2) with the boundary conditions

$$v(0) = \alpha, \qquad v(1) = \beta. \tag{5.1.9}$$

In this case we choose a trial value s for $v'(0)$ and solve the initial value problem

$$v'' = 3v + 10v^3 + x^2, \qquad v(0) = \alpha, \qquad v'(0) = s \tag{5.1.10}$$

up to $x = 1$. If the value of v at $x = 1$ is not sufficiently close to β, we adjust s and try again.

A key concern in the use of the shooting method is the adjustment of the parameter before the next shoot. We can address this question by recognizing that finding the right value of s is equivalent to finding a root of a nonlinear function. To see why this is so, consider (5.1.10). Let $v(x; s)$ be the solution of the initial-value problem with $v'(0) = s$ and define

$$f(s) = v(1; s) - \beta.$$

Then in the shooting method we need to find a value of s for which $f(s) = 0$. We can, in principle, use any number of numerical methods for finding solutions of equations; some of these methods will be discussed in the following section.

Systems of Equations

The shooting method can also be applied to two-point boundary-value problems for general first-order systems. Consider the system

$$\mathbf{u}' = R(\mathbf{u}, t), \qquad 0 \le t \le 1, \tag{5.1.11}$$

where $\mathbf{u}(t)$ is the n-vector with components $u_i(t)$, $i = 1, \ldots, n$. Assume that m of the functions $u_1, \ldots, u_n$ are prescribed at $t = 1$ and that $n-m$ are prescribed at $t = 0$ so that we have the correct number, n, of boundary conditions. We will denote the set of functions prescribed at $t = 0$ by U_0 and those prescribed at $t = 1$ by U_1. Note that these sets may overlap; for example, u_1 may be given at both $t = 0$ and $t = 1$, but u_2 may not be given at either end point. To apply the shooting method, we select initial values $s_1, \ldots, s_m$ for the m functions not prescribed at $t = 0$ and solve numerically the initial-value problem

$$\mathbf{u}' = R(\mathbf{u}, t), \qquad \mathbf{u}(0) \text{ given by boundary conditions or } \{s_1, \ldots, s_m\}.$$

Next, we compare the values of those $u_i \in U_1$ with the integrated values $\mathbf{u}(1; \mathbf{s})$, where $\mathbf{s} = (s_1, \ldots, s_m)$. To solve the boundary-value problem the initial values s_i must be such that

$$u_i(1; \mathbf{s}) = \text{given value}, \qquad u_i \in U_1.$$

This is a system of m nonlinear equations in the m unknowns $s_1, \ldots, s_m$. We will consider methods for the solution of systems of nonlinear equations in Section 5.3.

Instability

Although the shooting method is simple in concept, it can suffer from instabilities in the initial-value problems. Instabilities of this type were discussed in Chapter 2, and we give here another simple example similar to the one in Section 2.5. Consider the problem

$$u'' - 100u = 0, \tag{5.1.12}$$

with the boundary conditions

$$u(0) = 1, \qquad u(1) = 0. \tag{5.1.13}$$

It is easy to verify that the exact solution of this boundary-value problem is

$$u(t) = \frac{1}{1 - e^{-20}} e^{-10t} - \frac{e^{-20}}{1 - e^{-20}} e^{10t}. \tag{5.1.14}$$

Now we attempt to obtain the solution by the shooting method using

$$u'(0) = s. \tag{5.1.15}$$

The exact solution of the corresponding initial-value problem is

$$u(t; s) = \frac{10 - s}{20} e^{-10t} + \frac{10 + s}{20} e^{10t}, \tag{5.1.16}$$

and we see that the value $u(1; s)$ at the end point $t = 1$ is very sensitive to s. The value of s that will give the exact solution (5.1.14) of the boundary-value problem is

$$s = -10 \left(\frac{1 + e^{-20}}{1 - e^{-20}} \right) \doteq -10.$$

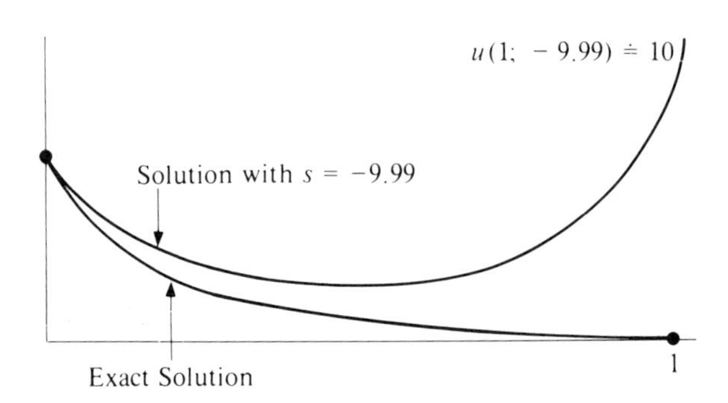

Figure 5.1: *Solutions of Exact and Nearby Problems*

If we solve the initial-value problem with the value of s correct to two decimal places, say $s = -9.99$, the solution of the initial-value problem is shown in Figure 5.1. The difficulty, of course, is that the solution of the initial-value problem grows like e^{10t}, and to suppress this fast-growing component, it is necessary to obtain a very accurate value of the initial condition. Even then, rounding and discretization error in the solution of the initial value problem will tend to cancel out this accuracy in s.

Supplementary Discussion and References: 5.1

The shooting method for two-point boundary-value problems is described in many books on numerical analysis. A particularly detailed treatment is given in Roberts and Shipman [1972]. See also Ascher et al. [1988].

Another approach to solving two-point boundary-value problems by means of initial-value problems is the method of invariant embedding. Here, however, the initial-value problems are for partial differential equations, rather than ordinary differential equations. For a thorough discussion of invariant embedding, see Meyer [1973].

EXERCISES 5.1

5.1.1. Solve the two-point boundary-value problem $y'' + y' + y = -(x^2 + x + 1)$, $y(0) = y(1) = 0$, by the shooting method using one of the methods of Chapter 2 for initial-value problems. Check your numerical results by finding the exact analytical solution to this problem. (*Hint:* For the analytical solution, try the method of undetermined coefficients for a quadratic polynomial.)

5.1.2. Solve the projectile boundary-value problem (5.1.4),(5.1.5),(5.1.7) numerically using one of the methods of Chapter 2 for initial-value problems.

5.1.3. Attempt to solve problem (5.1.12),(5.1.13) using the same method you used for Exercise 5.1.2. Compare your best result with the exact solution given by (5.1.16) and discuss the discrepancies. Also discuss any difficulties you encountered in obtaining your numerical solution.

5.2 Solution of a Single Nonlinear Equation

In the last section, we saw that the shooting method with one free parameter can be viewed as a problem of finding a solution of a nonlinear equation

$$f(x) = 0. \tag{5.2.1}$$

We also saw in Chapter 2 that the use of implicit methods required solving a nonlinear equation. Many other areas in scientific computing lead to the problem of finding roots of equations, or, more generally, solutions of a system of nonlinear equations, which we will discuss in the next section. In the present section we restrict our attention to functions of a single variable.

An important special case of (5.2.1) occurs when f is a polynomial:

$$f(x) = a_n x^n + \cdots + a_1 x + a_0. \tag{5.2.2}$$

In this case we know from the fundamental theorem of algebra that f has exactly n real or complex roots if we count multiplicities of the roots. For a general function f it is usually difficult to ascertain how many solutions equation (5.2.1) has: there may be none, one, finitely many, or infinitely many. A simple condition that ensures that there is at most one solution in a given interval (a, b) is that

$$f'(x) > 0 \qquad \text{for all} \quad x \in (a, b) \tag{5.2.3}$$

(or $f'(x) < 0$ in the interval), although this does not guarantee that a root exists in the interval. (The proof of these statements is left to Exercises 5.2.1 and 5.2.2.) If, however, f is continuous, and

$$f(a) < 0, \qquad f(b) > 0, \tag{5.2.4}$$

then it is intuitively clear (and rigorously proved by a famous theorem of the calculus) that f must have at least one root in the interval (a, b).

The Bisection Method

Let us now assume that (5.2.4) holds and, for simplicity, that there is just one root in the interval (a, b). We do not necessarily assume that (5.2.3) holds; the situation might be as shown in Figure 5.2. One of the simplest ways of approximating a root of f in this situation is the *bisection method*, which we now describe.

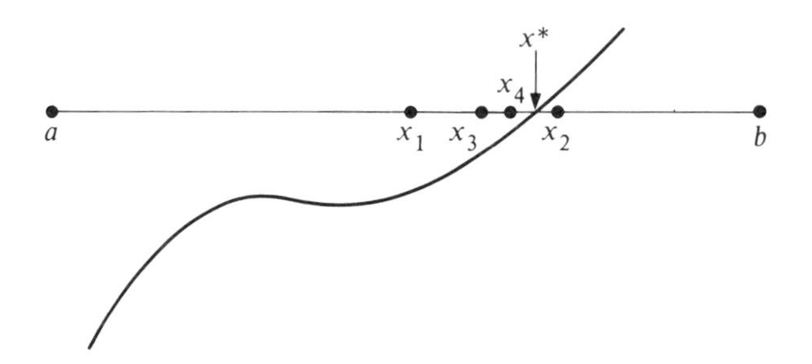

Figure 5.2: *The Bisection Method*

Let $x_1 = \frac{1}{2}(a + b)$ be the midpoint of the interval (a, b) and evaluate $f(x_1)$. If $f(x_1) > 0$, then the root, $x^\star$, must lie between a and x_1; if $f(x_1) < 0$, which is the situation shown in Figure 5.2, then $x^\star$ is between x_1 and b. We now continue this process, always keeping the interval in which $x^\star$ is known to lie and evaluating f at its midpoint to obtain the next interval. For the function shown in Figure 5.2, the steps would be as follows:

$$
\begin{aligned}
&f(x_1) < 0. \quad \text{Hence, } x^\star \in (x_1, b). \quad \text{Set } x_2 = \tfrac{1}{2}(x_1 + b).\\
&f(x_2) > 0. \quad \text{Hence, } x^\star \in (x_1, x_2). \ \text{Set } x_3 = \tfrac{1}{2}(x_1 + x_2).\\
&f(x_3) < 0. \quad \text{Hence, } x^\star \in (x_3, x_2). \ \text{Set } x_4 = \tfrac{1}{2}(x_2 + x_3).\\
&f(x_4) < 0. \quad \text{Hence, } x^\star \in (x_4, x_2). \ \text{Set } x_5 = \tfrac{1}{2}(x_2 + x_4).\\
&\qquad\qquad\qquad\vdots
\end{aligned}
$$

Clearly, each step of the bisection procedure reduces the length of the interval known to contain $x^\star$ by a factor of 2. Therefore after m steps the length of the interval will be $(b-a)2^{-m}$, and this provides a bound on the error in our current approximation to the root; that is,

$$|x_m - x^\star| \le \frac{|b-a|}{2^m}. \tag{5.2.5}$$

This bound has been obtained under the tacit assumption that the function values $f(x_i)$ are computed exactly. Of course, on a computer this will not be the case because of the rounding error (and possibly also discretization error – recall that the evaluation of the function f for the shooting method requires the solution of an initial-value problem). However, the bisection method does not use the value of $f(x_i)$ but only the sign of $f(x_i)$; therefore the bisection method is impervious to errors in evaluating the function f as long as the sign of $f(x_i)$ is evaluated correctly. One might think that the round-off error could not be so severe as to change the sign of the function, but this is not the case when the function values become sufficiently small. If the sign of $f(x_i)$ is incorrect, a wrong decision will be made in choosing the next subinterval, and the error bound (5.2.5) does not necessarily hold.

It is clear that if one makes a maximum error of E in evaluating f at any point in the interval (a, b), then the sign of f will be correctly evaluated as long as

$$|f(x)| > |E|.$$

Since the function f will be close to zero near the root $x^\star$, we can also argue the converse: there will be an *interval of uncertainty*, say, $(x^\star - \varepsilon, x^\star + \varepsilon)$, about the root in which the sign of f may not be correctly evaluated (see Figure 5.3). When our approximations reach this interval, their further progress toward the root is at best problematical. Unfortunately, it is extremely difficult to determine this interval in advance. It depends on the unknown root $x^\star$, the "flatness" of f in the neighborhood of the root, and the magnitude of the errors made in evaluating f. On the other hand, the interval is usually detectable during the course of the computation by an erratic behavior of the iterates; when this occurs, there is no longer any point in continuing the computation.

The fact that the sign of the function f may not be evaluated correctly near the root affects not only the bisection method but also the other methods we shall discuss later in the section.

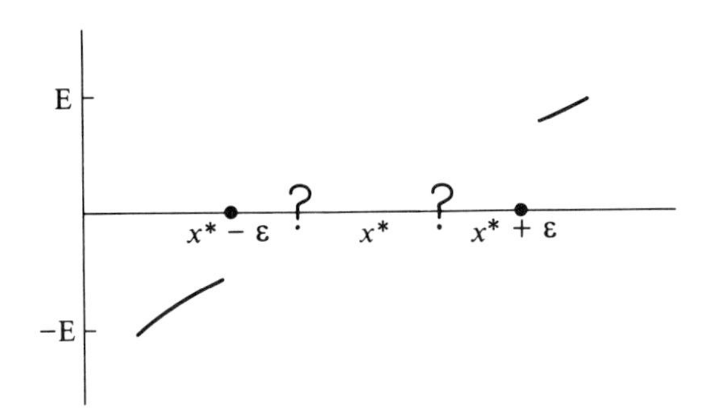

Figure 5.3: *The Interval of Uncertainty*

The Secant Method

One drawback of the bisection method is that it may be rather slow. To reduce the initial interval by a large factor, say 10^6, which may correspond to about six-decimal-digit accuracy, we would expect to require, from the error bound (5.2.5),

$$m = \frac{6}{\log_{10} 2} \doteq 20$$

evaluations of f. When each evaluation is expensive, as in the case of the shooting method, we would like to keep the number of evaluations as small as possible.

One possible way to speed up the bisection method is to use the values of the function f (instead of only its signs), and the simplest way to utilize this information is to choose the next point x_{i+1} as the zero of the linear function that interpolates f at x_{i-1} and x_i. This is shown in Figure 5.4. In the somewhat favorable situation shown in the figure, it is clear that x_{i+1} is a considerably better approximation to the root than would be the midpoint of the interval (x_{i-1}, x_i).

The linear interpolating function is given by (see Section 2.3, although it is easily checked directly)

$$l(x) = \frac{(x - x_{i-1})}{(x_i - x_{i-1})} f(x_i) - \frac{(x - x_i)}{(x_i - x_{i-1})} f(x_{i-1}), \tag{5.2.6}$$

and the root of this linear function is

$$x_{i+1} = \frac{x_{i-1} f(x_i) - x_i f(x_{i-1})}{f(x_i) - f(x_{i-1})}. \tag{5.2.7}$$

We may now proceed as in the bisection method, retaining x_{i+1} and either x_i or x_{i-1} so that the function values at the two retained points have different signs. This is the *regula falsi method.* Alternatively, in the *secant method*, we simply carry out (5.2.7) sequentially as indicated, keeping the last two iterates regardless of whether their function values have different signs.

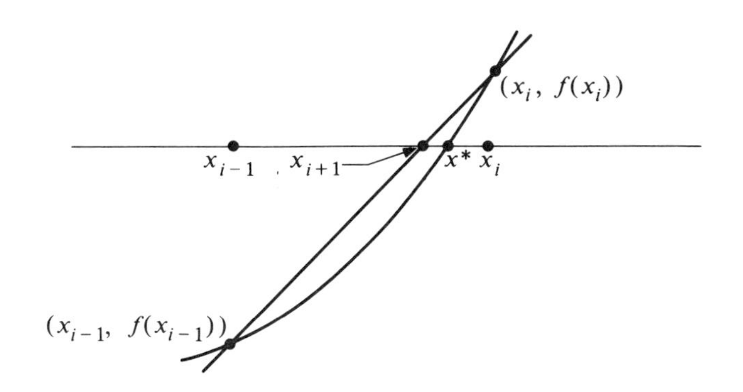

Figure 5.4: *The Secant Method*

It is better to rewrite (5.2.7) as

$$x_{i+1} = x_i - \frac{f(x_i)}{d_i}, \qquad d_i = \frac{f(x_i) - f(x_{i-1})}{x_i - x_{i-1}}, \tag{5.2.8}$$

which is easily verified as mathematically identical to (5.2.7). This form is preferable to (5.2.7) for computation since there is less cancellation.

Newton's Method

We can consider the quantity d_i in (5.2.8) to be a difference approximation to $f'(x_i)$, and, thus (5.2.8) may be viewed as a "discrete form" of the iterative method

$$x_{i+1} = x_i - \frac{f(x_i)}{f'(x_i)}. \tag{5.2.9}$$

This is known as *Newton's method* and is the most famous iterative method for obtaining roots of equations (as well as for solving systems of nonlinear equations, as we shall see in the next section). Geometrically, Newton's method can be interpreted as approximating the function f by the linear function

$$l_i(x) = f(x_i) + (x - x_i)f'(x_i),$$

which is tangent to f at x_i, and then taking the next iterate x_{i+1} to be the zero of $l_i(x)$; this is shown in Figure 5.5.

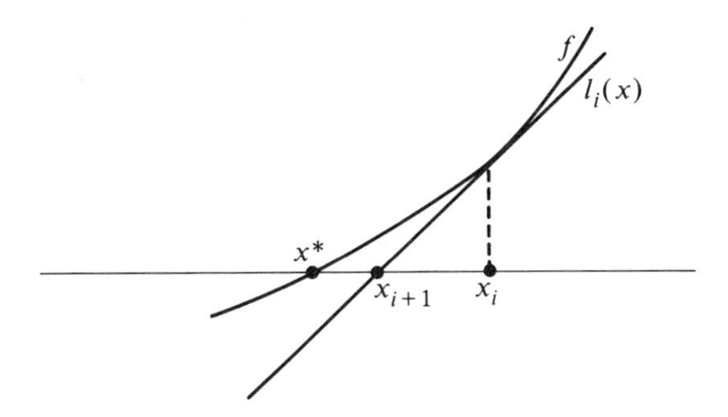

Figure 5.5: *Newtons's Method*

Iteration Functions and Convergence

The Newton iteration (5.2.9) can be written in the form

$$x_{i+1} = g(x_i), \tag{5.2.10}$$

where

$$g(x) = x - \frac{f(x)}{f'(x)}. \tag{5.2.11}$$

Many other iterative methods may also be written in the general form (5.2.10) for some *iteration function* g. For example, a very simple method is given by defining g to be

$$g(x) = x - \alpha f(x) \tag{5.2.12}$$

for some scalar α. This is sometimes called the *chord method* and is illustrated in Figure 5.6.

Iterative methods of the form (5.2.10) are called *one-step methods* since x_{i+1} depends only on the previous iterate x_i. On the other hand, the secant method (5.2.8) depends on both x_i and x_{i-1} and is an example of a *multistep method.* (Note the analogy with one-step and multistep methods for initial-value problems.)

To be useful the iteration function g must have the property

$$x^\star = g(x^\star) \tag{5.2.13}$$

for a root $x^\star$ of f. This is clearly the case for (5.2.11) and (5.2.12). (This is true for (5.2.11) even when $f'(x^\star) = 0$, although in this case we must define $g(x^\star)$ as the limit as $x \to x^\star$; see Exercise 5.2.3.) A value of $x^\star$ that satisfies (5.2.13) is called a *fixed point* of the function g. Assuming that the function g has a fixed point x^*, an important question is the convergence of the iterates x_i to x^*.

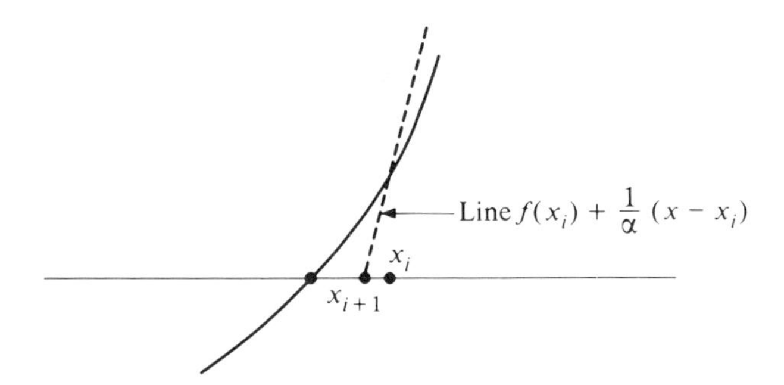

Figure 5.6: *The Chord Method*

We now discuss a basic property that ensures convergence of the iterates (5.2.10), at least when the starting iterate is sufficiently close to $x^\star$. We assume that g is continuously differentiable in a neighborhood of $x^\star$, that (5.2.13) holds, and that

$$|g'(x)| \leq \gamma < 1, \qquad \text{if } |x - x^\star| \leq \beta. \tag{5.2.14}$$

By the mean-value theorem of the calculus, we can write

$$g(x) - g(x^\star) = g'(\xi)(x - x^\star), \tag{5.2.15}$$

where ξ is between x and $x^\star$. Therefore if $|x - x^\star| \leq \beta$, then we can apply (5.2.14) to conclude that

$$|g(x) - g(x^\star)| \leq \gamma |x - x^\star|, \qquad \text{if } |x - x^\star| \leq \beta. \tag{5.2.16}$$

Suppose now that $|x_0 - x^\star| \leq \beta$. Then, using (5.2.10) and (5.2.13), we see from (5.2.16) that

$$|x_1 - x^\star| = |g(x_0) - g(x^\star)| \leq \gamma |x_0 - x^\star|.$$

Since $\gamma < 1$, this shows that x_1 is closer to $x^\star$ than x_0. Thus, $|x_1 - x^\star| \leq \beta$, and we can do the same thing again to obtain

$$|x_2 - x^\star| \leq \gamma |x_1 - x^\star| \leq \gamma^2 |x_0 - x^\star|,$$

and, in general,

$$|x_n - x^\star| \leq \gamma |x_{n-1} - x^\star| \leq \cdots \leq \gamma^n |x_0 - x^\star|. \tag{5.2.17}$$

Since $\gamma < 1$, this shows that $x_n \to x^\star$ as $n \to \infty$ (assuming no rounding or other errors).

It will be argued that (5.2.14) is an uncheckable condition since it requires knowing something about g' near $x^\star$, which is unknown. Surprisingly, however, we can obtain valuable information from the preceding analysis even without knowing $x^\star$. As a first illustration of this we consider an analysis of the second-order Adams-Moulton formula described in Section 2.4 for the solution of the ordinary differential equation $y' = f(y)$, where, for simplicity, we have dropped the dependence of f on x. The implicit formula is then given in (2.4.12) as

$$y_{k+1} = y_k + \frac{h}{2}[f(y_{k+1}) + f_k]. \tag{5.2.18}$$

This is a nonlinear equation for y_{k+1}, although it was used in Section 2.4 only as a "corrector formula"; that is, a predicted value $y_{k+1}^{(0)}$ was computed by an explicit method and then used in (5.2.18) to obtain a new estimate of y_{k+1} by

$$y_{k+1}^{(1)} = y_k + \frac{h}{2}[f(y_{k+1}^{(0)}) + f_k]. \tag{5.2.19}$$

Now we can correct this value again by using it in place of $y_{k+1}^{(0)}$ in (5.2.19). If we do this repeatedly, we obtain the sequence defined by

$$y_{k+1}^{(i+1)} = y_k + \frac{h}{2}[f(y_{k+1}^{(i)}) + f_k], \qquad i = 0, 1, \dots . \tag{5.2.20}$$

Clearly, (5.2.20) is just the iteration process $y_{k+1}^{(i+1)} = g(y_{k+1}^{(i)})$, where

$$g(y) = y_k + \frac{h}{2}[f(y) + f_k].$$

If y_{k+1} is the exact solution of (5.2.18), we can apply the previous analysis to conclude that the sequence of (5.2.20) will converge to y_{k+1} provided that $y_{k+1}^{(0)}$ (the predicted value) is sufficiently close to y_{k+1} and that

$$|g'(y)| = \left|\frac{h}{2}f'(y)\right| < 1$$

in a neighborhood of y_{k+1}; this will hold if h is sufficiently small.

As another illustration of the use of the convergence analysis, we consider Newton's method. Assume that $f'(x^\star) \neq 0$ and that f is twice continuously differentiable in a neighborhood of $x^\star$. Thus, by continuity, $f'(x) \neq 0$ in some neighborhood of $x^\star$, and we can differentiate the Newton iteration function (5.2.11) to obtain

$$g'(x) = 1 - \frac{[f'(x)]^2 - f(x)f''(x)}{[f'(x)]^2} = \frac{f(x)f''(x)}{[f'(x)]^2}.$$

Hence $g'(x^\star) = 0$, since $f(x^\star) = 0$. Therefore, by continuity, (5.2.14) must hold in a neighborhood of $x^\star$, and we conclude that the Newton iterates converge if x_0 is sufficiently close to $x^\star$. This shows that, under rather mild assumptions, the Newton iterates *must* converge to a root provided that x_0 (or any iterate x_k) is sufficiently close to $x^\star$. Although this type of convergence theorem, known as a *local convergence theorem*, does not help one decide if the iterates will converge from a given x_0, it gives an important intrinsic property of the iterative method.

When an iterate is not sufficiently close to a solution, various types of "bad" behavior can occur with Newton's method, as shown in Figure 5.7. Figure 5.7(a) illustrates that if $f'(x_i) = 0$, the next Newton iterate is not defined and the tangent line to f at x_i is horizontal. Figure 5.7(b) indicates the possibility of "cycling," in which $x_{i+2} = x_i$, and this cycle then repeats (see Exercise 5.2.5); thus there is no convergence but no divergence either. Cycles of order higher than 2 are also possible. Figure 5.7.(c) shows divergence to infinity, as would be the case if x_i is outside the domain of convergence to the solution of interest and the function behaves like, for example, e^{-x} as $x \to \infty$.

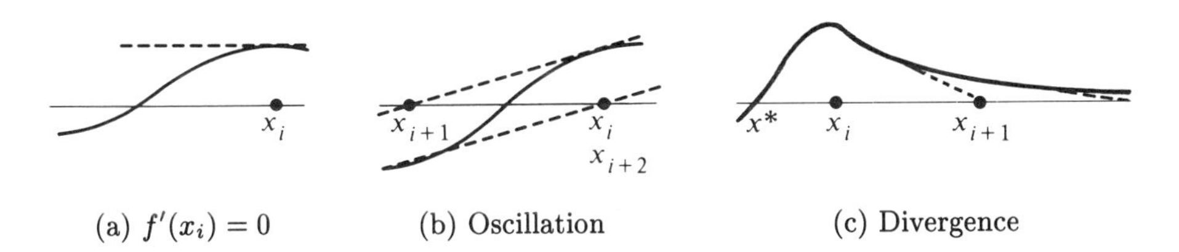

Figure 5.7: *Possible "Bad" Behavior of Newton's Method*

Convexity

In contrast to the above instances of bad behavior, there are situations in which Newton's method will converge for any starting approximation, no matter how far from the solution. In this case we speak of *global convergence.* Perhaps the simplest functions for which global convergence is obtained are those that are convex. A function is *convex* if it satisfies any one of the following equivalent properties, depending upon the differentiability of the function:

$$f''(x) \geq 0, \qquad \text{for all } x, \tag{5.2.21a}$$

$$f'(y) \geq f'(x), \qquad \text{if } y \geq x, \tag{5.2.21b}$$

$$f(\alpha x + (1-\alpha)y) \leq \alpha f(x) + (1-\alpha) f(y), \tag{5.2.21c}$$

where (5.2.21c) holds for any $\alpha \in (0,1)$ and all x, y.

A linear function $f(x) = ax + b$ is always convex, as is easily checked by any of the definitions of (5.2.21). However, we are mostly interested in functions that actually "bend upwards," as illustrated in Figure 5.8. Such functions are *strictly convex* and satisfy (5.2.21b,c) with strict inequality whenever $x \neq y$. Strict inequality in (5.2.21a) is also sufficient for strict convexity, but not necessary; the function $f(x) = x^4$ is strictly convex although $f''(0) = 0$.

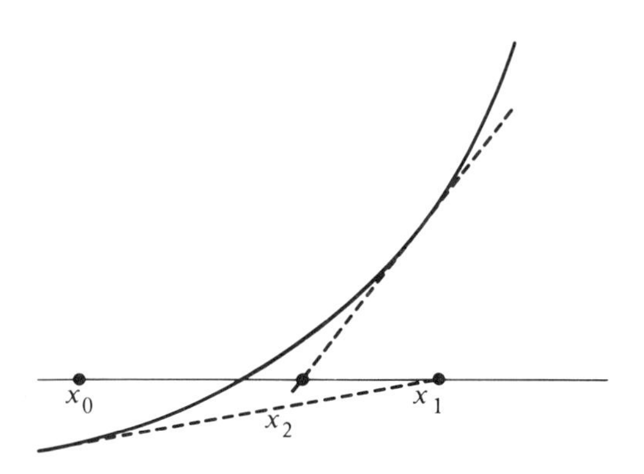

Figure 5.8: *Convergence of Newton's Method for a Convex Function*

A convex function may have infinitely many roots ($f(x) \equiv 0$) and even a strictly convex function may have no roots (for example, $f(x) = e^{-x}$). In the sequel, we will assume that f is strictly convex, $f'(x) > 0$ for all x, and $f(x) = 0$ has a solution; as noted previously, the condition on f' ensures that the solution is unique. These assumptions are illustrated by the function in Figure 5.8. In this case if x_0 is to the right of the solution, the Newton iterates converge monotonically to the solution, as is intuitively clear by drawing the tangent lines to the curve (see also Exercise 5.2.10). If x_0 is to the left of the solution, as shown in Figure 5.8, then the next Newton iterate is to the right of the solution, and thereafter the Newton iterates again converge monotonically to the solution. Figure 5.8 shows a function for which $f'(x) > 0$. If $f'(x) < 0$, the corresponding situation holds, but monotone convergence is now from left to right (Exercise 5.2.11). Similar convergence statements can be made if f is *concave*, that is, if $-f$ is convex.

The above discussion assumes that the properties of f hold for all x, in which case we obtain global convergence. They may, however, hold only in a neighborhood of a solution, and this will again ensure monotone convergence of the Newton iterates for suitable starting approximations x_0. For example, in Figure 5.7(c) there will be an interval $[x^\star, b]$ for which the Newton iterates will converge monotonically to $x^\star$ if $x_0 \in [x^\star, b]$. See also Exercises 5.2.5 and 5.2.6.

Rate of Convergence

From the standpoint of economical computation, the rate at which iterates converge to a root is almost as important as whether they converge at all. Suppose, in analogy to the estimate (5.2.17), that the errors behave as

$$|x^\star - x_{i+1}| = \gamma |x^\star - x_i|,$$

where γ is very close to 1, say $\gamma = 0.999$. Then, reducing the error in a given iterate by a factor of 10 would require well over two thousand iterations. Clearly, we wish the γ in the estimate (5.2.17) to be as small as possible, and from the derivation of (5.2.16) we see that it can be no smaller than $|g'(x^\star)|$. If $g'(x^\star) \neq 0$, then the rate of convergence is said to be *linear* or *geometric*, and $|g'(x^\star)|$ is the *asymptotic convergence factor*. Recall, however, that for Newton's method we showed that $g'(x^\star) = 0$ under the assumption that $f'(x^\star) \neq 0$. This does *not* imply, of course, that the iterates converge in one step, but it signals that the rate of convergence is faster than linear. In particular, it can be shown (see Exercise 5.2.4) that close to the solution the errors in Newton's method satisfy

$$|x^\star - x_{i+1}| \leq c|x^\star - x_i|^2, \tag{5.2.22}$$

where c depends on the ratio of f'' to f' near $x^\star$. The relation (5.2.22) defines *quadratic convergence*; the iterates converge very rapidly once they begin to get close to a root. For example, suppose $c = 1$ and $|x^\star - x_i| \doteq 10^{-3}$. Then $|x^\star - x_{i+1}| \doteq 10^{-6}$, so that the number of correct decimal places has been doubled in one iteration. It is this property of quadratic convergence that makes Newton's method of central importance. Quadratic convergence is lost, however, when $f'(x^\star) = 0$, so that $x^\star$ is a multiple root (see Exercise 5.2.12).

As an illustration of quadratic convergence in Newton's method, consider the problem of finding the zeros of $f(x) = 1/x + \ln x - 2$. This function is defined for all positive values of x and has two zeros: one between $x = 0$ and $x = 1$ and the other between $x = 6$ and $x = 7$, as illustrated in Figure 5.9. Table 5.1 contains a summary of the first six iterations of Newton's method using the starting value of $x = 0.1$. Note that once an approximation is "close enough" (in this case, after three iterations), the number of correct digits doubles in each iteration, which shows the quadratic convergence.

Rounding Error

So far the discussion of Newton's method has been predicated upon exact computation of the iterates, but rounding or other errors will inevitably cause

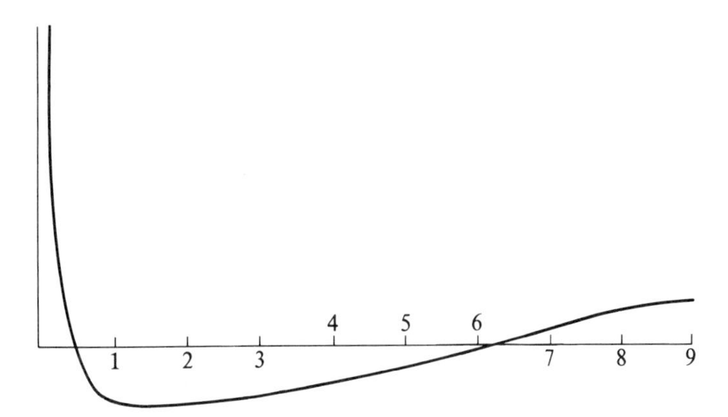

Figure 5.9: *The Function* $f(x) = 1/x + lnx - 2$

Table 5.1: *Convergence of Newton's Method for* $f(x) = 1/x + lnx - 2$

Iteration	x_{i-1}	$f(x_{i-1})$	x_i	Number of Correct Digits
1	0.1	5.6974149	0.16330461	0
2	0.16330461	2.3113878	0.23697659	0
3	0.23697659	0.7800322	0.29438633	1
4	0.29438633	0.1740346	0.31576121	2
5	0.31576121	0.0141811	0.31782764	4
6	0.31782764	0.0001134	0.31784443	8

the iterates to be computed inaccurately. For example, if ε_i and ε_i' are the errors made in computing $f(x_i)$ and $f'(x_i)$, respectively, then the computed next iterate $\hat{x}_{i+1}$ is

$$\hat{x}_{i+1} = x_i \ominus (f(x_i) + \varepsilon_i) \oplus\!\!\!\!\div (f'(x_i) + \varepsilon_i'),$$

where the circled operations indicate that rounding errors are also made in the subtraction and division. A full analysis of the effects of these errors is difficult, if even possible, and we content ourselves with the following remarks. If the errors ε_i and ε_i' are small, we can expect the computed iterates to behave roughly as the exact iterates would, at least as long as we are not close to the root. However, when $f(x_i)$ becomes so small that it is comparable in size to ε_i, then the computed iterates no longer behave like the exact ones. In particular, we saw in the case of the bisection method that when the sign of f can no longer be evaluated correctly, the method breaks down in the sense that a wrong decision may be made as to which interval the root is in. An analogous thing happens with Newton's method: if the sign of $f(x_i)$ is evaluated incorrectly, but that of $f'(x_i)$ correctly (a reasonable assumption if $f'(x^\star)$ is not particularly small), then the computed value of $f(x_i)/f'(x_i)$ has the wrong sign, and the computed next iterate moves in the wrong direction.

As with the bisection method, the notion of an interval of uncertainty about the root $x^\star$ applies equally well to Newton's method (as well as to essentially all iterative methods).

Ill-conditioning

In Chapter 4 we discussed ill-conditioning of a solution of a system of linear equations; an analogous problem can occur with roots of nonlinear equations. The simplest example of this is given by the trivial polynomial equation

$$x^n = 0,$$

which has an n-fold root equal to zero, and the polynomial equation

$$x^n = \varepsilon, \qquad \varepsilon > 0,$$

whose n roots are $\varepsilon^{1/n}$ times the nth roots of unity and therefore all have absolute value of $\varepsilon^{1/n}$. If, for example, $n = 10$ and $\varepsilon = 10^{-10}$, the roots of the second polynomial have absolute value 10^{-1}; thus, a change of 10^{-10} in one coefficient (the constant term) of the original polynomial has caused changes 10^9 times as great in the roots.

This simple example is a special case of the general observation that if a root $x^\star$ of a polynomial f is of multiplicity m, then small changes of order ε in the coefficients of f may cause a change of order $\varepsilon^{1/m}$ in $x^\star$; an analogous result holds for functions other than polynomials by expanding in a Taylor series about $x^\star$.

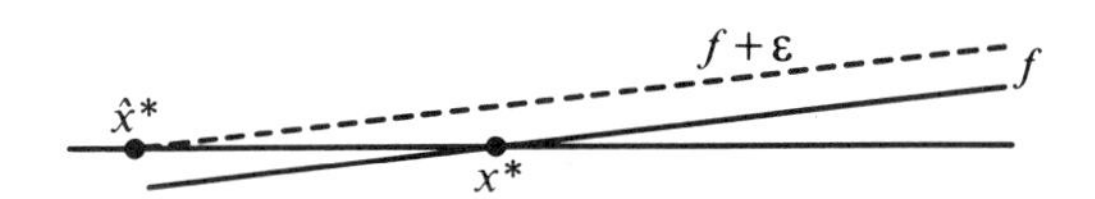

Figure 5.10: *A Large Change in $x^\star$ Due to a Small Change in f*

A necessary condition for a multiple root at $x^\star$ is that $f'(x^\star) = 0$. If $f'(x^\star) \neq 0$ but $f'(x)$ is small in the neighborhood of $x^\star$, then small changes in f can still cause large changes in $x^\star$, as Figure 5.10 illustrates. Perhaps the most famous example of how ill-conditioned nonmultiple roots can be is given by the following. Let f be the polynomial of degree 20 with roots $1, \ldots, 20$, and let $\hat{f}$ be the same polynomial but with the coefficient of x^{19} changed by $2^{-23} \doteq 10^{-7}$. Then the roots of $\hat{f}$ to one decimal place are given by

1.0	2.0	3.0	4.0	5.0	6.0	7.0	8.0	8.9

10.1 ± 0.6i	11.8 ± 1.7i	14.0 ± 2.5i	16.7 ± 2.8i	19.5 ± 1.9i	20.8

Since the coefficient of x^{19} in $f(x)$ is 210, we see that a change of about $10^{-7}\%$ in this one coefficient has caused such large changes in the roots that some have even become complex!

The Shooting Method

We end this section with a discussion of the shooting method introduced in Section 5.1. Consider the boundary value problem

$$v''(x) = g(x, v(x)), \qquad v(0) = \alpha, \qquad v(1) = \beta. \tag{5.2.23}$$

The equation (5.1.2) is a special case of (5.2.23) in which $g(x, v) = x^2 + 3v + 10v^3$. To apply the shooting method we solve the initial value problem

$$v''(x) = g(x, v(x)), \qquad v(0) = \alpha, \qquad v'(0) = s, \tag{5.2.24}$$

and denote the solution by $v(x; s)$. Then, as in Section 5.1, we define the function

$$f(s) = v(1; s) - \beta, \tag{5.2.25}$$

and we wish to solve the equation $f(s) = 0$.

Consider first the bisection method, in which we only need to evaluate the function f. Each such evaluation requires that we solve the initial value problem (5.2.24) so as to obtain $v(1; s)$. In general, the solution of (5.2.24) can be accomplished by any of the methods of Chapter 2 (after first converting the second order equation to a system of two first-order equations). To begin the bisection method we would need to find s_0 and s_1 so that $f(s_0)$ and $f(s_1)$ have different signs. This might require a little trial and error but then the bisection method proceeds systematically. Note that, in general, the computed $f(s)$ will be inaccurate because of both discretization error in solving the initial value problem as well as rounding error, and the bisection method will break down when the sign of f can no longer be evaluated correctly.

For this type of problem the evaluation of $f(s)$ can be time consuming since it requires the solution of the initial value problem (5.2.24). Therefore we would like to utilize the potentially rapid convergence of Newton's method. For Newton's method we need $f'(s)$ and we differentiate (5.2.25) to obtain

$$f'(s) = v_s(1; s), \tag{5.2.26}$$

where $v_s(1; s)$ is the partial derivative of $v(x; s)$ with respect to s and evaluated at $x = 1$. In order to obtain a way of computing $v_s(1; s)$, we differentiate

$$v''(x; s) = g(x, v(x; s))$$

with respect to s to obtain

$$\frac{\partial}{\partial s}(v''(x;s)) = g_v(x, v(x;s))v_s(x;s), \tag{5.2.27}$$

where $g_v(x, v)$ is the partial derivative of $g(x, v)$ with respect to v. Assuming that differentiation with respect to s and x can be interchanged on the left side of (5.2.27), we then have

$$v_s''(x;s) = g_v(x, v(x,s))v_s(x;s), \tag{5.2.28}$$

which is called the *adjoint equation* for v_s. In (5.2.28) we are assuming that s is held fixed so that this is a differential equation for $v_s(x;s)$ considered only as a function of x. The initial conditions for (5.2.28) are obtained by differentiating those of (5.2.24) with respect to s; thus

$$v_s(0;s) = 0, \qquad v_s'(0;s) = 1. \tag{5.2.29}$$

If we knew the exact solution $v(x;s)$ of (5.2.24), we could put this into g_v in (5.2.28) to obtain a known function of x, and then (5.2.28), (5.2.29) would be a linear initial-value problem for $v_s(x;s)$. Upon solving this initial value problem we obtain $f'(s) = v_s(1;s)$. Of course, we do not know the exact solution of (5.2.24), but we solve this initial-value problem approximately to obtain values $v_i \doteq v(x_i;s)$ at the grid points x_i. We can then use these approximate values to evaluate g_v in (5.2.28) and in this way we can obtain an approximate solution to (5.2.28) at the same time we obtain the approximate solution to (5.2.24). Thus we will be able to carry out Newton's method, at least approximately, for $f(s) = 0$. We could also approximate Newton's method by the secant or regula falsi methods in which only values of $f(s)$ are required.

We should caution, however, that the shooting method is not always viable, as discussed in Section 5.1, and the finite difference method to be discussed in Section 5.3 may be preferable for two-point boundary value problems.

Supplementary Discussion and References: 5.2

For a thorough treatment of the theory of iterative methods for roots of equations, see Traub [1964], and for an excellent discussion of rounding error, see Wilkinson ([1963], [1965]). In particular, the example of the ill-conditioned polynomial of degree 20 is due to Wilkinson.

For roots of polynomials there are a number of specialized methods, such as Bairstow's method for polynomials with real coefficients but complex roots, and Laguerre's method, which has the property of cubic convergence (that is, an error estimate of the form (5.2.22) holds with $|x^\star - x_i|^3$ on the right-hand side). Also, roots of polynomials are eigenvalues of the corresponding companion matrix and can be obtained in principle by the methods of Chapter 7.

When to stop the iteration is a problem for which there is still no definitive solution. The usual simplest tests are $|f(x_i)| < \varepsilon$ or $|x_{i+1} - x_i| < \varepsilon$, where ε is some given tolerance. The first can be misleading when the function f is very "flat" near the root, as is the case of a multiple root, and the second can fail in a variety of situations depending on the iterative method. For example, for Newton's method it can fail when the derivative is very large at the current iterate. These two possibilities are depicted in the following figure:

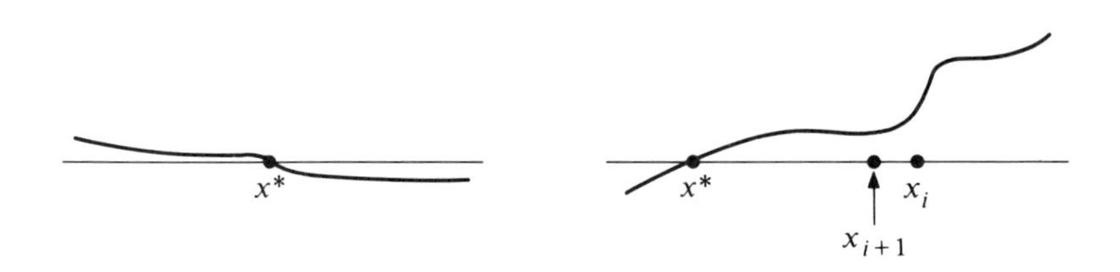

EXERCISES 5.2

5.2.1. If f is a continuously differentiable function, use the mean-value theorem (see Appendix 1) to show that if (5.2.3) holds then f has at most one root in the interval (a, b).

5.2.2. Let $f(x) = e^x$. Show that $f'(x) > 0$ for all x but f does not have any finite roots.

5.2.3. Let f be twice continuously differentiable and suppose that $f'(x^\star) = 0$ at a root $x^\star$ of f but $f'(x) \neq 0$ in a neighborhood of $x^\star$. Show that the limit of the iteration function g of (5.2.11) exists and equals $x^\star$ as $x \to x^\star$.

5.2.4. Let f be twice continuously differentiable and suppose that $f'(x^\star) \neq 0$ at a root $x^\star$ of f. Show that the error relation (5.2.22) holds for Newton's method in some interval about $x^\star$. *Hint:* Expand $0 = f(x^\star) = f(x) + f'(x)(x^\star - x) + \frac{1}{2}f''(\xi)(x^\star - x)^2$, solve this for $x^\star$, and then use (5.2.9).

5.2.5. Consider the function $f(x) \equiv x - x^3$ with roots at 0 and ± 1.

a. Show that Newton's method is locally convergent to each of the three roots.

b. Carry out several steps of Newton's method starting with the initial approximation $x_0 = 2$. Discuss the rate of convergence that you observe in your computed iterates.

c. Carry out several steps of both the bisection and secant methods starting with the interval $(\frac{3}{4}, 2)$. Compare the rate of convergence of the iterates from these methods with that of the Newton iterates.

d. Determine the set of points S for which the Newton iterates will converge (in the absence of rounding errors) to the root 1 for any starting approximation x_0 in S. Do the same for the roots 0 and -1.

5.2.6. Consider the equation $x - 2\sin x = 0$.

a. Show graphically that this equation has precisely three roots: 0, and one in each of the intervals $(\pi/2, 2)$ and $(-2, -\pi/2)$.

b. Show that the iterates $x_{i+1} = 2\sin x_i$, $i = 0, 1, \ldots$, converge to the root in $(\pi/2, 2)$ for any x_0 in this interval.

c. Apply the Newton iteration to this equation and ascertain for what starting values the iterates will converge to the root in $(\pi/2, 2)$. Compare the rate of convergence of the Newton iterates with those of part **b**.

5.2.7. Let n be a positive integer and α a positive number. Show that Newton's method for the equation $x^n - \alpha = 0$ is

$$x_{k+1} = \frac{1}{n}\left[(n-1)x_k + \frac{\alpha}{x_k^{n-1}}\right], \qquad k = 0, 1, \cdots,$$

and that this Newton sequence converges for any $x_0 > 0$. Discuss the case $n = 2$.

5.2.8. Ascertain whether the following statements are true or false and prove your assertions:

a. Let $\{x_k\}$ be a sequence of Newton iterates for a continuously differentiable function f. If for some i, $|f(x_i)| \leq 0.01$ and $|x_{i+1} = x_i| \leq 0.01$, then x_{i+1} is within 0.01 of a root of $f(x) = 0$.

b. The Newton iterates converge to the unique solution of $x^2 - 2x + 1 = 0$ for any $x_0 \neq 1$. (Ignore rounding error.)

5.2.9. Consider the equation $x^2 - 2x + 2 = 0$. What is the behavior of the Newton iterates for various real starting values?

5.2.10. Show that the Newton iterates converge to the unique solution of $e^{2x} + 3x + 2 = 0$ for any starting value x_0.

5.2.11. Assume that f is differentiable, convex, and $f'(x) < 0$ for all x. If $f(x) = 0$ has a solution $x^\star$, show that $x^\star$ is unique and that the Newton iterates converge monotonically upward to $x^\star$ if $x_0 < x^\star$. What happens if $x_0 > x^\star$?

5.2.12. Show that the Newton iterates for $f(x) \equiv x^p$ converge to the solution $x^\star = 0$ only linearly with an asymptotic convergence factor of $(p-1)/p$.

5.2.13. Newton's method can be used for determining the reciprocal of numbers when division is not available.

a. Show how Newton's method can be applied to the equation

$$f(x) = \frac{1}{x} - a,$$

without using division.

b. Give an equation for the error term, $e_k = x_k - a^{-1}$, and show that the convergence is quadratic.

c. Give conditions on the initial approximation so that $x_k \to a^{-1}$ as $k \to \infty$. If $0 < a < 1$, give a numerical value of x_0 which will guarantee convergence.

5.3 Systems of Nonlinear Equations

We mentioned in Section 5.1 that the shooting method applied to a system of ordinary differential equations can lead to the problem of solving a system of nonlinear algebraic equations in order to supply the missing initial conditions. As we shall see later, the application of the finite difference method to nonlinear boundary-value problems also leads to nonlinear systems of equations. Therefore we consider in this section how the methods discussed in the previous section for a single equation can be extended to systems of equations.

The problem is to obtain an approximate solution to the system of equations

$$f_i(x_1, x_2, \ldots, x_n) = 0, \qquad i = 1, \ldots, n, \tag{5.3.1}$$

where $f_1, f_2, \ldots, f_n$ are given functions of the n variables $x_1, \ldots, x_n$. We shall usually use vector notation and write (5.3.1) as

$$\mathbf{F}(\mathbf{x}) = \mathbf{0}, \tag{5.3.2}$$

where, as usual, $\mathbf{x}$ is the vector with components $x_1, \ldots, x_n$, and $\mathbf{F}$ is the vector function with components $f_1, \ldots, f_n$. The special case of solving (5.3.1) when $n = 1$ is just the problem of finding roots of a single equation that was considered in the previous section. On the other hand, the special case in which

$$\mathbf{F}(\mathbf{x}) \equiv A\mathbf{x} - \mathbf{b},$$

where A is a given matrix and $\mathbf{b}$ a given vector, is that of solving a system of linear equations, which was treated in Chapter 4.

The problem of ascertaining when (5.3.2) has solutions, and how many, is generally very difficult. In the relatively simple case $n = 2$, it is easy to see the various possibilities geometrically, at least in principle. For example, if we plot in the x_1, x_2 plane the set of points for which $f_1(x_1, x_2) = 0$, and then the set of points for which $f_2(x_1, x_2) = 0$, the intersection of these sets is precisely the set of solutions of (5.3.2). (Here, and henceforth, we are restricting our attention to only real solutions.) Figure 5.11 illustrates a few possible situations. Later

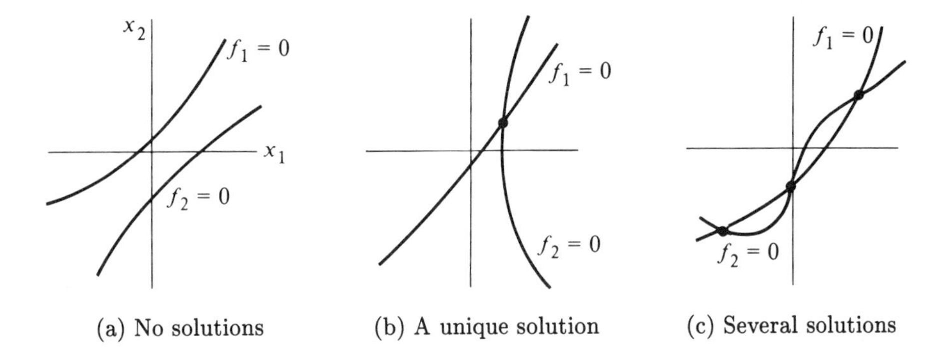

Figure 5.11: *Possible Solutions for* $n = 2$

we shall assume that (5.3.2) has a solution $\mathbf{x}^*$ that is the one of interest to us, although the system may have additional solutions.

Picard Iterations

In many situations the system (5.3.2) has the form

$$\mathbf{F}(\mathbf{x}) \equiv A\mathbf{x} + \mathbf{H}(\mathbf{x}) = 0, \qquad (5.3.3)$$

where A is a given nonsingular matrix and $\mathbf{H}$ is a given vector of nonlinear functions. In this case a somewhat natural (although not necessarily good) iterative procedure is

$$\mathbf{x}^{i+1} = -A^{-1}\mathbf{H}(\mathbf{x}^i), \qquad i = 0, 1, \ldots\ , \qquad (5.3.4)$$

where the superscript indicates iteration number. Here, as well as later, we mean by (5.3.4) that at each step of the iteration the linear system of equations

$$A\mathbf{x}^{i+1} = -\mathbf{H}(\mathbf{x}^i)$$

is to be solved to obtain the next iterate. The iteration (5.3.4) is known as a *Picard iteration.* It may be considered a special case of the extension of the chord method of the previous section to n equations; this would take the form

$$\mathbf{x}^{i+1} = \mathbf{x}^i - B\mathbf{F}(\mathbf{x}^i), \qquad i = 0, 1, \ldots, \qquad (5.3.5)$$

for a given nonsingular matrix B. It is easy to see (Exercise 5.3.2) that (5.3.5) reduces to (5.3.4) if $\mathbf{F}$ is of the form (5.3.3) and $B = A^{-1}$.

Convergence

When will the iterations (5.3.5) or (5.3.4) converge? The situation is precisely analogous to that in the scalar case but complicated by the need to work with vector-valued functions. Consider the general one-step iteration

$$\mathbf{x}^{i+1} = \mathbf{G}(\mathbf{x}^i), \qquad i = 0, 1, \dots, \tag{5.3.6}$$

where $\mathbf{G}$ is a given iteration function; for example, for (5.3.5)

$$\mathbf{G}(\mathbf{x}) \equiv \mathbf{x} - B\mathbf{F}(\mathbf{x}). \tag{5.3.7}$$

We shall assume that the solution $\mathbf{x}^*$ of $\mathbf{F}(\mathbf{x}) = 0$ satisfies $\mathbf{x}^* = \mathbf{G}(\mathbf{x}^*)$ and, conversely, that if $\mathbf{x}^* = \mathbf{G}(\mathbf{x}^*)$, then $\mathbf{F}(\mathbf{x}^*) = 0$; it is clear that this is the case for (5.3.7) if B is nonsingular.

In the previous section, the convergence theory was based on $|g'(x)| < 1$ in a neighborhood of the solution. For systems of equations the corresponding result is the following. If

$$||\mathbf{G}'(\mathbf{x})|| \le \gamma < 1, \qquad \text{for } ||\mathbf{x} - \mathbf{x}^*|| \le \beta, \tag{5.3.8}$$

then the iterates (5.3.6) converge if $||\mathbf{x}^0 - \mathbf{x}^*|| \le \beta$ (or if $||\mathbf{x}^k - \mathbf{x}^*|| \le \beta$ for any k). Here, as in Section 4.4, $||\ ||$ denotes a vector norm or the corresponding matrix norm, and $\mathbf{G}'(\mathbf{x})$ is the Jacobian matrix (see Appendix 1) of $\mathbf{G}$ evaluated at $\mathbf{x}$. We shall not prove this convergence statement but only note that it can be rather easily proven after a proper extension of the mean-value theorem to n dimensions.

If we apply the criterion (5.3.8) to the iteration (5.3.5), we obtain (see Exercise 5.3.4)

$$||I - B\mathbf{F}'(\mathbf{x})|| \le \gamma < 1 \qquad \text{for } ||\mathbf{x} - \mathbf{x}^*|| < \beta \tag{5.3.9}$$

and, in particular, for the iteration (5.3.4),

$$||A^{-1}\mathbf{H}'(\mathbf{x})|| \le \gamma < 1 \qquad \text{for } ||\mathbf{x} - \mathbf{x}^*|| \le \beta. \tag{5.3.10}$$

Intuitively, (5.3.10) says that the iteration (5.3.4) will converge provided that $A^{-1}\mathbf{H}'(\mathbf{x})$ is "small" when $\mathbf{x}$ is close to $\mathbf{x}^*$. Similarly, the iteration (5.3.5) will converge if $B\mathbf{F}'(\mathbf{x})$ is close to the identity, or, equivalently, if B^{-1} is close to $\mathbf{F}'(\mathbf{x})$. Since $\mathbf{x}^*$ is not known, these criteria are not meant to be used to check whether a given iteration will converge, but rather to give some insight as to what factors govern the convergence.

Newton's Method

Analogously to the previous section, the size of $||\mathbf{G}'(\mathbf{x})||$ will tend to determine the rate of convergence, and we would like this quantity to be as small as possible, at least near the solution $\mathbf{x}^*$. Suppose that for the iteration (5.3.5) we could choose $B = [\mathbf{F}'(\mathbf{x}^*)]^{-1}$; then $\mathbf{G}'(\mathbf{x}^*) = 0$, and the rate of convergence will be rapid near the solution. Of course, this choice of B is essentially impossible since $\mathbf{x}^*$ is not known, but we can achieve this effect by the following *Newton iteration:*

$$\mathbf{x}^{i+1} = \mathbf{x}^i - [\mathbf{F}'(\mathbf{x}^i)]^{-1}\mathbf{F}(\mathbf{x}^i), \qquad i = 0, 1, \ldots . \tag{5.3.11}$$

Here, we are assuming, of course, that the matrices $\mathbf{F}'(\mathbf{x}^i)$ are nonsingular, and we would carry out (5.3.11) by the following steps:

$$\begin{array}{l} \text{1. Solve the linear system } \mathbf{F}'(\mathbf{x}^i)\mathbf{y}^i = -\mathbf{F}(\mathbf{x}^i). \\ \text{2. Set } \mathbf{x}^{i+1} = \mathbf{x}^i + \mathbf{y}^i. \end{array} \tag{5.3.12}$$

The iteration (5.3.12) can be derived as follows. We approximate the functions f_i at $\mathbf{x}^k$ by a first-order Taylor expansion:

$$f_i(\mathbf{x}) \doteq l_i(\mathbf{x}) \equiv f_i(\mathbf{x}^k) + f_i'(\mathbf{x}^k)(\mathbf{x} - \mathbf{x}^k), \quad i = 1, \cdots, n. \tag{5.3.13}$$

Here, $f_i'(\mathbf{x}^k)$ is the ith row of the Jacobian matrix $\mathbf{F}'(\mathbf{x}^k)$ and (5.3.13) can be written as

$$\mathbf{F}(\mathbf{x}) \doteq L(\mathbf{x}) \equiv \mathbf{F}(\mathbf{x}^k) + \mathbf{F}'(\mathbf{x}^{\mathbf{k}})(\mathbf{x} - \mathbf{x}^{\mathbf{k}}).$$

The solution of the linear system $L(\mathbf{x}) = 0$ then gives the next Newton iterate $\mathbf{x}^{k+1}$. Geometrically, $l_i(\mathbf{x}) = 0$ is the equation of the "hyperplane" tangent to f_i at $\mathbf{x}^k$, and $\mathbf{x}^{k+1}$ is the intersection of the n sets $\{\mathbf{x} : l_i(\mathbf{x}) = 0\}$. This generalizes to n dimensions the property of Newton's method in a single variable that the next Newton iterate is the intersection of the x-axis with the tangent line to f at x_i.

Clearly, the iteration (5.3.11) reduces to Newton's method of the previous section if $n = 1$. We would hope that (5.3.11) retains the basic property of quadratic convergence. This is true, and we state the following result without proof.

> THEOREM 5.3.1 (Newton Convergence) *If $\mathbf{F}$ is two times continuously differentiable in a neighborhood of $\mathbf{x}^*$, and if $\mathbf{F}'(\mathbf{x}^*)$ is nonsingular, then the iterates (5.3.11) will converge to $\mathbf{x}^*$ provided that $\mathbf{x}^0$ is sufficiently close to $\mathbf{x}^*$ (local convergence theorem), and they will have the property of quadratic convergence:*

$$||\mathbf{x}^{i+1} - \mathbf{x}^*|| \le c||\mathbf{x}^i - \mathbf{x}^*||^2. \tag{5.3.14}$$

As an example of Newton's method (5.3.12), we give in Table 5.2 the first four iterations for the system of nonlinear equations

$$x_1^2 + x_2^2 - 1 = 0, \qquad x_1^2 - x_2 = 0, \tag{5.3.15}$$

Table 5.2: *Convergence of Newton's Method for* (5.3.15)

Iteration	x_1	x_2	Number of Correct Digits
0	0.5	0.5	0,0
1	0.87499999	0.62499999	0,1
2	0.79067460	0.61805555	1,4
3	0.78616432	0.61803399	4,8
4	0.78615138	0.61803399	8,8

using the starting values $x_1 = 0.5$ and $x_2 = 0.5$. Note that we can observe the approximate quadratic convergence by the number of correct digits in the iterates.

The quadratic convergence property (5.3.14) (which is lost if $\mathbf{F}'(\mathbf{x}^*)$ is singular) is highly desirable and makes Newton's method of central importance in the solution of nonlinear systems of equations. But there are three obstacles to its successful use. The first is the need to compute the Jacobian matrix at each step, and this requires evaluation of the n^2 partial derivatives $\partial f_i / \partial x_j$. If n is large and/or the functions f_i are complicated, it can be drudgery to work out by hand – and then convert to computer code – the expressions for these derivatives; this can sometimes be mitigated by the use of symbolic differentiation techniques, as discussed in Chapter 1. Another commonly used approach is to approximate the partial derivatives by finite differences; for example,

$$\frac{\partial f_i}{\partial x_j}(\mathbf{x}) \doteq \frac{1}{h}[f_i(x_1, \ldots, x_{j-1}, x_j + h, x_{j+1}, \ldots, x_n) - f_i(\mathbf{x})]. \tag{5.3.16}$$

This has the advantage of requiring only the expressions for the f_i, which are needed in any case. But the actual numerical evaluation of the Jacobian matrix, either by expressions for the partial derivatives or by approximations such as (5.3.16), can be costly in computer time. This leads to a frequently used modification of Newton's method in which the Jacobian matrix is reevaluated only periodically rather than at each iteration. For example, the iteration might be:

1. Evaluate $\mathbf{F}'(\mathbf{x}^0)$.
2. Compute $\mathbf{x}^{i+1} = \mathbf{x}^i - [\mathbf{F}'(\mathbf{x}^0)]^{-1}\mathbf{F}(\mathbf{x}^i), \quad i = 0, 1, \ldots, k.$
3. Evaluate $\mathbf{F}'(\mathbf{x}^{k+1})$. (5.3.17)
4. Compute $\mathbf{x}^{i+1} = \mathbf{x}^i - [\mathbf{F}'(\mathbf{x}^{k+1})]^{-1}\mathbf{F}(\mathbf{x}^i), \quad i = k+1, \ldots, 2k.$

$\vdots$

The modified Newton iteration (5.3.17) can also be useful in alleviating the second disadvantage of Newton's method: the need to solve a system of linear equations at each step. The advantage of (5.3.17) in this regard is that the actual implementation involves, of course, solving a number of linear systems (as in step 2),

$$\mathbf{F}'(\mathbf{x}^0)\mathbf{y}^i = -\mathbf{F}(\mathbf{x}^i), \qquad i = 0, 1, \ldots, k,$$

where the coefficient matrix $\mathbf{F}'(\mathbf{x}^0)$ is the same. Hence, as discussed in Section 4.2, the LU factors of $\mathbf{F}'(\mathbf{x}^0)$ from the Gaussian elimination process can be retained and used for all $k+1$ right-hand sides.

The third – and most troublesome – difficulty with Newton's method is that the iterates may not converge from a given starting approximation $\mathbf{x}^0$; the local convergence theorem only insures convergence once $\mathbf{x}^0$ (or some other iterate) is "sufficiently close" to $\mathbf{x}^\star$. One remedy for this difficulty is to obtain the best possible first approximation using any physical or other knowledge about the problem. However, this is not always sufficient. An approach that often – but certainly not always – works is the *continuation method*, which we describe briefly in the Supplementary Discussion.

Nonlinear Boundary Value Problems

We discuss now the extension to nonlinear two-point boundary-value problems of the finite difference method presented in Chapter 3. We shall consider the equation

$$v'' = g(x, v), \qquad 0 \le x \le 1, \tag{5.3.18}$$

with the boundary conditions

$$v(0) = \alpha, \qquad v(1) = \beta. \tag{5.3.19}$$

Here g is a given function of two variables, and α and β are given constants.

We proceed exactly as in Section 3.1. The interval $[0, 1]$ is partitioned by grid points

$$0 = x_0 < x_1 < \cdots < x_n < x_{n+1} = 1$$

with spacing h. At each interior grid point x_i we approximate the second derivative by central differences and use these approximations in (5.3.18). This leads to the system of equations (corresponding to (3.1.8))

$$-v_{i+1} + 2v_i - v_{i-1} + h^2 g(x_i, v_i) = 0, \qquad i = 1, \ldots, n, \tag{5.3.20}$$

where $v_0 = \alpha$ and $v_{n+1} = \beta$ are known by the boundary conditions (5.3.18). This is a system of n equations in the n unknowns $v_1, \ldots, v_n$ and is nonlinear if the function g is nonlinear in v. A solution $v_1^\star, \ldots, v_n^\star$ of (5.3.19), if it exists, is an approximation to the corresponding solution v of (5.3.18) at the grid points $x_1, \ldots, x_n$.

We can write the system (5.3.19) in matrix-vector form as

$$\mathbf{F}(\mathbf{v}) \equiv A\mathbf{v} + \mathbf{H}(\mathbf{v}) = 0, \tag{5.3.21}$$

where $\mathbf{v}$ is the vector with components $v_1, \ldots, v_n$, A is the $(2, -1)$ tridiagonal matrix of (3.1.10) and

$$\mathbf{H}(\mathbf{v}) = h^2 \begin{bmatrix} g(x_1, v_1) \\ \\ \vdots \\ \\ g(x_n, v_n) \end{bmatrix} - \begin{bmatrix} \alpha \\ 0 \\ \vdots \\ 0 \\ \beta \end{bmatrix}. \tag{5.3.22}$$

As an example, consider the problem

$$v''(x) = 3v(x) + x^2 + 10[v(x)]^3, \qquad 0 \le x \le 1, \qquad v(0) = v(1) = 0. \tag{5.3.23}$$

Here

$$g(x, v) = 3v + x^2 + 10v^3, \tag{5.3.24}$$

and with $h = 1/(n+1)$ and

$$x_i = ih, \qquad i = 0, 1, \ldots, n+1, \tag{5.3.25}$$

the difference equations (5.3.20) are

$$-v_{i+1} + 2v_i - v_{i-1} + h^2(3v_i + i^2h^2 + 10v_i^3) = 0, \qquad i = 1, \ldots, n, \tag{5.3.26}$$

where, from the boundary conditions, $v_0 = v_{n+1} = 0$. Hence the ith component of the function $\mathbf{H}(\mathbf{v})$ of (5.3.22) is $h^2(3v_i + i^2h^2 + 10v_i^3)$.

We now consider some numerical methods for the system (5.3.21). The Picard iteration discussed earlier is

$$A\mathbf{v}^{k+1} = -\mathbf{H}(\mathbf{v}^k). \tag{5.3.27}$$

The time to carry out one of these iterative steps depends almost entirely on the complexity of $\mathbf{H}$ since the solution of tridiagonal linear systems is very rapid, as we saw in Section 3.2. Moreover, in this case the LU decomposition of A can be done once and for all. Whether the iteration (5.3.26) even converges, however, will depend upon the properties of H.

Next consider Newton's method for (5.3.21). The Jacobian matrix will be (see Exercise 5.3.8)

$$\mathbf{F}'(\mathbf{v}) = A + \mathbf{H}'(\mathbf{v}). \tag{5.3.28}$$

Since the ith component, $H_i(\mathbf{v}) = h^2 g(x_i, v_i)$, of $\mathbf{H}$ depends only on v_i, we have that $\frac{\partial H_i}{\partial v_j} = 0$, $j \neq i$. Thus the matrix $\mathbf{H}'(\mathbf{v})$ is diagonal and $\mathbf{F}'(\mathbf{v})$ is tridiagonal with a typical row given by

$$-1 \qquad 2 + h^2 \frac{\partial g}{\partial v}(x_i, v_i) \qquad -1.$$

The Newton iteration is then

$$\begin{aligned} &1. \text{ Solve } [A + \mathbf{H}'(\mathbf{v}^k)]\mathbf{y}^k = -[A\mathbf{v}^k + \mathbf{H}(\mathbf{v}^k)], \\ &2. \text{ Set } \mathbf{v}^{k+1} = \mathbf{v}^k + \mathbf{y}^k, \end{aligned} \tag{5.3.29}$$

so that at each iteration a tridiagonal linear system is to be solved. If the function g is complicated, a major portion of the work of each Newton iteration will be the evaluation of $\mathbf{H}(\mathbf{v}^k)$ and $\mathbf{H}'(\mathbf{v}^k)$.

For the boundary-value problem (5.3.23) g is given by (5.3.24), so that

$$\frac{\partial g}{\partial v}(x, v) = 3 + 30v^2,$$

and the ith diagonal element of the Jacobian matrix (5.3.28) is $2 + h^2(3 + 30v_i^2)$. Since the $(2, -1)$ tridiagonal matrix A is diagonally dominant, it is clear that the addition of the positive terms $h^2(3 + 30v_i^2)$ to the diagonal only enhances the diagonal dominance. More generally, whenever (Exercise 5.3.11)

$$\frac{\partial g}{\partial v}(x, v) \geq 0, \qquad 0 \leq x \leq 1, \qquad -\infty < v < \infty, \tag{5.3.30}$$

and A is the $(2, -1)$ matrix, then

$$A + \mathbf{H}'(\mathbf{v}) \text{ is diagonally dominant,} \tag{5.3.31}$$

$$A + \mathbf{H}'(\mathbf{v}) \text{ is symmetric positive-definite.} \tag{5.3.32}$$

As we saw in Section 4.3, either of these properties is sufficient to ensure that the solution of the tridiagonal systems (5.3.29) of Newton's method can be carried out by Gaussian elimination without any need for interchanging rows to preserve numerical stability.

It is also true (but beyond the scope of this book to prove) that either of the conditions (5.3.31) or (5.3.32) ensures that the system (5.3.21) has a unique solution. On the other hand, if (5.3.30) does not hold the differential equation (5.3.18) need not have a unique solution, and this will be reflected in the discrete system (5.3.21). For example, if $g(x, v) = v^4$ there will be two solutions of both (5.3.18) and (5.3.21). This is explored further in Exercise 5.3.13.

For the difference equations (5.3.26) we tabulate in Table 5.3 the results of Newton's method at the grid points $0.1, 0.2, \ldots, 0.9$ for $h = 0.1$, 0.01, and 0.001 ($n = 9$, 99, and 999). In all cases the initial approximation for Newton's method was taken to be $\mathbf{v}^0 = 0$, and the iteration terminated when all components of the Newton correction vector $\mathbf{y}^k$ of (5.3.28) were less than 10^{-6} in magnitude.

Table 5.3: *Newton's Method for the Difference Equations* (5.3.26)

x	$h = 0.1$	$h = 0.01$	$h = 0.001$
0.1	-0.0058	-0.0058	-0.0058
0.2	-0.0116	-0.0118	-0.0118
0.3	-0.0174	-0.0176	-0.0176
0.4	-0.0223	-0.0230	-0.0230
0.5	-0.0274	-0.0276	-0.0276
0.6	-0.0302	-0.0304	-0.0304
0.7	-0.0303	-0.0305	-0.0305
0.8	-0.0265	-0.0266	-0.0266
0.9	-0.0170	-0.0171	-0.0171

Supplementary Discussion and References: 5.3

For a more detailed discussion and analysis of a variety of methods for solving systems of nonlinear equations numerically, see Ortega and Rheinboldt [1970] and Dennis and Schnabel [1983]. In particular, these references contain various discrete forms of Newton's method where the Jacobian matrix is approximated in some fashion. Certain of these approximations lead to natural generalizations of the secant method to systems of equations, and others give what are known as *quasi-Newton* methods, which are among the most promising methods for nonlinear systems. For a review of quasi-Newton methods, see also Dennis and Moré [1977].

An attractive alternative to symbolic differentiation or approximation of the partial derivatives in the Jacobian matrix by finite differences is *automatic differentiation.* See Griewank [1989] for a review and Griewank [1990] for application to Newton's method.

Many systems of equations arise in the attempt to minimize (or maximize) a function g of n variables. From the calculus we know that if g is continuously differentiable, then a necessary condition for a local minimum is that the gradient vector vanishes:

$$\left(\frac{\partial g}{\partial x_1}, \ldots, \frac{\partial g}{\partial x_n}\right) = 0.$$

Thus by solving this system of equations, one obtains a possible local minimizer of g, and in many situations it will be known that this vector must indeed minimize g. Alternatively, if we are given an arbitrary system of equations $f_i(\mathbf{x}) = 0$, $i = 1, \ldots, n$, we can convert the solution of this system to a minimization problem by defining a function

$$g(\mathbf{x}) = \sum_{i=1}^{n} [f_i(\mathbf{x})]^2.$$

Clearly, g takes on a minimum value of zero only when all $f_i(\mathbf{x})$ are zero. This conversion, however, is usually not recommended for obtaining a numerical solution of the system since the ill-conditioning of the problem will be increased.

In many problems the equations are to be solved for various values of one or more parameters. Suppose there is a single parameter α and we write the system of equations as

$$\mathbf{F}(\mathbf{x}; \alpha) = \mathbf{0}. \tag{5.3.33}$$

Assume that we wish solutions $\mathbf{x}_0^\star, \cdots, \mathbf{x}_N^\star$ for values $\alpha_0 < \alpha_1 < \cdots < \alpha_N$, where α_0 corresponds to a trivial, or at least an easy, problem; for example, the equations for α_0 may be linear. If $\mathbf{x}_0^\star$ can be computed and if $|\alpha_1 - \alpha_0|$ is small, then we hope that $\mathbf{x}_0^\star$ is sufficiently close to $\mathbf{x}_1^\star$ so that $\mathbf{x}_0^\star$ is a suitable starting approximation for the equation $\mathbf{F}(\mathbf{x}; \alpha_1) = 0$. Continuing in this way, we use each previous solution as a starting approximation for the next problem. This is called the *continuation method.*

If the equations to be solved do not contain a parameter, we can always introduce one artificially. For example, let $\mathbf{F}(\mathbf{x}) = 0$ be the system and let $\mathbf{x}^0$ be our best approximation to the solution (but not good enough for the Newton iteration to converge). Define a new set of equations depending on a parameter α by

$$\hat{\mathbf{F}}(\mathbf{x}; \alpha) = \mathbf{F}(\mathbf{x}) + (\alpha - 1)\mathbf{F}(\mathbf{x}^0) = 0, \qquad 0 \le \alpha \le 1. \tag{5.3.34}$$

Then $\hat{\mathbf{F}}(\mathbf{x}; 0) = \mathbf{F}(\mathbf{x}) - \mathbf{F}(\mathbf{x}^0) = 0$, for which $\mathbf{x}^0$ is a solution, and $\hat{\mathbf{F}}(\mathbf{x}; 1) = \mathbf{F}(\mathbf{x}) = 0$, which are the equations to be solved. Hence, we proceed as in the previous paragraph for parameters $0 = \alpha_0 < \alpha_1 < \cdots < \alpha_N = 1$.

The continuation method is closely related to *Davidenko's method.* Consider (5.3.33) and assume that for each $\alpha \in [0, 1]$ the equation defines a solution $\mathbf{x}(\alpha)$ that is continuously differentiable in α. Then if we differentiate

$$\mathbf{F}(\mathbf{x}(\alpha)) + (\alpha - 1)\mathbf{F}(\mathbf{x}^0) = 0$$

with respect to α, we obtain by the chain rule

$$\mathbf{F}'(\mathbf{x}(\alpha))\mathbf{x}'(\alpha) + \mathbf{F}(\mathbf{x}^0) = 0,$$

or, assuming that the Jacobian matrix $\mathbf{F}'(\mathbf{x}(\alpha))$ is nonsingular,

$$\mathbf{x}'(\alpha) = -[\mathbf{F}'(\mathbf{x}(\alpha))]^{-1}\mathbf{F}(\mathbf{x}^0),$$

with the initial condition $\mathbf{x}(0) = \mathbf{x}^0$. The solution $\mathbf{x}(\alpha)$ of this initial-value problem at $\alpha = 1$ will, we hope, be the desired solution of the original system of equations $\mathbf{F}(\mathbf{x}) = 0$. In practice we will have to solve the differential equations numerically, and we can, in principle, use any of the methods of Chapter 2. Although Davidenko's method and the continuation method are attractive possibilities, their reliability in practice has been less than desired. In particular, it is possible that the Jacobian matrix will become singular for some $\mathbf{x}(\alpha)$ with $\alpha < 1$, or even that the solution curve itself will blow up prematurely. For a review of possible ways of overcoming some of these difficulties, see Allgower and Georg [1990].

The proof that the system (5.3.21) has a unique solution under the conditions (5.3.31) or (5.3.32) can be found, for example, in Ortega and Rheinboldt [1970, Section 4.4].

EXERCISES 5.3

5.3.1. Show graphically that the system of equations $x_1^2 + x_2^2 = 1$, $x_1^2 - x_2 = 0$ has precisely two solutions.

5.3.2. Show that (5.3.5) reduces to (5.3.4) when $\mathbf{F}$ is of the form (5.3.3) and $B = A^{-1}$.

5.3.3. Compute the Jacobian matrix $\mathbf{G}'(\mathbf{x})$ for

$$\mathbf{G}(\mathbf{x}) = \begin{bmatrix} x_1^2 + x_1x_2x_3 + x_3^3 \\ x_1^3x_2 + x_2x_3^2 \\ x_1/x_2^3 \end{bmatrix}.$$

5.3.4. If $\mathbf{G}(\mathbf{x}) = \mathbf{x} - B\mathbf{F}(\mathbf{x})$, show that $\mathbf{G}'(\mathbf{x}) = I - B\mathbf{F}'(\mathbf{x})$ and conclude that (5.3.9) and (5.3.10) follow from (5.3.8).

5.3.5. For the functions of Exercise 5.3.1, compute the tangent planes at $x_1 = 2$, $x_2 = 2$.

5.3.6. Give the Newton iteration for the equations of Exercise 5.3.1. For what points $\mathbf{x}$ is the Jacobian matrix nonsingular?

5.3.7. Write a program for Newton's method to solve n equations in n unknowns. Use Gaussian elimination with partial pivoting to solve the linear equations.

5.3.8. If $\mathbf{F}(\mathbf{v}) = A\mathbf{v} + \mathbf{H}(\mathbf{v})$ for some matrix A, verify that $\mathbf{F}'(\mathbf{v}) = A + \mathbf{H}'(\mathbf{v})$. Apply this to obtaining the Jacobian matrices used in (5.3.9) and (5.3.10).

5.3.9. Write out the difference equations (5.3.20) and the corresponding Jacobian matrices for:

a. $g(x, v) = v + v^2$

b. $g(x, v) = xv^3$

5.3.10 Write out the Newton iteration (5.3.29) explicitly for the difference equations (5.3.20) with g given by Exercise 5.3.9.

5.3.11. Let D be a diagonal matrix with non-negative elements.

a. If A is symmetric positive definite, show that $A + D$ is positive definite.

b. If A is diagonally dominant and has positive diagonal elements, show that $A + D$ is diagonally dominant.

c. Apply parts **a.** and **b.** to show that (5.3.31) and (5.3.32) follow from (5.3.30).

5.3.12. Consider the two-point boundary-value problem $v'' = e^v + 2 - e^{x^2}$, $0 \leq x \leq 1$, $v(0) = 0$, $v(1) = 1$.

a. Write the finite difference equations for this problem in matrix-vector form for $h = 0.01$.

b. Discuss in detail how you would solve the system of equations in part **a** on a computer. The discussion should include a clear description of the method, what, if any, problems you expect the method to have, how much computer time you would expect the method to use, and so on.

5.3.13. Consider the two-point boundary-value problem $v'' = v^4$, $0 \leq x \leq 1$, $v(0) = 1$, $v(1) = \frac{1}{2}$.

a. Find an approximation to a solution by a third degree polynomial obtained by using the data $v(0)$, $v(1)$, $v''(0)$, $v''(1)$.

b. Obtain an approximate solution by the shooting method using both bisection and a chord method for the resulting single nonlinear equation.

c. Use the finite difference method and obtain a solution to the discrete system by the Picard method and Newton's method. For an initial approximation for these iterative methods, use the approximate solution of part **a**.

d. As mentioned in the text, a boundary value problem need not have a unique solution. Can you find a second approximate solution for this problem?

5.3.14. Consider the initial value problem

$$\mathbf{y}' = \mathbf{f}(x, \mathbf{y}), \quad \mathbf{y}(0) = \mathbf{y}_0,$$

and the backward Euler method

$$\mathbf{y}_{k+1} = \mathbf{y}_k + h\mathbf{f}(x_{k+1}, \mathbf{y}_{k+1}), \quad k = 0, 1, \ldots.$$

a. Discuss Newton's method applied to the backward Euler system for $\mathbf{y}_{k+1}$. How might you obtain an initial approximation for Newton's method?

b. Describe the Picard iteration for this system, and give conditions under which the Picard iterates will converge.

Chapter 6

Is There More Than Finite Differences?

6.1 Introduction to Projection Methods

In the previous chapters we have studied in some detail the application of finite difference methods to the approximate solution of differential equations. In this chapter we will consider another approach which has several variants known by such names as the finite element method, Galerkin's method, and the Rayleigh-Ritz method. The underlying theme of all these methods is that one attempts to approximate the solution of the differential equation by a finite linear combination of known functions. These known functions, usually called the *basis functions*, have the common property that they are relatively simple: polynomials, trigonometric functions, and, most importantly, spline functions, which will be studied in the following section. Conceptually, we regard the solution as lying in some appropriate (infinite-dimensional) function space, and we attempt to obtain an approximate solution that lies in the finite-dimensional subspace that is determined by the basis functions. The "projection" of the solution onto the finite-dimensional subspace is the approximate solution.

We will illustrate these general ideas with the linear two-point boundary-value problem

$$v''(x) + q(x)v = f(x), \qquad 0 \le x \le 1, \tag{6.1.1}$$

with

$$v(0) = 0, \qquad v(1) = 0, \tag{6.1.2}$$

where, for simplicity, we have taken the interval to be $[0, 1]$ and the boundary conditions to be zero (see Exercises 3.1.1 and 6.1.6).

Suppose that we look for an approximate solution of (6.1.1), (6.1.2) of the form

$$u(x) = \sum_{j=1}^{n} c_j \phi_j(x), \qquad (6.1.3)$$

where the basis functions ϕ_j satisfy the boundary conditions:

$$\phi_j(0) = \phi_j(1) = 0, \qquad j = 1, \ldots, n. \qquad (6.1.4)$$

If (6.1.4) holds, then the approximate solution u given by (6.1.3) satisfies the boundary conditions. A classical example of a set of basis functions that satisfy (6.1.4) is

$$\phi_j(x) = \sin j\pi x, \qquad j = 1, \ldots, n. \qquad (6.1.5)$$

Another example is the set of polynomials

$$\phi_j(x) = x^j(1-x), \qquad j = 1, \ldots, n. \qquad (6.1.6)$$

In the latter case the approximate solution (6.1.3) is of the form

$$u(x) = x(1-x)(c_1 + c_2 x + \cdots + c_n x^{n-1}),$$

which is a polynomial of degree $n+1$ with the property that it vanishes at 0 and 1. Our main example of a set of basis functions, however, is spline functions, which, as mentioned previously, will be studied in the following section.

Given a set of basis functions, we need to specify in what sense (6.1.3) is to be an approximate solution; that is, what is the criterion for determining the coefficients c_k in the linear combination? There are several possible approaches, and we will discuss here only two, both of which are generally applicable and widely used.

Collocation

The first criterion is that of *collocation*. Let $x_1, \ldots, x_n$ be n (not necessarily equally spaced) grid points in the interval $[0, 1]$. We then require that the approximate solution satisfy the differential equation at these n points. Thus for the equation (6.1.1) and the approximation (6.1.3) we require that

$$\frac{d^2}{dx^2}\left(\sum_{j=1}^{n} c_j \phi_j(x)\right)\Bigg|_{x_i} + q(x_i)\sum_{j=1}^{n} c_j \phi_j(x_i) = f(x_i), \quad i = 1, \ldots, n, \quad (6.1.7)$$

and we assume, of course, that the basis functions are twice differentiable. If we carry out the differentiation in (6.1.7) and collect coefficients of the c_j, we have

$$\sum_{j=1}^{n} c_j[\phi_j''(x_i) + q(x_i)\phi_j(x_i)] = f(x_i), \quad i = 1, \ldots, n. \qquad (6.1.8)$$

This is a system of n linear equations in the n unknowns $c_1, \ldots, c_n$. The computational problem is first to evaluate the coefficients

$$a_{ij} \equiv \phi_j''(x_i) + q(x_i)\phi_j(x_i), \tag{6.1.9}$$

and then solve the system of linear equations

$$A\mathbf{c} = \mathbf{f}, \tag{6.1.10}$$

where A is the $n \times n$ matrix (a_{ij}), $\mathbf{c} = (c_1, \ldots, c_n)^T$, and $\mathbf{f} = (f(x_1), \ldots, f(x_n))^T$.

We give a simple example. Consider the problem

$$v''(x) + x^2 v(x) = x^3, \qquad 0 \le x \le 1, \tag{6.1.11}$$

with the boundary conditions (6.1.2). Here $f(x) = x^3$ and $q(x) = x^2$. With the basis functions (6.1.5), we have

$$\phi_j'(x) = j\pi \cos j\pi x, \qquad \phi_j''(x) = -(j\pi)^2 \sin j\pi x,$$

so that the coefficients (6.1.9) are

$$a_{ij} = -(j\pi)^2 \sin j\pi x_i + x_i^2 \sin j\pi x_i.$$

The coefficients of the right-hand side of the system (6.1.10) are $f(x_i) = x_i^3$. If we use the basis functions (6.1.6), then

$$\phi_j'(x) = x^{j-1}[j - (j+1)x], \qquad \phi_j''(x) = jx^{j-2}[j - 1 - (j+1)x],$$

so that the coefficients (6.1.9) are now

$$a_{ij} = jx_i^{j-2}[j - 1 - (j+1)x_i] + x_i^{j+2}(1 - x_i).$$

Again, the components of the right-hand side are x_i^3. In both cases the system (6.1.10) is easily constructed once the grid points $x_1, \ldots, x_n$ are specified.

Galerkin's Method

We will return to a discussion of the collocation method after we consider another approach to determining the coefficients $c_1, \ldots, c_n$. This is known as *Galerkin's method* and is based on the concept of orthogonality of functions. Recall that two vectors $\mathbf{f}$ and $\mathbf{g}$ are orthogonal if the inner product satisfies

$$(\mathbf{f}, \mathbf{g}) \equiv \mathbf{f}^T \mathbf{g} = \sum_{j=1}^{n} f_j g_j = 0. \tag{6.1.12}$$

Now suppose that the components of the vectors $\mathbf{f}$ and $\mathbf{g}$ are the values of two functions f and g at n equally spaced grid points in the interval $[0, 1]$; that is,

$$\mathbf{f} = (f(h), f(2h), \ldots, f(nh)),$$

where $h = (n+1)^{-1}$ is the grid-point spacing, and similarly for $\mathbf{g}$. Then the orthogonality relation (6.1.12) is

$$\sum_{j=1}^{n} f(jh)g(jh) = 0,$$

and this relation is unchanged if we multiply by h:

$$h\sum_{j=1}^{n} f(jh)g(jh) = 0. \tag{6.1.13}$$

Now let $n \to \infty$ (or, equivalently, let $h \to 0$). Then, assuming that the functions f and g are integrable, the sum in (6.1.13) will tend to the integral

$$\int_0^1 f(x)g(x)dx = 0. \tag{6.1.14}$$

With this motivation, we define two functions f and g to be *orthogonal* on the interval $[0, 1]$ if the relation (6.1.14) holds.

The rationale for the Galerkin approach is as follows. Let the *residual function* for $u(x)$ be defined by

$$r(x) = u''(x) + q(x)u(x) - f(x), \qquad 0 \le x \le 1. \tag{6.1.15}$$

If $u(x)$ were the exact solution of (6.1.1), then the residual function would be identically zero. Obviously, the residual would then be orthogonal to every function, and, in particular, it would be orthogonal to the set of basis functions. However, we cannot expect $u(x)$ to be the exact solution because we restrict $u(x)$ to be a linear combination of the basis functions. The Galerkin criterion is to choose $u(x)$ so that its residual is orthogonal to all of the basis functions $\phi_1, \ldots, \phi_n$:

$$\int_0^1 [u''(x) + q(x)u(x) - f(x)]\phi_i(x)dx = 0, \quad i = 1, \ldots, n. \tag{6.1.16}$$

If we put (6.1.3) into (6.1.16) and interchange the summation and integration, we obtain

$$\sum_{j=1}^{n} c_j \int_0^1 [\phi_j''(x) + q(x)\phi_j(x)]\phi_i(x)dx = \int_0^1 f(x)\phi_i(x)dx, \quad i = 1, \ldots, n.$$

Again, this is a system of linear equations of the form (6.1.10) with

$$f_i = \int_0^1 f(x)\phi_i(x)dx, \qquad i = 1, \ldots, n, \tag{6.1.17}$$

and

$$a_{ij} = \int_0^1 [\phi_j''(x) + q(x)\phi_j(x)]\phi_i(x)dx.$$

If we integrate the first term in this integral by parts,

$$\int_0^1 \phi_j''(x)\phi_i(x)dx = \phi_j'(x)\phi_i(x)\mid_0^1 - \int_0^1 \phi_j'(x)\phi_i'(x)dx,$$

and note that the first term vanishes because ϕ_i is zero at the end points, we can rewrite a_{ij} as

$$a_{ij} = -\int_0^1 \phi_i'(x)\phi_j'(x)dx + \int_0^1 q(x)\phi_i(x)\phi_j(x)dx. \tag{6.1.18}$$

Thus the system of equations to solve for the coefficients $c_1, \ldots, c_n$ in Galerkin's method is $A\mathbf{c} = \mathbf{f}$, with the elements of A given by (6.1.18) and those of $\mathbf{f}$ by (6.1.17). An example of the evaluation of the a_{ij} of (6.1.18) is left to Exercise 6.1.5.

Comparison of Methods

We now make several comments regarding the finite difference, collocation, and Galerkin methods as applied to (6.1.1). In each case the central computational problem is to solve a system of linear equations. In the finite difference and collocation methods these linear systems are determined by n grid points in the interval, although the nature of the linear systems is quite different: the finite difference method gives an approximation to the solution of the differential equation at the grid points, whereas the collocation method gives the coefficients of the representation (6.1.3) of the approximate solution. With the collocation (or Galerkin) method, the value of the approximate solution at any point $\bar{x}$ in the interval is obtained by the additional evaluation

$$u(\bar{x}) = \sum_{j=1}^{n} c_j\phi_j(\bar{x}).$$

Although the finite difference method requires no additional work to obtain the approximate solution at the grid points, it is defined *only* at the grid points, and obtaining an approximation at other points in the interval necessitates an interpolation process. The collocation and Galerkin methods, on the other hand, give an approximate solution on the whole interval.

As we saw in Chapter 3, the linear system of equations of the finite difference method is easily obtained and has the important property (for the second-order difference approximations used there) that the coefficient matrix is tridiagonal; thus, the solution of the linear system requires relatively little computation, the number of arithmetic operations required being proportional to n (Section 3.2). For the collocation method, the elements of the coefficient matrix are also evaluated relatively easily by (6.1.9), provided that the basis functions ϕ_j are suitably simple. However, the coefficient matrix will now generally be full, which means not only that all n^2 elements need to be evaluated but also that the solution time will be proportional to n^3. One of the very important properties of the spline basis functions – to be discussed in the next section – is that ϕ_i will be identically zero except in a subinterval about x_i. In the cases considered in Section 6.4, this subinterval will extend only from x_{i-2} to x_{i+2}, and the coefficient matrix will be tridiagonal.

These same comments apply to Galerkin's method: the coefficient matrix will in general be full, but the use of an appropriate spline function basis will allow us to recover a tridiagonal matrix. However, there is now another complication. The evaluation of the matrix coefficients (6.1.18) and elements of the right-hand side (6.1.17) requires integration over the whole interval. Only if the functions q and f are very simple will one be able to evaluate these integrals explicitly in closed form. Usually they must be approximated, and this leads us to the topic of numerical integration, which we consider in Section 6.3. Symbolic computation systems may also be used under certain circumstances. Finally, one advantage of the Galerkin method is that it always yields a symmetric matrix, as can be seen by (6.1.18), whereas collocation does not. No method has a clear advantage over the others. For each method there are problems for which it is best. Given a particular problem, analyses of the three methods applied to the problem may be necessary to evaluate their relative effectivenesses.

Nonlinear Problems

We end this section by indicating briefly how the collocation and Galerkin methods can be applied to nonlinear problems. For this purpose we will consider the equation

$$v'' = g(v), \qquad v(0) = v(1) = 0, \tag{6.1.19}$$

where g is a given nonlinear function of a single variable. For the collocation method applied to (6.1.19) we substitute the approximate solution (6.1.3) and evaluate at the grid points $x_1, \ldots, x_n$ as before. This then leads to the nonlinear system of equations

$$\sum_{j=1}^{n} c_j \phi_j''(x_i) = g\left(\sum_{j=1}^{n} c_j \phi_j(x_i)\right), \qquad i = 1, \ldots, n, \tag{6.1.20}$$

for the coefficients $c_1, \ldots c_n$.

Similarly, for the Galerkin method the residual function (6.1.15) now becomes

$$r(x) = \sum_{j=1}^{n} c_j \phi_j''(x) - g\left(\sum_{j=1}^{n} c_j \phi_j(x)\right),$$

so that the system of equations corresponding to (6.1.16) is

$$\int_0^1 \left[\sum_{j=1}^n c_j \phi_j''(x) - g\left(\sum_{j=1}^n c_j \phi_j(x) \right) \right] \phi_i(x) dx = 0, \quad i = 1, \ldots, n. \quad (6.1.21)$$

As before, we can integrate the first term by parts to put (6.1.21) in the form

$$-\sum_{j=1}^n c_j \int_0^1 \phi_i'(x)\phi_j'(x) dx = \int_0^1 g\left(\sum_{j=1}^n c_j \phi_j(x) \right) \phi_i(x) dx, \quad (6.1.22)$$

which is again a nonlinear system for $c_1, \ldots, c_n$. The methods of the previous chapter can be applied, in principle, to approximate solutions of both (6.1.20) and (6.1.22).

Supplementary Discussion and References: 6.1

A related approach to projection methods is by means of a *variational principle.* Consider the problem

$$\underset{v \in V}{\text{Minimize}} \int_0^1 \{[v'(x)]^2 - q(x)[v(x)]^2 - 2f(x)v(x)\} dx, \quad (6.1.23)$$

where V is a set of suitably differentiable functions that vanish at the end points $x = 0$ and $x = 1$. By results in the calculus of variations, the solution of (6.1.23) is also the solution of the differential equation (6.1.1), which is known as the *Euler equation* for (6.1.23). Thus we can solve (6.1.1) by solving (6.1.23), and we can attempt to approximate a solution to (6.1.23) in a manner analogous to the Galerkin method. This is known as the *Rayleigh-Ritz method.*

Let $\phi_1, \ldots, \phi_n$ be a set of basis functions such that $\phi_i(0) = \phi_i(1) = 0$, $i = 1, \ldots, n$. Then we wish to minimize

$$\int_0^1 \left\{ \left[\sum_{i=1}^n c_i \phi_i'(x) \right]^2 - q(x) \left[\sum_{i=1}^n c_i \phi_i(x) \right]^2 - 2f(x) \sum_{i=1}^n c_i \phi_i(x) \right\} dx \quad (6.1.24)$$

over the coefficients $c_1, \ldots, c_n$. If $c_1^\star, \ldots, c_n^\star$ is the solution of the minimization problem, then

$$u(x) = \sum_{i=1}^n c_i^\star \phi_i(x)$$

is taken as an approximate solution for (6.1.23). If we use the same basis functions for the Galerkin method applied to (6.1.1), we will obtain the same approximate solution. A good reference for the Rayleigh-Ritz and Galerkin methods is Strang and Fix [1973]. A good reference for collocation methods is Ascher et al. [1988].

An important question is when does the system of linear equations obtained by the discretization methods of this section have a unique solution. This is generally easier to ascertain in the case of the Rayleigh-Ritz method since the question reduces to when the functional (6.1.24) has a minimum. For an introduction to these existence and uniqueness theorems as well as the important question of discretization error for the Galerkin and collocation methods, see, for example, Prenter [1975] and Hall and Porsching [1990].

EXERCISES 6.1

6.1.1. **a.** For the two-point boundary-value problem $y''(x) = y(x) + x^2$, $0 \le x \le 1$, $y(0) = y(1) = 0$, write out explicitly the system of equations (6.1.8) for $n = 3$, $\phi_j(x) = \sin j\pi x$, and $x_i = i/3$, $i, j = 1, 2, 3$.

b. Repeat part **a** with $\phi_j(x) = x^j(1-x)$, $j = 1, 2, 3$.

c. Write out the system for general n in matrix form using both (6.1.5) and (6.1.6) as the basis functions.

6.1.2. Show that the functions $\sin k\pi x$ are mutually orthogonal on the interval $[0, 1]$, that is, $\int_0^1 \sin k\pi x \sin j\pi x \, dx = 0$, $j, k = 0, 1, \ldots, j \ne k$.

6.1.3. Repeat Exercise 6.1.1 for the Galerkin equations $A\mathbf{c} = \mathbf{f}$, where $\mathbf{f}$ is given by (6.1.17) and A by (6.1.18).

6.1.4. Let $g(v) = e^v$.

a. For $n = 3$ and $\phi_j(x) = \sin j\pi x$, $j = 1, 2, 3$, write out explicitly the equations (6.1.20) and (6.1.22) for the two-point boundary-value problem (6.1.19).

b. Repeat part **a** for the basis functions $\phi_j(x) = x^j(1-x)$, $j = 1, 2, 3$.

6.1.5. If $q(x) = x^2$, evaluate the coefficients a_{ij} of (6.1.18) for the basis functions (6.1.5) and (6.1.6).

6.1.6. Show that the boundary value problem

$$v''(x) + p(x)v'(x) + q(x)v(x) = f(x), \quad v(0) = \alpha, \ v(1) = \beta$$

can be converted to a problem with zero boundary conditions as follows. Let $u(x) = v(x) - (\beta - \alpha)x - \alpha$. Show that $u(0) = u(1) = 0$ and that u satisfies the differential equation

$$u''(x) + p(x)u'(x) + q(x)u(x) = f(x) - \alpha q(x) - [p(x) + q(x)](\beta - \alpha)x.$$

6.2 Spline Approximation

In Section 2.3 we considered the problem of approximating a function by polynomials or piecewise polynomials. In the present section we will extend this to piecewise polynomials that have additional properties.

Piecewise Quadratic Functions

Let $a \le x_1 < x_2 < \cdots < x_n \le b$ be nodes subdividing the interval $[a, b]$, and let $y_1, \ldots, y_n$ be corresponding function values. In Section 2.3, we used piecewise polynomials that matched at certain grid points. For example, the function of (2.3.8) was a piecewise quadratic that agreed with given data at seven nodes; it was composed of three quadratics and was continuous but failed to be differentiable at the nodes where the different quadratics met. Now suppose that we wish to approximate by piecewise quadratics, but we require that the approximating function be differentiable everywhere. Then we need a different approach than that of Section 2.3. To illustrate this approach let $n = 4$, $I_i = [x_i, x_{i+1}]$, $i = 1, 2, 3$, and

$$q_i(x) = a_{i2}x^2 + a_{i1}x + a_{i0}, \qquad i = 1, 2, 3. \tag{6.2.1}$$

We will define a piecewise quadratic function q such that $q(x) = q_i(x)$ if $x \in I_i$, $i = 1, 2, 3$, as illustrated in Figure 6.1. For q to be continuous and take on the prescribed values y_i at the nodes, we require that

$$\begin{aligned} &q_1(x_1) = y_1, \qquad && q_1(x_2) = y_2, \qquad && q_2(x_2) = y_2, \\ &q_2(x_3) = y_3, && q_3(x_3) = y_3, && q_3(x_4) = y_4. \end{aligned} \tag{6.2.2}$$

If we also wish that q be differentiable at the nodes, then q_1' must equal q_2' at x_2, and q_2' must equal q_3' at x_3:

$$q_1'(x_2) = q_2'(x_2), \qquad q_2'(x_3) = q_3'(x_3). \tag{6.2.3}$$

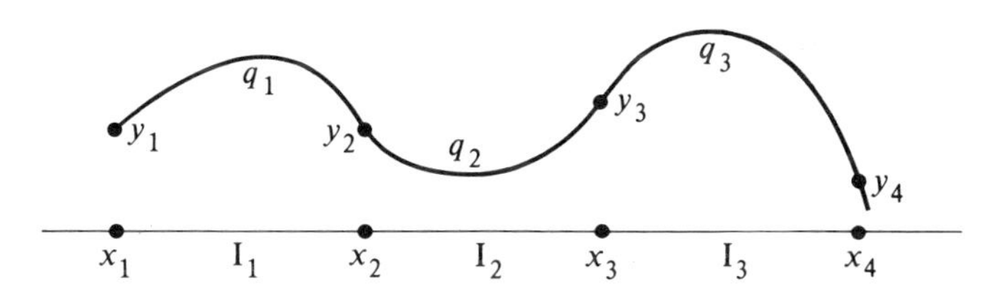

Figure 6.1: *A Piecewise Quadratic Function*

The function q is determined by the nine coefficients in (6.2.1) that define q_1, q_2, and q_3. The relations (6.2.2) and (6.2.3) give only eight conditions that these nine coefficients must satisfy, and hence another relation must be specified to determine q uniquely. Usually a value of q' at some node is specified, for example,

$$q_1'(x_1) = d_1, \tag{6.2.4}$$

where d_1 is some given value. The nine relations (6.2.2), (6.2.3), and (6.2.4) are then a system of nine linear equations for the coefficients of the q_i.

This approach is easily extended to an arbitrary number, n, of nodes. In this case there will be $n-1$ intervals I_i and $n-1$ quadratics q_i defined on these intervals. The conditions (6.2.2) and (6.2.3) become

$$q_i(x_i) = y_i \qquad q_i(x_{i+1}) = y_{i+1}, \qquad i = 1, \ldots, n-1, \tag{6.2.5}$$

and

$$q_i'(x_{i+1}) = q_{i+1}'(x_{i+1}), \qquad i = 1, \ldots, n-2. \tag{6.2.6}$$

These relations give $3n-4$ linear equations for the $3n-3$ unknown coefficients of the polynomials $q_1, \ldots, q_{n-1}$. Again, one additional condition is needed, and we can use, for example, (6.2.4). Thus, to determine the piecewise quadratic, we need to solve $3n-3$ linear equations.

Writing out equations (6.2.5), (6.2.6), and (6.2.4) for the quadratics q_i, we have

$$\begin{aligned}
&a_{i2}x_i^2 + a_{i1}x_i + a_{i0} = y_i, && i = 1, \ldots, n-1,\\
&a_{i2}x_{i+1}^2 + a_{i1}x_{i+1} + a_{i0} = y_{i+1}, && i = 1, \ldots, n-1,\\
&2a_{i2}x_{i+1} + a_{i1} = 2a_{i+1,2}x_{i+1} + a_{i+1,1}, && i = 1, \ldots, n-2,\\
&2a_{12}x_1 + a_{11} = d_1.
\end{aligned}$$

If we order the unknowns as $a_{12}, a_{11}, a_{10}, a_{22}, a_{21}, a_{20}$ and so on, the coefficient matrix of this system has the structure

$$\begin{bmatrix}
x_1^2 & x_1 & 1 &&&&&&& \\
x_2^2 & x_2 & 1 &&&&&&& \\
&&& x_2^2 & x_2 & 1 &&&& \\
&&& x_3^2 & x_3 & 1 &&&& \\
&&&&&& \ddots &&& \\
&&&&&&& x_{n-1}^2 & x_{n-1} & 1 \\
&&&&&&& x_n^2 & x_n & 1 \\
2x_2 & 1 & 0 & -2x_2 & -1 &&&&& \\
&&& 2x_3 & 1 & 0 & -2x_3 & -1 && \\
&&&&&& \ddots &&& \\
&&&& 2x_{n-1} & 1 & 0 & -2x_{n-1} & -1 & 0 \\
2x_1 & 1 &&&&&&&&
\end{bmatrix}. \tag{6.2.7}$$

Cubic Splines

For the purpose of approximating solutions to differential equations – as well as for many other situations – it is necessary that the approximating functions be at least twice continuously differentiable. This is not possible with piecewise quadratics unless the data are such that a single quadratic will suffice. We are thus led to consider a piecewise cubic polynomial $c(x)$ with the following properties:

$$c \text{ is twice continuously differentiable} \tag{6.2.8}$$

$$\text{In each interval } I_i = [x_i, x_{i+1}], c \text{ is a cubic polynomial.} \tag{6.2.9}$$

Such a function is called a *cubic spline*, the name being derived from a flexible piece of wood used by draftsmen for drawing curves.

The function c will be represented by

$$c(x) = c_i(x) = a_{i3}x^3 + a_{i2}x^2 + a_{i1}x + a_{i0}, \quad x \in I_i, \quad i = 1, \ldots, n-1. \tag{6.2.10}$$

The condition (6.2.8) implies that both c and c' are also continuous on the whole interval I. Hence we must have

$$c_{i-1}(x_i) = c_i(x_i) \qquad c'_{i-1}(x_i) = c'_i(x_i) \qquad c''_{i-1}(x_i) = c''_i(x_i), \tag{6.2.11}$$

for $i = 2, \ldots, n-1$, which are $3n - 6$ conditions. Since there are $4n - 4$ unknown coefficients a_{ij} to be obtained for the function c of (6.2.10), we need $n + 2$ additional conditions. Especially for the purpose of interpolation or approximation, we will require that c take on the prescribed values

$$c(x_i) = y_i, \qquad i = 1, \ldots, n, \tag{6.2.12}$$

which gives another n conditions. We still need two more conditions, and there are various possibilities for this. The *natural cubic spline* satisfies the additional conditions

$$c''(x_1) = c''(x_n) = 0. \tag{6.2.13}$$

It can be shown that if $\hat{c}$ is any other cubic spline that satisfies (6.2.8), (6.2.9), and (6.2.12), then

$$\int_a^b [c''(x)]^2 dx \leq \int_a^b [\hat{c}''(x)]^2 dx, \tag{6.2.14}$$

so that the natural cubic spline has "minimum curvature."

We could determine c by solving the system of linear equations given by (6.2.11) – (6.2.13) for the unknown coefficients a_{ij}. In this case the coefficient matrix would have a somewhat unwieldy structure similar to (6.2.7). However, for the natural cubic spline there is another approach that will lead to a simple tridiagonal system of equations in which the unknowns are the values of the second derivatives of c at the nodes. Then by integration we can determine c

itself. Obtaining this tridiagonal system requires a good deal of manipulation, which we now begin.

Computation of the Natural Cubic Spline

We first note that c_i'' is linear since c_i is a cubic. Therefore the formula for linear interpolation yields

$$c_i''(x) = c_i''(x_i) + \frac{(x - x_i)}{h_i}[c_i''(x_{i+1}) - c_i''(x_i)], \tag{6.2.15}$$

where we have set $h_i = x_{i+1} - x_i$, $i = 1, \ldots, n-1$. We now integrate (6.2.15) twice to obtain an expression for $c(x)$:

$$\begin{aligned} c_i'(x) &= c_i'(x_i) + \int_{x_i}^{x} c_i''(t)dt = c_i'(x_i) + c_i''(x_i)(x - x_i) \\ &+ \frac{[c_i''(x_{i+1}) - c_i''(x_i)]}{2h_i}(x - x_i)^2, \end{aligned} \tag{6.2.16}$$

$$\begin{aligned} c_i(x) &= c_i(x_i) + \int_{x_i}^{x} c_i'(t)dt = c_i(x_i) + c_i'(x_i)(x - x_i) \\ &+ c_i''(x_j)\frac{(x - x_i)^2}{2} + \frac{[c_i''(x_{i+1}) - c_i''(x_i)]}{6h_i}(x - x_i)^3. \end{aligned} \tag{6.2.17}$$

For convenience we will henceforth use the notation

$$\begin{aligned} y_i = c_i(x_i) = c_{i-1}(x_i), \qquad y_i' = c_i'(x_i) = c_{i-1}'(x_i), \\ y_i'' = c_i''(x_i) = c_{i-1}''(x_i), \end{aligned} \tag{6.2.18}$$

where we have invoked the conditions (6.2.11). Now replace i by $i - 1$ in (6.2.16), and then set $x = x_i$ to obtain

$$y_i' = y_{i-1}' + (y_i'' + y_{i-1}'')\frac{h_{i-1}}{2}. \tag{6.2.19}$$

Next, set $x = x_{i+1}$ in (6.2.17) and solve for y_i':

$$y_i' = \frac{y_{i+1} - y_i}{h_i} - y_{i+1}''\frac{h_i}{6} - y_i''\frac{h_i}{3}. \tag{6.2.20}$$

Equating the right-hand sides of (6.2.19) and (6.2.20) gives

$$y_{i-1}' + (y_i'' + y_{i-1}'')\frac{h_{i-1}}{2} = \frac{y_{i+1} - y_i}{h_i} - y_{i+1}''\frac{h_i}{6} - y_i''\frac{h_i}{3}. \tag{6.2.21}$$

We next wish to eliminate y'_{i-1} from (6.2.21). To do this, replace i by $i-1$ in (6.2.20) and substitute the resulting expression for y'_{i-1} into (6.2.21):

$$\frac{y_i - y_{i-1}}{h_{i-1}} - y''_i \frac{h_{i-1}}{6} - y''_{i-1}\frac{h_{i-1}}{3} + (y''_i + y''_{i-1})\frac{h_{i-2}}{2}$$
$$= \frac{y_{i+1} - y_i}{h_i} - y''_{i+1}\frac{h_i}{6} - y''_i\frac{h_i}{3}.$$

After rearranging, this becomes

$$y''_{i-1}h_{i-1} + 2y''_i(h_i + h_{i-1}) + y''_{i+1}h_i = \gamma_i, \qquad i = 2, \cdots, n-1, \tag{6.2.22}$$

where

$$\gamma_i = 6\left[\frac{y_{i+1} - y_i}{h_i} - \frac{y_i - y_{i-1}}{h_{i-1}}\right], \qquad i = 2, \ldots, n-1. \tag{6.2.23}$$

Since $y''_1 = y''_n = 0$ by the condition (6.2.13), (6.2.22) is a system of $n-2$ linear equations in the $n-2$ unknowns $y''_2, \ldots, y''_{n-1}$; it is of the form $H\mathbf{y} = \boldsymbol{\gamma}$, where $\mathbf{y} = (y''_2, \ldots, y''_{n-1})$, $\boldsymbol{\gamma} = (\gamma_2, \ldots, \gamma_{n-1})$, and

$$H = \begin{bmatrix} 2(h_1+h_2) & h_2 & & & \\ h_2 & 2(h_2+h_3) & h_3 & & \\ & h_3 & \ddots & \ddots & \\ & & \ddots & & h_{n-2} \\ & & & h_{n-2} & 2(h_{n-2}+h_{n-1}) \end{bmatrix}. \tag{6.2.24}$$

The matrix H is tridiagonal and clearly diagonally dominant. (It is also symmetric and positive-definite.) Hence, the system $H\mathbf{y} = \boldsymbol{\gamma}$ can be easily and efficiently solved by Gaussian elimination with no interchanges.

After the y''_i are computed, we still need to obtain the polynomials $c_1, \ldots, c_{n-1}$. The first derivatives y'_i at the node points can be obtained from (6.2.20) since the y_i are also known; thus

$$\begin{aligned} y'_i &= c'_i(x_i) = c'_{i-1}(x_i) \\ &= \frac{y_{i+1} - y_i}{h_i} - y''_{i+1}\frac{h_i}{6} - y''_i\frac{h_i}{3}, \qquad i = 1, \ldots, n-1. \end{aligned} \tag{6.2.25}$$

Then, the c_i themselves can be computed from (6.2.17), which we write in terms of y_i, y'_i, and y''_i for $i = 1, \cdots, n-1$:

$$c_i(x) = y_i + y'_i(x - x_i) + y''_i\frac{(x - x_i)^2}{2} + (y''_{i+1} - y''_i)\frac{(x - x_i)^3}{6h_i}. \tag{6.2.26}$$

Note that if we wish to evaluate c for some particular value $\hat{x}$, we first need to ascertain the interval I_i in which the point $\hat{x}$ lies and thus select the correct c_k for the evaluation.

An Example

We now give a simple example of the computation of a natural cubic spline. Suppose that we have the following nodes and function values:

$$\begin{array}{lllll} x_1 = 0 & x_2 = \frac{1}{4} & x_3 = \frac{1}{2} & x_4 = \frac{3}{4} & x_5 = 1 \\ y_1 = 1 & y_2 = 2 & y_3 = 1 & y_4 = 0 & y_5 = 1. \end{array}$$

Here, $n = 5$ and the h_i are all equal to $\frac{1}{4}$. The matrix H of (6.2.24) and the vector γ whose components are given by (6.2.23) are

$$H = \frac{1}{4}\begin{bmatrix} 4 & 1 & 0 \\ 1 & 4 & 1 \\ 0 & 1 & 4 \end{bmatrix}, \qquad \gamma = \begin{bmatrix} -48 \\ 0 \\ 48 \end{bmatrix}.$$

Hence, the quantities y_2'', y_3'', and y_4'' are obtained as the solution of the linear system

$$\begin{aligned} 4y_2'' + y_3'' \qquad &= -192 \\ y_2'' + 4y_3'' + y_4'' &= \quad 0 \\ y_3'' + 4y_4'' &= \quad 192, \end{aligned}$$

which is easily solved by Gaussian elimination. Since $y_1'' = y_5'' = 0$ by the condition (6.2.13), we have

$$y_1'' = 0, \qquad y_2'' = -48, \qquad y_3'' = 0, \qquad y_4'' = 48, \qquad y_5'' = 0.$$

Using these values of the y_i'', we next compute the y_i' from (6.2.25):

$$y_1' = 6, \quad y_2' = 0, \quad y_3' = -6, \quad y_4' = 0,$$

and with these we obtain the cubic polynomials c_1, c_2, c_3, and c_4 from (6.2.26). This then gives the cubic spline

$$c(x)\begin{cases} = c_1(x) = 1 + 6x - 32x^3, & 0 \le x \le \frac{1}{4}, \\ = c_2(x) = 2 - 24(x - \frac{1}{4})^2 + 32(x - \frac{1}{4})^3, & \frac{1}{4} \le x \le \frac{1}{2}, \\ = c_3(x) = 1 - 6(x - \frac{1}{2}) + 32(x - \frac{1}{2})^3, & \frac{1}{2} \le x \le \frac{3}{4}, \\ = c_4(x) = 24(x - \frac{3}{4})^2 - 32(x - \frac{3}{4})^3, & \frac{3}{4} \le x \le 1. \end{cases}$$

If we wish to evaluate c at some point, say $x = 0.35$, we note that $0.35 \in [\frac{1}{4}, \frac{1}{2}]$, and thus use c_2 for the evaluation:

$$c(0.35) = c_2(0.35) = 2 - 24(0.1)^2 + 32(0.1)^3 = 1.792.$$

In Section 6.4, we will return to the collocation and Galerkin methods using cubic splines as the basis functions. We stress again, however, that spline functions are extremely useful in many types of approximation applications.

Supplementary Discussion and References: 6.2

For further reading on spline functions, see Prenter [1975] and de Boor [1978]. In particular, splines using polynomials of degree higher than cubic are sometimes very useful and are developed in these references.

EXERCISES 6.2

6.2.1. Assume that f is a given function for which the following values are known: $f(1) = 2$, $f(2) = 3$, $f(3) = 5$, $f(4) = 3$, $f(4) = 3$. For these data:

a. Find the interpolating polynomial of degree 3 and write it in the form $a_0 + a_1 x + a_2 x^2 + a_3 x^3$.

b. Find the quadratic spline function that satisfies the condition $q'(1) = 0$. (*Hint:* Start from the left.)

c. Find the cubic spline function that satisfies $c''(1) = 6$, $c''(4) = -9$. (*Hint:* Try the polynomial of part **a.**)

6.2.2. Reorder the unknowns in the system of equations (6.2.7) so as to obtain a coefficient matrix with as small a bandwidth as you can.

6.2.3. Use (6.2.11) – (6.2.13) to write out the system of equations for the unknown coefficients a_{ij} of the cubic spline (6.2.10).

6.2.4. For the function of Exercise 6.2.1, find the cubic spline c that satisfies $c'(1) = 1$, $c'(4) = -1$, rather than the condition (6.2.13). (*Hint:* Think.)

6.2.5. Write a computer program to obtain the natural cubic spline for a given set of nodes $x_1 < \cdots < x_n$ and corresponding function values $y_1, \ldots, y_n$ by first solving the tridiagonal system with the coefficient matrix (6.2.24) and then using (6.2.25) and (6.2.26). Also write a program for evaluating this cubic spline at a given value x. Test your program on the example given in the text.

6.3 Numerical Integration

The Galerkin method described in Section 6.1 requires the evaluation of definite integrals of the form

$$I(f) = \int_a^b f(x)dx,$$

and the need to evaluate such integrals also arises in a number of other problems in scientific computing. The integrand, $f(x)$, may be given in one of three ways:

1. An explicit formula for $f(x)$ is given; for example, $f(x) = (\sin x)e^{-x^2}$.

2. The function $f(x)$ is not given explicitly but can be computed for any value of x in the interval $[a, b]$, usually by means of a computer program.

3. A table of values $\{x_i, f(x_i)\}$ is given for a fixed, finite set of points x_i in the interval.

Functions in the first category are sometimes amenable to methods of symbolic computation, either by hand or by computer systems, although many integrands will not have a "closed form" integral. The integrals of functions that fall into the second and third categories – as well as the first category if symbolic methods are not used – are usually approximated by numerical methods; such methods are called *quadrature rules* and are derived by approximating the function $f(x)$ by some other function, $\tilde{f}(x)$, whose integral is relatively easy to evaluate. Any class of simple functions may be used to approximate $f(x)$, such as polynomials, piecewise polynomials, and trigonometric, exponential, or logarithmic functions. The choice of the class of functions used may depend on some particular properties of the integrand, but the most common choice, which we will use here, is polynomials or piecewise polynomials.

The Newton-Cotes Formulas

The simplest polynomial is a constant. In the *rectangle rule*, f is approximated by its value at the end point a (or, alternatively, at b) so that

$$I(f) \doteq R(f) = (b-a)f(a). \tag{6.3.1}$$

We could also approximate f by another constant obtained by evaluating f at a point interior to the interval; the most common choice is $(a+b)/2$, the center of the interval, which gives the *midpoint rule*

$$I(f) \doteq M(f) = (b-a)f\left(\frac{a+b}{2}\right). \tag{6.3.2}$$

The rectangle and midpoint rules are illustrated in Figure 6.2.

The next simplest polynomial is a linear function. If it is chosen so that it agrees with f at the end points a and b, then a trapezoid is formed, as illustrated in Figure 6.3. The area of this trapezoid – the integral of the linear function – is the approximation to the integral of f and is given by

$$I(f) \doteq T(f) = \frac{(b-a)}{2}[f(a) + f(b)]. \tag{6.3.3}$$

This is known as the *trapezoid rule.*

To obtain one further formula, we next approximate f by an interpolating quadratic polynomial that agrees with f at the end points a and b and the

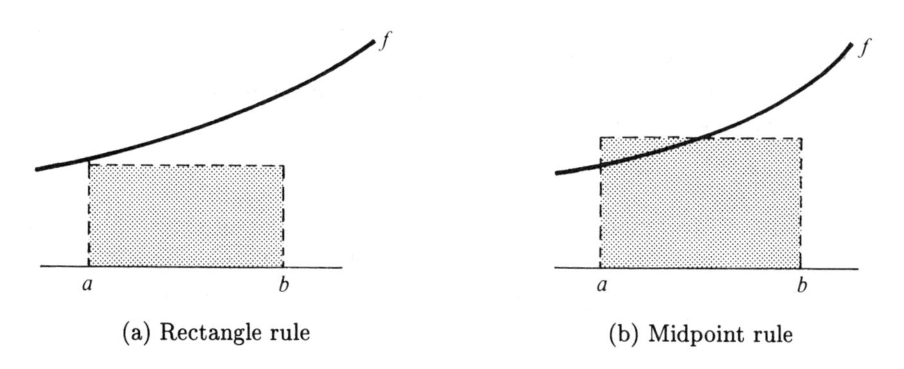

(a) Rectangle rule (b) Midpoint rule

Figure 6.2: *Integration Approximations*

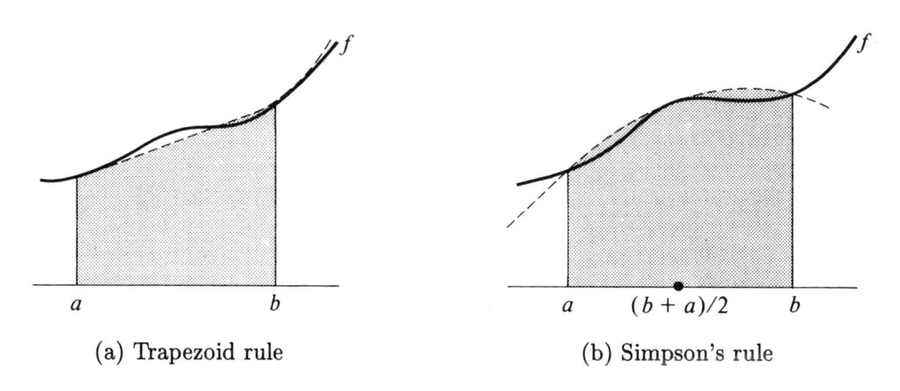

(a) Trapezoid rule (b) Simpson's rule

Figure 6.3: *More Approximations*

midpoint $(a+b)/2$. The integral of this quadratic is given by (see Exercise 6.3.1)

$$I(f) \doteq S(f) = \frac{(b-a)}{6}\left[f(a) + 4f\left(\frac{a+b}{2}\right) + f(b)\right], \tag{6.3.4}$$

which is *Simpson's rule* and is illustrated in Figure 6.3. We note that Simpson's rule may also be viewed as a linear combination of the trapezoid rule and the midpoint rule since

$$\frac{1}{6}\left[f(a) + 4f\left(\frac{a+b}{2}\right) + f(b)\right] = \frac{1}{3}\left[\frac{f(a)+f(b)}{2}\right] + \frac{2}{3}f\left(\frac{a+b}{2}\right).$$

We can continue the preceding method of generating quadrature formulas by using polynomials of still higher degree. The interval $[a,b]$ is divided by m equally spaced points, an interpolating polynomial of degree $m+1$ is constructed to agree with f at these m points plus the two end points, and this polynomial is then integrated from a to b to give an approximation to the integral. Such quadrature formulas are called the *Newton-Cotes formulas* (See the Supplementary Discussion.)

Error Formulas

We consider next the error made in using the quadrature rules that have been described. In all cases f is approximated by an interpolating polynomial p of degree n over the interval $[a, b]$, and the integral of p is the approximation to the integral. Hence the error in this approximation is

$$E = \int_a^b [f(x) - p(x)]dx. \tag{6.3.5}$$

By the Interpolation Error Theorem 2.3.2, this can be written as

$$E = \frac{1}{(n+1)!} \int_a^b (x - x_0) \cdots (x - x_n) f^{(n+1)}(z(x))dx, \tag{6.3.6}$$

where $x_0, x_1, \ldots, x_n$ are the interpolation points, and $z(x)$ is a point in the interval $[a, b]$ that depends on x. We now apply (6.3.6) to some specific cases.

For the rectangle rule (6.3.1), $n = 0$ and $x_0 = a$; hence (6.3.6) becomes

$$|E_R| = \left| \int_a^b (x - a) f'(z(x))dx \right| \leq M_1 \int_a^b (x - a)dx = \frac{M_1}{2}(b - a)^2, \tag{6.3.7}$$

where M_1 is a bound for $|f'(x)|$ over the interval $[a, b]$. Note that the bound (6.3.7) will not be small unless M_1 is small, which means that f is close to constant, or the length of the interval is small; we shall return to this point later when we discuss the practical use of these quadrature formulas. For the trapezoid rule (6.3.3), $n = 1$, $x_0 = a$, and $x_1 = b$. Hence, again applying (6.3.6), we have

$$|E_T| = \frac{1}{2} \left| \int_a^b (x - a)(x - b) f''(z(x))dx \right| \leq \frac{M_2}{12}(b - a)^3, \tag{6.3.8}$$

where M_2 is a bound on $|f''(x)|$ over $[a, b]$.

Consider next the midpoint rule (6.3.2), in which $n = 0$ and $x_0 = (a+b)/2$. If we apply (6.3.6) and proceed as in (6.3.7), we obtain

$$|E_M| = \left| \int_a^b \left[x - \frac{(a+b)}{2} \right] f'(z(x))dx \right| \leq \frac{M_1}{4}(b - a)^2. \tag{6.3.9}$$

This, however, is not the best bound we can obtain. We shall instead expand the integrand of (6.3.5) in a Taylor series about $m = (a+b)/2$. Since the interpolating polynomial is simply the constant $p(x) = f(m)$, this gives

$$f(x) - p(x) = f'(m)(x-m) + \tfrac{1}{2}f''(z(x))(x-m)^2,$$

where z is a point in the interval and depends on x. Thus the error in the midpoint rule is

$$\begin{aligned} |E_M| &= \left| \int_a^b [f'(m)(x-m)dx + \tfrac{1}{2}f''(z(x))(x-m)^2]dx \right| \qquad (6.3.10) \\ &\leq \left| f'(m)\int_a^b (x-m)dx \right| + \frac{1}{2}\left| \int_a^b f''(z(x))(x-m)^2 dx \right| \\ &\leq \frac{M_2}{24}(b-a)^3, \end{aligned}$$

since

$$\int_a^b (x-m)dx = 0, \qquad \int_a^b (x-m)^2 dx = \frac{(b-a)^3}{12}.$$

In a similar way we can derive the following bound for the error in Simpson's rule (6.3.4), which we state without proof (M_4 is a bound for the fourth derivative):

$$|E_S| \leq \frac{M_4}{2880}(b-a)^5. \qquad (6.3.11)$$

Composite Formulas

The above error bounds all involve powers of the length $b-a$ of the interval, and unless this length is small the bounds will not, in general, be small. However, in practice, we will only apply these quadrature formulas to sufficiently small intervals which we obtain by subdividing the given interval $[a,b]$. Thus we partition the interval $[a,b]$ into n subintervals $[x_{i-1}, x_i]$, $i = 1, \ldots, n$, where $x_0 = a$ and $x_n = b$. Then

$$I(f) = \int_a^b f(x)dx = \sum_{i=1}^n \int_{x_{i-1}}^{x_i} f(x)dx.$$

If we apply the rectangle rule to each subinterval $[x_{i-1}, x_i]$, we obtain the *composite rectangle rule*

$$I(f) \doteq I_{CR}(f) = \sum_{i=1}^n h_i f(x_{i-1}), \qquad (6.3.12)$$

where $h_i = x_i - x_{i-1}$. The *composite midpoint, trapezoid,* and *Simpson's rules* are obtained in the same way by applying the basic rule to each subinterval; they are given by

$$I_{CM}(f) = \sum_{i=1}^{n} h_i f\left(\frac{x_i + x_{i-1}}{2}\right), \tag{6.3.13}$$

$$I_{CT}(f) = \sum_{i=1}^{n} \frac{h_i}{2}[f(x_{i-1}) + f(x_i)], \tag{6.3.14}$$

$$I_{CS}(f) = \frac{1}{6}\sum_{i=1}^{n} h_i \left[f(x_{i-1}) + 4f\left(\frac{x_{i-1} + x_i}{2}\right) + f(x_i)\right]. \tag{6.3.15}$$

We note that the composite rules may all be viewed as approximating the integrand f on the interval $[a, b]$ by a piecewise polynomial function (see Section 2.3) and then integrating the piecewise polynomial to obtain an approximation to the integral. For the midpoint and rectangle rules, the approximating function is piecewise constant; for the trapezoid rule it is piecewise linear, and for Simpson's rule it is piecewise quadratic.

We can now apply the previous error bounds on each subinterval. For example, for the rectangle rule, we use (6.3.7) to obtain the following bound on the error in the composite rule:

$$E_{CR} \leq \frac{M_1}{2}\sum_{i=1}^{n} h_i^2. \tag{6.3.16}$$

Note that we have used the maximum M_1 of $|f'(x)|$ on the whole interval $[a, b]$, although a better bound in (6.3.16) could be obtained if we used the maximum of $|f'(x)|$ separately on each subinterval.

In the special case that the subintervals are all of the same length, $h_i = h = (b - a)/n$, (6.3.16) becomes

$$E_{CR} \leq \frac{M_1}{2}(b - a)h \qquad \text{(Composite rectangle rule error)}, \tag{6.3.17}$$

which shows that the composite rectangle rule is a first-order method; that is, the error reduces only linearly in h. In a similar fashion, we can obtain bounds for the errors in the other composite rules by using (6.3.8), (6.3.10), and (6.3.11). The following bounds are given in the case that the intervals are all of the same length h:

$$E_{CM} \leq \frac{M_2}{24}(b - a)h^2 \qquad \text{(Composite midpoint rule error)}, \tag{6.3.18}$$

$$E_{CT} \leq \frac{M_2}{12}(b - a)h^2 \qquad \text{(Composite trapezoid rule error)}, \tag{6.3.19}$$

$$E_{CS} \leq \frac{M_4}{2880}(b-a)h^4 \qquad \text{(Composite Simpson's rule error)}. \qquad (6.3.20)$$

Thus the composite midpoint and trapezoid rules are both second order, whereas the composite Simpson's rule is fourth order. Because of its relatively high accuracy and simplicity, the composite Simpson's rule is an often-used method.

Supplementary Discussion and References: 6.3

A difficulty with quadrature rules, as well as with other numerical methods that we have discussed earlier, is that some choice of the step sizes, h_j, must be made. If the numerical integration schemes were to be used as described previously, the user would be required to specify h_j a priori. In practice, high-quality quadrature software will employ some automatic adaptive scheme that will vary the step size depending on estimates of the error obtained during the computation. The user will be required to specify an acceptable tolerance for the error, and the program will automatically specify the step size as it is computing.

The solution at $x = b$ of the initial-value problem

$$y'(x) = f(x), \qquad y(a) = 0, \qquad a \leq x \leq b, \qquad (6.3.21)$$

is $y(b) = \int_a^b f(x)dx$. Hence integration may be viewed as the "trivial" subcase of solving an initial-value problem in which the right-hand side is independent of y. Any of the methods discussed in Chapter 2 may be applied to (6.3.21), in principle. In fact, most of those methods reduce to some quadrature rule that we have discussed. For example, Euler's method is the composite rectangle rule, the second-order Runge-Kutta method is the composite trapezoid rule, and the fourth-order Runge-Kutta method is the composite Simpson's rule (see Exercise 6.3.7).

The Newton-Cotes formulas, mentioned in the text as being derived by integrating an interpolating polynomial of degree n, can be written in the form

$$I(f) \doteq \sum_{i=0}^{n} \alpha_i f(x_i), \qquad (6.3.22)$$

where the x_i are equally spaced points in the interval $[a, b]$, with $x_0 = a$, $x_n = b$; Simpson's rule is the case $n = 2$. For $n \leq 7$, the coefficients α_i are all positive, but beginning with $n = 8$ certain coefficients will be negative; this has a deleterious effect on rounding error since cancellations will occur. The Newton-Cotes formulas also have the unsatisfactory theoretical property that as $n \to \infty$, convergence to the integral will not necessarily occur, even for infinitely differentiable functions.

The representation (6.3.22) provides another approach to the derivation of quadrature formulas – the *method of undetermined coefficients.* Assume first that the x_i are given. If we seek to determine the α_i so that the formula is exact for polynomials of as high a degree as possible, then in particular it must be exact for $1, x, x^2, \ldots, x^m$, where m is to be as large as possible. This means that we must have

$$\sum_{i=0}^{n} \alpha_i x_i^j = \frac{b^{j+1} - a^{j+1}}{j+1}, \qquad j = 0, 1, \ldots, m, \tag{6.3.23}$$

where the right-hand sides of these relations are the exact integrals of the powers of x. The relations (6.3.23) constitute a system of linear equations for the unknown coefficients α_i. If $m = n$, then the coefficient matrix is the Vandermonde matrix discussed in Section 2.3. It is nonsingular if the x_i are all distinct, and hence the α_i are uniquely determined for $m = n$. If the x_i are equally spaced, then this approach again gives the Newton-Cotes formulas.

Now assume that we do not specify the points x_i in advance but consider them to be unknowns in the relations (6.3.23). Then if $m = 2n + 1$, (6.3.23) is a system of $2n + 2$ equations in the $2n + 2$ unknowns $\alpha_0, \alpha_1, \ldots, \alpha_n$ and $x_0, x_1, \ldots, x_n$. The solution of these equations for the α_i and x_i give the *Gaussian quadrature formulas.* For example, in the case $n = 1$ on the interval $[a, b] = [-1, 1]$, the formula is

$$\int_{-1}^{1} f(x)dx \doteq f\left(-\frac{1}{\sqrt{3}}\right) + f\left(\frac{1}{\sqrt{3}}\right).$$

In general, the abscissas x_i of these quadrature formulas are roots of certain orthogonal polynomials. Gaussian quadrature rules are popular because of their high-order accuracy, and the weights are always non-negative.

We remarked in the text that Simpson's rule can be viewed as a linear combination of the trapezoid and midpoint rules. By taking suitable linear combinations of the trapezoid rule for different spacings h, we can also derive higher-order quadrature formulas. This is known as *Romberg integration* and is a special case of Richardson extrapolation discussed earlier. The basis for the derivation of Romberg integration is that the trapezoid approximation can be shown to satisfy

$$T(h) = I(f) + C_2 h^2 + C_4 h^4 + \cdots + C_{2m} h^{2m} + 0(h^{2m+2}) \tag{6.3.24}$$

where the C_i depend on f and the interval but are independent of h. The expansion (6.3.24) holds provided that f has $2m + 2$ derivatives. Now define a new approximation to the integral by

$$T_1(h) = \tfrac{1}{3}\left[4T\left(\frac{h}{2}\right) - T(h)\right]. \tag{6.3.25}$$

The coefficients of this linear combination are chosen so that when the error in (6.3.25) is computed using (6.3.24), the coefficient of the h^2 term is zero. Thus

$$T_1(h) = I(f) + C_4^{(1)} h^4 + \cdots + 0(h^{2m+2}),$$

so that T_1 is a fourth-order approximation to the integral. One can continue the process by combining $T_1(h)$ and $T_1(h/2)$ in a similar fashion to eliminate the h^4 term in the error for T_1. More generally, we can construct the triangular array

$$\begin{array}{lll} T(h) & & \\ T(h/2) & T_1(h) & \\ T(h/4) & T_1(h/2) & T_2(h) \\ \vdots & \vdots & \ddots \end{array}$$

where

$$T_k\left(\frac{h}{2^{j-1}}\right) = \frac{4^j T_{k-1}\left(\frac{h}{2^j}\right) - T_{k-1}\left(\frac{h}{2^{j-1}}\right)}{4^j - 1}.$$

The elements in the ith column of this array converge to the integral at a rate depending on h^{2i}. Provided that f is infinitely differentiable, however, the elements on the diagonal of the array converge at a rate that is superlinear, that is, faster than any power of h.

We have not touched at all upon several other important topics in numerical integration: techniques for handling integrands with a singularity, integrals over an infinite interval, multiple integrals, and adaptive procedures that attempt to fit the grid spacing automatically to the integrand. For a discussion of these matters, as well as further reading on the topics covered in this section, see Davis and Rabinowitz [1984] and Stroud [1971].

EXERCISES 6.3

6.3.1. Write down explicitly the interpolating quadratic polynomial that agrees with f at the three points a, b, and $(a+b)/2$. Integrate this quadratic from a to b to obtain Simpson's rule (6.3.4).

6.3.2. Show that the trapezoid rule integrates any linear function exactly and that Simpson's rule integrates any cubic polynomial exactly. (*Hint:* Expand the cubic about the midpoint.)

6.3.3. Apply the rectangle, midpoint, trapezoid, and Simpson's rules to the function $f(x) = x^4$ on the interval $[0, 1]$. Compare the actual error in the approximations to the bounds given by (6.3.7), (6.3.8), (6.3.10), and (6.3.11).

6.3.4. Based on the bound (6.3.19), how small would h need to be to guarantee an error no larger than 10^{-6} in the composite trapezoid rule approximation for $f(x) = x^4$ on $[0, 1]$. How small for the composite Simpson's rule?

6.3.5. Write a computer program to carry out the composite trapezoid and Simpson's rules for an "arbitrary" function on the interval $[a, b]$ and with an arbitrary subdivision of $[a, b]$. Test your program on $f(x) = x^4$ on $[0, 1]$ and find the actual h needed, in the case of an equal subdivision, to achieve an error of less than 10^{-6} with the composite trapezoid rule. Do the same for $f(x) = e^{-x^2}$.

6.3.6. Derive the four-point quadrature formula based on interpolation of the integrand by a cubic polynomial at equally spaced points. (Hint: Think. The calculation can be simplified somewhat.)

6.3.7. Show that Euler's method applied to the initial value problem (6.3.21) is the composite rectangle rule, that the second-order Runge-Kutta method is the composite trapezoid rule, and that the fourth-order Runge-Kutta method is the composite Simpson rule.

6.4 The Discrete Problem Using Splines

We now return to the original problem (6.1.1), (6.1.2) of this chapter: for the two-point boundary-value problem

$$v''(x) + q(x)v(x) = f(x), \qquad 0 \le x \le 1, \tag{6.4.1}$$

and

$$v(0) = v(1) = 0, \tag{6.4.2}$$

we wish to find an approximate solution of the form

$$u(x) = \sum_{j=1}^{n} c_j \phi_j(x), \tag{6.4.3}$$

where $\phi_1, \ldots, \phi_n$ are given functions.

Collocation

Recall from Section 6.1 that the collocation method for (6.4.1) requires solving the linear system of equations

$$A\mathbf{c} = \mathbf{f}, \tag{6.4.4}$$

where the elements of the matrix A are

$$a_{ij} = \phi_j''(x_i) + q(x_i)\phi_j(x_i), \qquad i, j = 1, \ldots, n, \tag{6.4.5}$$

$\mathbf{c}$ is the vector of unknown coefficients $c_1, \ldots, c_n$, $\mathbf{f}$ is the vector of values $f(x_1), \ldots, f(x_n)$, and $x_1, \ldots, x_n$ are given points in the interval $[0, 1]$. In Section 6.1, we considered the choice of the basis functions ϕ_j as either polynomials or trigonometric functions and saw that, in general, the coefficient matrix A was dense – that is, it had few zero elements – in contrast to the tridiagonal coefficient matrix that was obtained in Chapter 3 using the finite difference method. In the present section we shall use spline functions, and we will see that in the simplest case this again leads to a tridiagonal coefficient matrix.

Since the coefficients a_{ij} use $\phi''(x_i)$, it is necessary for the basis functions to have a second derivative at the nodes $x_1, \ldots, x_n$. Thus linear and quadratic splines will not suffice; cubic splines, however, are twice differentiable, and we will consider them first as our basis functions. We will need to choose the basis functions so that they are linearly independent in a sense to be made clear shortly. We would also like to choose them so that the coefficient matrix A has as small a bandwidth as possible. To illustrate this last point, let us attempt to make the coefficient matrix tridiagonal. Assuming that the function q has no special properties, we see from (6.4.5) that this will only be achieved if we can choose the ϕ_j so that

$$\phi_j''(x_i) = \phi_j(x_i) = 0, \qquad |i-j| > 1. \tag{6.4.6}$$

This, in turn, will be true if we can choose ϕ_i such that it vanishes identically outside the interval $[x_{i-2}, x_{i+2}]$, and if

$$\phi_i''(x_{i-2}) = \phi_i(x_{i-2}) = \phi_i''(x_{i+2}) = \phi_i(x_{i+2}) = 0. \tag{6.4.7}$$

B-Splines

Now recall that a cubic spline was defined by the conditions (6.2.8) and (6.2.9). These conditions, together with a specification of the function values at the node points $x_1, \ldots, x_n$, give $4n-6$ relations to determine the $4n-4$ unknown coefficients that define the cubic spline. In Section 6.2 we used the additional two conditions (6.2.13), which determine a natural cubic spline; unfortunately, this natural cubic spline cannot satisfy the condition (6.4.6) unless it is identically zero. However, if we do not impose the additional conditions (6.2.13), we can obtain a cubic spline that does indeed satisfy the conditions (6.4.6). We denote this spine by $B_i(x)$ and define it explicitly by

$$\begin{array}{ll}
\frac{1}{4h^3}(x-x_{i-2})^3, & x_{i-2} \le x \le x_{i-1}, \\[2mm]
\frac{1}{4} + \frac{3}{4h}(x-x_{i-1}) + \frac{3}{4h^2}(x-x_{i-1})^2 - \frac{3}{4h^3}(x-x_{i-1})^3, & x_{i-1} \le x \le x_i, \\[2mm]
\frac{1}{4} + \frac{3}{4h}(x_{i+1}-x) + \frac{3}{4h^2}(x_{i+1}-x)^2 - \frac{3}{4h^3}(x_{i+1}-x)^3, & x_i \le x \le x_{i+1}, \\[2mm]
\frac{1}{4h^3}(x_{i+2}-x)^3, & x_{i+1} \le x \le x_{i+2}, \quad \text{otherwise,}
\end{array} \tag{6.4.8}$$

where we have now assumed that the node points $x_1, \ldots, x_n$ are equally spaced with spacing h. It is straightforward (Exercise 6.4.1) to verify that this function is a cubic spline with the function values

$$B_i(x_i) = 1, \qquad B_i(x_{i\pm 1}) = \tfrac{1}{4},$$

and zero at the other nodes. Moreover, if $\phi_i = B_i$ the conditions (6.4.6) and (6.4.7) are satisfied (Exercise 6.4.1). Such a spline function, which is illustrated in Figure 6.4, is called a *cubic basis spline*, or *cubic B-spline* for short, since any cubic spline on the interval $[a, b]$ may be written as a linear combination of B-splines. More precisely, we state the following theorem without proof:

Theorem 6.4.1 *Let $c(x)$ be a cubic spline for the equally spaced node points $x_1 < \cdots < x_n$. Then there are constants $\alpha_0, \alpha_1, \ldots, \alpha_{n+1}$ such that*

$$c(x) = \sum_{i=0}^{n+1} \alpha_i B_i(x). \tag{6.4.9}$$

Note that the functions B_0, B_1, B_n, and B_{n+1} used in (6.4.9) require the introduction of the auxiliary grid points x_{-2}, x_{-1}, x_0 and $x_{n+1}, x_{n+2}, x_{n+3}$.

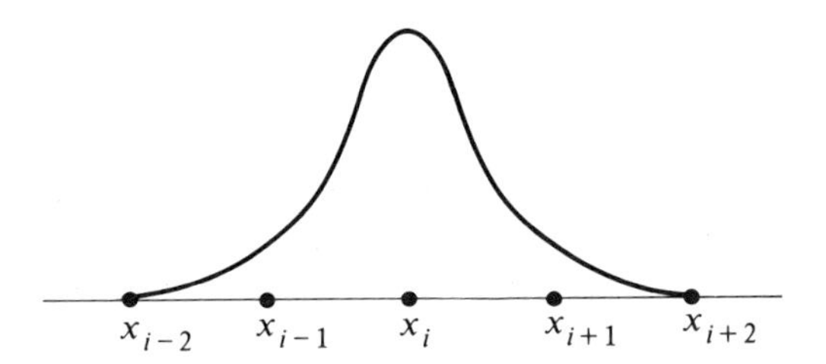

Figure 6.4: *A Cubic B-Spline*

Application to the Boundary-Value Problem

We now return to the boundary-value problem (6.4.1). We assume again that the node points are equally spaced with spacing h, and that $x_1 = 0$, $x_n = 1$. We wish to take the basis functions $\phi_1, \ldots, \phi_n$ to be the B-splines $B_1, \ldots, B_n$. However, although $B_3, \ldots, B_{n-2}$ satisfy the zero boundary conditions, B_1, B_2, B_{n-1}, and B_n do not. Therefore we define the ϕ_i to be

$$\begin{aligned} \phi_i(x) &= B_i(x), \qquad i = 3, \ldots, n-2 \\ \phi_1(x) &= B_1(x) - 4B_0(x), \quad \phi_2(x) = 4B_2(x) - B_1(x), \\ \phi_{n-1}(x) &= 4B_{n-1}(x) - B_n(x), \quad \phi_n(x) = B_n(x) - 4B_{n+1}(x). \end{aligned} \tag{6.4.10}$$

It is easy to verify from the definition (6.4.8) that $\phi_1(0) = \phi_2(0) = \phi_{n-1}(0) = \phi_n(0) = 0$. Moreover, it is clear that any linear combination (6.4.3) is a cubic spline and satisfies $u(0) = u(1) = 0$.

We next need to evaluate the coefficients (6.4.5); for this, we will need B_i, B_i', and B_i'' evaluated at the nodal points. This is easily done (Exercise 6.4.2), and we summarize the results in Table 6.1. Note that by the definition of the B_i, all values at node points not indicated in Table 6.1 are zero. This implies, in particular, that the coefficients a_{ij} of (6.4.5) are all zero unless $|i-j| \leq 1$ or, possibly, if $i = 1$ or n.

Table 6.1: Values of B_i, B_i', B_i''

	x_{i-1}	x_i	x_{i-1}
B_i	$1/4$	1	$1/4$
B_i'	$3/(4h)$	0	$-3/(4h)$
B_i''	$3/(2h^2)$	$-3/h^2$	$3/(2h^2)$

For the evaluation of the nonzero coefficients a_{ij} of (6.4.5), we set $q_i = q(x_i)$. Then using Table 6.1 we have

$$\begin{aligned} a_{ii} &= B_i''(x_i) + q_i B_i(x_i) = \frac{-3}{h^2} + q_i, \quad i = 3, \ldots, n-2, \\ a_{i,i+1} &= B_{i+1}''(x_i) + q_i B_{i+1}(x_i) = \frac{3}{2h^2} + \frac{q_i}{4}, \quad i = 2, \ldots, n-3, \\ a_{i,i-1} &= B_{i-1}''(x_i) + q_i B_{i-1}(x_i) = \frac{3}{2h^2} + \frac{q_i}{4}, \quad i = 4, \ldots, n-1. \end{aligned} \tag{6.4.11}$$

For the remaining coefficients we use the functions ϕ_1, ϕ_2, ϕ_{n-1} and ϕ_n of (6.4.10) and obtain

$$\begin{aligned} a_{11} &= -\frac{9}{h^2}, & a_{n-2,n-1} &= \frac{6}{h^2} + q_{n-2}, \\ a_{12} &= \frac{9}{h^2}, & a_{n-1,n-1} &= \frac{27}{2h^2} + \frac{15}{4} q_{n-1} \\ a_{21} &= \frac{3}{2h^2} - \frac{1}{4} q_2, & a_{n,n-1} &= \frac{3}{2h} + \frac{1}{4} q_{n-1}, \\ a_{22} &= -\frac{27}{2h^2} + \frac{15}{4} q_2, & a_{n,n-1} &= \frac{9}{h^2}, \\ a_{32} &= \frac{6}{h^2} + q_3, & a_{nn} &= -\frac{9}{h^2}. \end{aligned} \tag{6.4.12}$$

The components of the right-hand side $\mathbf{f}$ of the system (6.4.4) are $f(x_1), \ldots, f(x_n)$. Then the solution of the system (6.4.4) with the coefficients of the tridiagonal matrix A defined by (6.4.11) and (6.4.12) yields the coefficients $c_1, \ldots, c_n$ of the linear combination (6.4.3).

We note that in the special case $q(x) \equiv 0$, so that (6.4.1) is $v''(x) = f(x)$, the coefficient matrix A reduces to

$$A = -\frac{3}{2h^2} \begin{bmatrix} 6 & -6 & & & & & & \\ -1 & 9 & -1 & & & & & \\ & -4 & 2 & -1 & & & & \\ & & -1 & 2 & \ddots & & & \\ & & & \ddots & \ddots & & & \\ & & & & & -1 & & \\ & & & & -1 & 2 & -4 & \\ & & & & & -1 & 9 & -1 \\ & & & & & & -6 & 6 \end{bmatrix},$$

which, aside from the factor $-3/2(h^2)$ and the elements of the first and last three rows and columns, is the $(2, -1)$ coefficient matrix obtained by the finite difference method of Section 3.1.

It can be shown that if the solution v of (6.4.1) is sufficiently differentiable, the preceding method has a discretization error of order h^2, the same as the finite difference method of Section 3.1 using centered differences. Higher-order accuracy can be obtained by using splines of higher degree. For example, fourth-order accuracy can be obtained with quintic splines, functions that are four times continuously differentiable and that reduce to fifth-degree polynomials on each subinterval $[x_i, x_{i+1}]$.

The Galerkin Method

We now turn to the Galerkin method. We will assume again that we wish to approximate a solution of (6.4.1),(6.4.2) by a linear combination of the form (6.4.3), where now the coefficients $c_1, \ldots, c_n$ are determined by the Galerkin criterion discussed in Section 6.1. This leads us to the solution of the linear system (6.4.4) where from (6.1.18)

$$a_{ij} = -\int_0^1 \phi_i'(x)\phi_j'(x)dx + \int_0^1 q(x)\phi_i(x)\phi_j(x)dx, \quad i, j = 1, \ldots, n, \quad (6.4.13)$$

and from (6.1.17)

$$f_i = \int_0^1 f(x)\phi_i(x)dx, \qquad i = 1, \ldots, n. \qquad (6.4.14)$$

For the basis functions ϕ_i we will again use piecewise polynomials. As the simplest possibility, let us first consider piecewise linear functions. In particular, assuming that the grid points $0 = x_0, x_1, \ldots, x_n, x_{n+1} = 1$, are equally spaced with spacing h, we will take the basis functions ϕ_i, $i = 1, \ldots, n$, to be

$$\begin{aligned}\phi_i(x) &= \frac{1}{h}(x - x_{i-1}), & x_{i-1} \le x \le x_i, \\ &= -\frac{1}{h}(x - x_{i+1}), & x_i \le x \le x_{i+1}, \\ &= 0 & x < x_{i-1},\ x > x_{i+1}.\end{aligned} \tag{6.4.15}$$

These particular piecewise linear functions are called *hat functions*, or *linear B-splines*, and are illustrated in Figure 6.5. It is intuitively clear, and easily shown (Exercise 6.4.4), that any piecewise linear function that is defined on the nodes $x_0, x_1, \ldots, x_{n+1}$ and that vanishes at x_0 and x_{n+1} can be expressed as a linear combination of these $\phi_1, \ldots, \phi_n$.

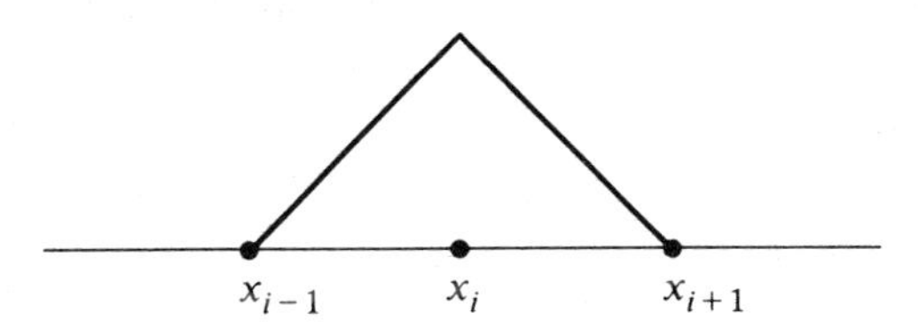

Figure 6.5: *A Hat Function*

We now wish to use the basis functions (6.4.15) in Galerkin's method. At first glance, it would seem that there is a difficulty in using the ϕ_i in the computation of the a_{ij} since this requires ϕ_i', which does not exist at the points x_{i-1}, x_i, and x_{i+1}. Note, however, that ϕ_i' is simply the piecewise constant function

$$\begin{aligned}\phi_i'(x) &= \frac{1}{h}, & x_{i-1} < x < x_i, \\ &= -\frac{1}{h}, & x_i < x < x_{i+1}, \\ &= 0, & x < x_{i-1}, x > x_{i+1}.\end{aligned} \tag{6.4.16}$$

There are discontinuities in this function at the points x_{i-1}, x_i, and x_{i+1}, but these do not affect the integration, and the integrations in (6.4.13) can be carried out on each subinterval to give

$$a_{ij} = \sum_{k=0}^{n} \int_{x_k}^{x_{k+1}} [-\phi_i'(x)\phi_i'(x) + q(x)\phi_i(x)\phi_j(x)]dx. \tag{6.4.17}$$

By the definition of the ϕ_i, the products $\phi_i\phi_j$ and $\phi_i'\phi_j'$ vanish identically unless $i - 1 \le j \le i + 1$. Thus

$$a_{ij} = 0, \qquad \text{if } |i - j| > 1. \tag{6.4.18}$$

To evaluate the other a_{ij} we first introduce the quantities

$$R_i = \int_{x_i}^{x_{i+1}} q(x)(x-x_{i+1})^2 dx, \qquad Q_i = \int_{x_{i-1}}^{x_i} q(x)(x-x_{i-1})^2 dx, \qquad (6.4.19)$$

$$S_i = \int_{x_{i-1}}^{x_i} q(x)(x-x_{i-1})(x-x_i)dx,$$

and note that

$$\int_{x_i}^{x_{i+1}} \phi_i'(x)\phi_i'(x)dx = \frac{1}{h}, \qquad \int_{x_{i-1}}^{x_i} \phi_i'(x)\phi_i'(x)dx = \frac{1}{h},$$

$$\int_{x_i}^{x_{i+1}} \phi_i'(x)\phi_{i+1}'(x)dx = -\frac{1}{h}, \qquad \int_{x_i}^{x_{i+1}} q(x)\phi_i(x)\phi_i(x)dx = \frac{1}{h^2}R_i,$$

$$\int_{x_{i-1}}^{x_i} q(x)\phi_i(x)\phi_i(x)dx = \frac{1}{h^2}Q_i, \qquad \int_{x_{i-1}}^{x_i} q(x)\phi_{i-1}(x)\phi_i(x)dx = -\frac{1}{h^2}S_i.$$

Therefore, from (6.4.17),

$$\begin{aligned} a_{ii} &= \frac{1}{h^2}(-2h+Q_i+R_i), \qquad & i &= 1,\ldots,n, \\ a_{i,i+1} &= \frac{1}{h^2}(h-S_{i+1}), & i &= 1,\ldots,n-1, \\ a_{i,i-1} &= \frac{1}{h^2}(h-S_i), & i &= 2,\ldots,n, \end{aligned} \qquad (6.4.20)$$

and the right-hand side components of (6.4.4) are given for $i = 1,\cdots,n$ by

$$f_i = \frac{1}{h}\int_{x_{i-1}}^{x_i} f(x)(x-x_{i-1})dx + \frac{1}{h}\int_{x_i}^{x_{i+1}} f(x)(x_{i+1}-x)dx. \qquad (6.4.21)$$

Thus the linear system (6.4.4) to be solved for the coefficients $c_1,\ldots,c_n$ of (6.4.3) consists of the tridiagonal matrix A whose components are given by (6.4.20), and the right-hand side $\mathbf{f}$ with components given by (6.4.21). We note that unless $q(x)$ and $f(x)$ are such that the integrals in (6.4.19) and (6.4.21) can be evaluated exactly, we would use the numerical integration techniques of the previous section to approximate these integrals. In the special case $q(x) \equiv 0$, all Q_i, R_i, and S_i are zero; hence

$$a_{ii} = \frac{-2}{h} \qquad a_{i,i+1} = \frac{1}{h} \qquad a_{i-1,i} = \frac{1}{h}.$$

If we then multiply the equations (6.4.4) by $-h^{-1}$, the new coefficient matrix will be exactly the $(2,-1)$ tridiagonal matrix that arose in Chapter 3 from the finite difference approximation to $v'' = f$. The right-hand side of the Galerkin equations will be different, however, involving the integrals of f given in (6.4.21).

Provided that the solution of (6.4.1) is sufficiently differentiable, it can be shown that the Galerkin procedure using the piecewise linear functions (6.4.15) is second-order accurate; that is, the discretization error is $0(h^2)$. By using cubic splines it is possible to increase the order of accuracy by two, so as to make the discretization error $0(h^4)$.

Comparison of Methods

We now compare the three methods that we have discussed for two-point boundary-value problems: finite differences, collocation, and Galerkin. The finite difference method is conceptually simple, easy to implement, and yields second-order accuracy with the centered differences that we used in Chapter 3. The collocation method with cubic splines is slightly more difficult to implement but still relatively easy. For the Galerkin method, however, we must evaluate the integrals of (6.4.19) and (6.4.21), and generally this will require the use of numerical integration or symbolic computation systems. In all three cases the system of linear equations to be solved has a tridiagonal coefficient matrix. All three methods have higher-order versions, which are, naturally, more complicated. It is probably fair to say that for ordinary differential equations the simplicity of the finite difference and collocation methods allows them to be preferred in most cases. The power of the Galerkin method becomes more apparent for partial differential equations.

Supplementary Discussion and References: 6.4

The books by Ascher et.al. [1988], deBoor [1978] and Prenter [1975] are good sources for further reading on the material of this section and for a proof of Theorem 6.4.1. See also Strang and Fix [1973] and Hall and Porsching [1990] for further discussion of the Galerkin method.

EXERCISES 6.4

6.4.1. Show that the function defined by (6.4.8) is a cubic spline on the interval $[0, 1]$ and satisfies the conditions (6.4.6) and (6.4.7).

6.4.2. Show that the values of B_i, B_i', and B_i'' at x_{i-1}, x_i, x_{i+1} are as given in Table 6.1.

6.4.3. Consider the two-point boundary-value problem

$$-v'' + (1 + x^2)v = x^2, \quad v(0) = 0, \quad v(1) = 0.$$

a. Let $h = \frac{1}{4}$ and write out the coefficients a_{ij} of (6.4.11) and (6.4.12) for the collocation method and then the complete system of linear equations (6.4.4). Ascertain whether the coefficient matrix is symmetric positive definite and diagonally dominant. Solve the system and express the approximate solution in the form (6.4.3), where the basis functions are given by (6.4.10).

b. Repeat part **a** for the Galerkin method using the basis functions (6.4.15).

6.4.4. Let $f(x)$ be a piecewise linear function with nodes $x_i = ih$, $i = 0, 1, \ldots, n+1$, $h = (n+1)^{-1}$, and that vanishes at x_0 and x_{n+1}. Show that there are constants $\alpha_1, \ldots, \alpha_n$ such that $f(x) = \sum_{i=1}^{n} \alpha_i \phi_i(x)$, where the ϕ_i are defined by (6.4.15).

6.4.5. Consider the boundary-value problem

$$v''(x) - 3v(x) = x^2, \ 0 \le x \le 1, \ v(0) = v(1) = 0.$$

a. Derive the system of tridiagonal equations to be solved to carry out the collocation method using the basis functions of (6.4.10) with the points $x_i = (i-1)h, i = 1, \cdots, n, h = 1/(n-1)$.

b. Derive the system of tridiagonal equations to be solved to carry out Galerkin's method using the basis functions (6.4.15).

c. Add the nonlinear term $10[v(x)]^3$ to the right hand side of the differential equation and repeat parts **a.** and **b.** Then compute the Jacobian matrices for these systems and discuss how to carry out Newton's method.

Chapter 7

N Important Numbers

7.1 Eigenvalue Problems

In this chapter we shall consider the numerical solution of eigenvalue problems. Such problems occur frequently in engineering, physics, chemistry, economics, and statistics, as well as other areas. In this section we shall discuss a few example problems, classify different types of eigenvalue problems, and provide some mathematical background.

In the *matrix eigenvalue problem* we wish to find a real or complex number λ, an eigenvalue, and a corresponding nonzero vector $\mathbf{x}$, an *eigenvector*, that satisfy the equation

$$A\mathbf{x} = \lambda\mathbf{x}, \tag{7.1.1}$$

where A is a given real or complex $n \times n$ matrix. As discussed in Appendix 2, a solution λ of (7.1.1) is a root of the characteristic polynomial $\det(A - \lambda I) = 0$. This is a polynomial of degree n and therefore has exactly n real or complex roots, $\lambda_1, \cdots, \lambda_n$, provided that the multiplicity of each root is counted. Once an eigenvalue λ_i is known, a corresponding eigenvector $\mathbf{x}_i$ can be determined, in principle, as a solution of the homogeneous system of equations

$$(A - \lambda_i I)\mathbf{x} = 0. \tag{7.1.2}$$

Note that even if the matrix A is real, its eigenvalues – and consequently also its eigenvectors – may be complex. For example (Exercise 7.1.1), the matrix

$$\begin{bmatrix} 2 & -1 \\ 1 & 2 \end{bmatrix}$$

has eigenvalues $2 \pm i$.

The preceding mathematical procedure – form the characteristic polynomial, compute its roots, and solve the homogeneous equations (7.1.2) – is not a viable computational procedure except for the most trivial problems. The main purpose of this chapter is to give alternative computational methods.

Differential Equations

As an example of how eigenvalue problems arise, consider the system of ordinary differential equations

$$\frac{d\mathbf{y}}{dt} = A\mathbf{y} \tag{7.1.3}$$

for a given constant $n \times n$ matrix A. If we try a solution of (7.1.3) of the form

$$\mathbf{y}(t) = e^{\lambda t}\mathbf{x} \tag{7.1.4}$$

for some constant unknown vector $\mathbf{x}$ and unknown parameter λ, then we must have

$$\frac{d\mathbf{y}}{dt} = \lambda e^{\lambda t}\mathbf{x} = A(e^{\lambda t}\mathbf{x}),$$

or, since $e^{\lambda t}$ is always nonzero, $A\mathbf{x} = \lambda\mathbf{x}$; that is, (7.1.4) is a solution of (7.1.3) if and only if λ and $\mathbf{x}$ are an eigenvalue and a corresponding eigenvector of A.

An important type of matrix is one that has n linearly independent eigenvectors (see Theorem A.2.1 in Appendix 2 for the definition of linear independence). If this is the case, and $\lambda_1, \ldots, \lambda_n$ and $\mathbf{x}_1, \ldots, \mathbf{x}_n$ are the eigenvalues and corresponding eigenvectors, then

$$\mathbf{y}_1(t) = e^{\lambda_1 t}\mathbf{x}_1, \qquad \mathbf{y}_2(t) = e^{\lambda_2 t}\mathbf{x}_2, \qquad \cdots, \qquad \mathbf{y}_n(t) = e^{\lambda_n t}\mathbf{x}_n \tag{7.1.5}$$

is a complete set of linearly independent solutions of the differential equation (7.1.3). Hence, any solution of (7.1.3) may be written in the form

$$\mathbf{y}(t) = \sum_{i=1}^{n} c_i \mathbf{y}_i(t) = \sum_{i=1}^{n} c_i e^{\lambda_i t}\mathbf{x}_i, \tag{7.1.6}$$

where the constants $c_1, \ldots, c_n$ may be determined by initial or other conditions. Thus, the general solution of (7.1.3) may be obtained by solving the eigenvalue problem for the matrix A. If A does not have n linearly independent eigenvectors, a similar but more complicated representation of the solution may be given.

Linearly Independent Eigenvectors

We now discuss in more detail the property of a matrix having n linearly independent eigenvectors. As in Appendix 2, a similarity transformation of the matrix A is of the form PAP^{-1}, where P is any nonsingular matrix. A similarity transformation of A arises from a change of variables; for example, consider the system of equations $A\mathbf{x} = \mathbf{b}$ and make the change of variables $\mathbf{y} = P\mathbf{x}$ and $\mathbf{c} = P\mathbf{b}$, where P is a nonsingular matrix. In the new variables the system of equations is $AP^{-1}\mathbf{y} = P^{-1}\mathbf{c}$ or, upon multiplying through by P, $PAP^{-1}\mathbf{y} = \mathbf{c}$. Thus, the coefficient matrix of the system in the new variables is the similarity transform PAP^{-1}.

An important property of similarity transformations is that they preserve the eigenvalues of A: the matrices A and PAP^{-1} have the same eigenvalues. This is easily seen by considering the characteristic polynomial and using the fact that the determinant of a product of matrices is the product of the determinants. Thus

$$\det(A - \lambda I) = \det(PP^{-1})\det(A - \lambda I) = \det(P)\det(A - \lambda I)\det(P^{-1})$$
$$= \det(PAP^{-1} - \lambda I),$$

which shows that the characteristic polynomials, and hence the eigenvalues, of A and PAP^{-1} are identical. However, the eigenvectors change under a similarity transformation. Indeed,

$$PAP^{-1}\mathbf{y} = \lambda\mathbf{y} \quad \text{or} \quad AP^{-1}\mathbf{y} = \lambda P^{-1}\mathbf{y}$$

shows that the eigenvector $\mathbf{y}$ of PAP^{-1} is related to the eigenvector $\mathbf{x}$ of A by $P^{-1}\mathbf{y} = \mathbf{x}$ or $\mathbf{y} = P\mathbf{x}$.

An important question is how "simple" the matrix A may be made under a similarity transformation. A basic result in this regard, which brings us back to linear independence of the eigenvectors, is the following:

> THEOREM 7.1.1 *A matrix A is similar to a diagonal matrix if and only if A has n linearly independent eigenvectors.*

The proof of this theorem is both simple and illustrative. Let $\mathbf{x}_1, \ldots, \mathbf{x}_n$ be n linearly independent eigenvectors of A with corresponding eigenvalues $\lambda_1, \ldots, \lambda_n$, and let P be the matrix with columns $\mathbf{x}_1, \ldots, \mathbf{x}_n$: then P is nonsingular since its columns are linearly independent. By the basic definition $A\mathbf{x}_i = \lambda_i\mathbf{x}_i$ applied to each column of P, we have

$$AP = A(\mathbf{x}_1, \mathbf{x}_2, \ldots \mathbf{x}_n) = (\lambda_1\mathbf{x}_1, \ldots, \lambda_n\mathbf{x}_n) = PD, \tag{7.1.7}$$

where D is the diagonal matrix $\text{diag}(\lambda_1, \lambda_2, \ldots, \lambda_n)$. Thus (7.1.7) is equivalent to $A = PDP^{-1}$, which shows that A is similar to a diagonal matrix whose diagonal entries are the eigenvalues of A. Conversely, if A is similar to a diagonal matrix, then (7.1.7) shows that the columns of the similarity matrix P must be eigenvectors of A, and they are linearly independent by the nonsingularity of P.

Two important special cases of the preceding result are the following theorems, which we state without proof.

THEOREM 7.1.2 *If A has distinct eigenvalues, then A is similar to a diagonal matrix.*

THEOREM 7.1.3 *If A is a real symmetric matrix (*that is, $A = A^T$*), then A is similar to a diagonal matrix, and the similarity matrix may be taken to be orthogonal (*that is, $PP^T = I$*).*

Symmetric matrices are extremely important in applications. They also have many nice properties as regards their eigenvalues and eigenvectors. In particular, the eigenvalues of a symmetric matrix are always real and are positive if A is positive definite. Moreover, the last part of Theorem 7.1.3 can be rephrased to say that the eigenvectors can be chosen to be orthonormal (Appendix 2).

The Jordan Form

Theorem 7.1.2 shows that if a matrix A does not have n linearly independent eigenvectors, then necessarily it has multiple eigenvalues. (But note that a matrix may have n linearly independent eigenvectors even though it has multiple eigenvalues; the identity matrix is an example.) The matrix

$$A = \begin{bmatrix} 1 & 1 \\ 0 & 1 \end{bmatrix} \tag{7.1.8}$$

is a simple example of a matrix that does not have two linearly independent eigenvectors (see Exercise 7.1.4) and is not similar to a diagonal matrix. However, any $n \times n$ matrix may be made similar to a matrix of the form

$$J = \begin{bmatrix} \lambda_1 & \delta_1 & & & \\ & \lambda_2 & \delta_2 & & \\ & & \ddots & \ddots & \\ & & & & \delta_{n-1} \\ & & & & \lambda_n \end{bmatrix},$$

where the λ_i are the eigenvalues of A and the δ_i are either 0 or 1. If q is the number of δ_i that are nonzero, then A has $n - q$ linearly independent eigenvectors, and whenever a δ_i is nonzero, then the eigenvalues λ_i and λ_{i-1} are identical. Thus the matrix J can be partitioned as

$$J = \begin{bmatrix} J_1 & & \\ & \ddots & \\ & & J_p \end{bmatrix}, \tag{7.1.9a}$$

where p is the number of linearly independent eigenvectors, and each J_i is a matrix of the form

$$J_i = \begin{bmatrix} \lambda_i & 1 & & \\ & \ddots & \ddots & \\ & & & 1 \\ & & & \lambda_i \end{bmatrix} \tag{7.1.9b}$$

with identical eigenvalues and all 1's on the first superdiagonal. The matrix J of (7.1.9) is called the *Jordan canonical form* of A. Note that if A has n linearly independent eigenvectors, then $p = n$; in this case each J_i reduces to a 1×1 matrix, and J is diagonal.

The Jordan canonical form is for useful theoretical purposes but not very useful in practice. For many computational purposes it is very desirable to work with orthogonal or unitary matrices. (A *unitary* matrix U is a complex matrix that satisfies $U^\star U = I$, where $U^\star$ is the conjugate transpose of U; a real unitary matrix is an orthogonal matrix.) We next state without proof two basic results on similarity transformations with unitary or orthogonal matrices.

SCHUR'S THEOREM. *For an arbitrary $n \times n$ matrix A, there is a unitary matrix U such that $UAU^\star$ is triangular.*

MURNAGHAN-WINTNER THEOREM. *For a real $n \times n$ matrix A, there is an orthogonal matrix P so that*

$$PAP^T = \begin{bmatrix} T_{11} & & \cdots & T_{1m} \\ & T_{22} & & \\ & & \ddots & \vdots \\ & & & T_{mm} \end{bmatrix},$$

where each T_{ii} is either 2×2 or 1×1.

In the case of Schur's Theorem, the diagonal elements of $UAU^\star$ are the eigenvalues of A since $UAU^\star$ is a similarity transformation. If A is real but has some complex eigenvalues, then U is necessarily complex. The Murnaghan-Winter Theorem comes as close to a triangular form as possible with a real orthogonal matrix. In this case, if T_{ii} is 1×1, then it is a real eigenvalue of A, whereas if T_{ii} is 2×2, its two eigenvalues are a complex conjugate pair of eigenvalues of A.

Other Differential Equations

We now return to further examples of eigenvalue problems. Many applications lead in certain simple cases to the ordinary differential equation

$$-y''(x) = \lambda y(x), \qquad y(0) = 0, \qquad y(1) = 0. \tag{7.1.10}$$

Here we wish to find values – again called eigenvalues – of the scalar λ so that (7.1.10) has corresponding nonzero solutions – called eigenfunctions – that satisfy the given zero boundary conditions. This particularly simple problem can be solved explicitly. There are infinitely many eigenvalues and corresponding eigenfunctions that are given by

$$\lambda_k = k^2\pi^2 \qquad y_k(x) = \sin k\pi x, \qquad k = 1, 2, \ldots, \tag{7.1.11}$$

as is easily checked by substitution into (7.1.10).

Next suppose that (7.1.10) is modified by adding a nonconstant coefficient of y; that is,

$$-y''(x) = \lambda c(x)y(x), \qquad y(0) = 0, \qquad y(1) = 0, \tag{7.1.12}$$

where c is a given positive function. Now it is no longer possible, in general, to obtain the eigenvalues and eigenfunctions of (7.1.12) explicitly, but we can approximate them numerically by the following procedure. Just as in the treatment of boundary-value problems in Chapter 3, we discretize the interval $[0, 1]$ with grid points $x_i = ih$, $i = 0, 1, \ldots, n+1$, $h = 1/(n+1)$, and replace the second derivative in (7.1.12) by the corresponding difference quotient. This gives the discrete equations

$$\frac{1}{h^2}(-y_{i+1} + 2y_i - y_{i-1}) = \lambda c_i y_i, \qquad i = 1, \ldots, n, \tag{7.1.13}$$

where $c_i = c(x_i)$, $y_0 = y_{n+1} = 0$, and y_i is an approximation to $y(x_i)$.

The equations (7.1.13) constitute a matrix eigenvalue problem of the form

$$A\mathbf{y} = \lambda B\mathbf{y}, \tag{7.1.14}$$

where A is the $(2, -1)$ tridiagonal matrix of (3.1.10) and B is a diagonal matrix with elements $h^2 c_i$. Equation (7.1.14) is an example of a *generalized eigenvalue problem* in which the matrix B on the right-hand side of the equation is not the identity matrix. In the present case we have assumed that the function $c(x)$ is positive; therefore B is non-singular and we can multiply (7.1.14) by B^{-1} to convert it to the standard eigenvalue problem $B^{-1}A\mathbf{y} = \lambda\mathbf{y}$.

It is not always advisable to convert (7.1.14) back to a standard eigenvalue problem even if B is non-singular (see the Supplementary Discussion of Section 7.2). Moreover, if A and B are symmetric, the product $B^{-1}A$ is not symmetric, in general. However, if B is also positive definite, we can convert (7.1.14) to a standard eigenvalue problem for a symmetric matrix as follows. First, compute the Cholesky decomposition $B = LL^T$ (Section 4.5). Then, multiply (7.1.14) by L^{-1} so that (7.1.14) becomes

$$L^{-1}AL^{-T}L^Ty = \lambda L^Ty \quad \text{or} \quad \hat{A}\mathbf{z} = \lambda\mathbf{z},$$

where $\mathbf{z} = L^T\mathbf{y}$ and $\hat{A} = L^{-1}AL^{-T}$ is symmetric.

Although the main purpose of this chapter is to describe methods for computing eigenvalues and eigenvectors, it is important to note that many problems require information only about the location of eigenvalues and not their precise values. As an example of this we return to the system (7.1.3) of ordinary differential equations. An important property of this system is whether all solutions tend to zero as t tends to infinity. If so, the zero solution is said to be *asymptotically stable*. Assuming again that the matrix A has n linearly independent eigenvectors, all solutions will go to zero as t goes to infinity if and only if each of the solutions (7.1.5) does, and since the vectors $\mathbf{x}_i$ are constant, this will be the case if and only if $e^{\lambda_i t}$ approaches zero as t approaches infinity for each i. If λ_i is real, this will be the case if and only if $\lambda_i < 0$, and if λ_i is complex, the real part of λ_i, denoted by $\text{Re}(\lambda_i)$, must be negative. Thus the zero solution of (7.1.3) is asymptotically stable if and only if

$$\text{Re}(\lambda_i) < 0, \qquad i = 1, \ldots, n, \tag{7.1.15}$$

so that all the eigenvalues lie in the left-half of the complex plane. A related example is an iterative method of the form

$$\mathbf{x}^{k+1} = A\mathbf{x}^k + \mathbf{d}, \qquad k = 0, 1, \ldots . \tag{7.1.16}$$

As we will see in Chapter 9, the iterates $\mathbf{x}^k$ will converge for any starting vector $\mathbf{x}^0$ if and only if all the eigenvalues of A satisfy $|\lambda_i| < 1$. This will be true if $||A|| < 1$ for some norm, but the following approach is sometimes more easily applied.

Gerschgorin's Theorem

Let $A = (a_{ij})$ be a real or complex $n \times n$ matrix and let

$$r_i = \sum_{\substack{j=1 \\ j \neq i}}^{n} |a_{ij}|, \qquad i = 1, \ldots, n.$$

That is, r_i is the sum of the absolute values of the off-diagonal elements in the ith row of A. Next define disks in the complex plane centered at a_{ii} and with radius r_i:

$$\Lambda_i = \{z : |z - a_{ii}| \leq r_i\}, \qquad i = 1, \ldots, n.$$

We then have the following:

GERSCHGORIN'S THEOREM *All the eigenvalues of A lie in the union of the disks $\Lambda_1, \ldots, \Lambda_n$. Moreover, if S is a union of m disks such that S is disjoint from all the other disks, then S contains exactly m eigenvalues of A (counting multiplicities).*

As a simple example of the use of Gerschgorin's theorem, consider the matrix

$$A = \frac{1}{16}\begin{bmatrix} -8 & -2 & 4 \\ -1 & -4 & 2 \\ 2 & 2 & -10 \end{bmatrix}, \tag{7.1.17}$$

for which the Gerschgorin disks are illustrated in Figure 7.1. Note that we can immediately conclude that all eigenvalues of A have negative real part; hence if A were the coefficient matrix of the system of differential equations (7.1.3), the zero solution of that system would be asymptotically stable. Similarly, we can immediately conclude that the eigenvalues of A are all less than 1 in absolute value, so the vectors $\mathbf{x}^k$ defined by (7.1.16) converge.

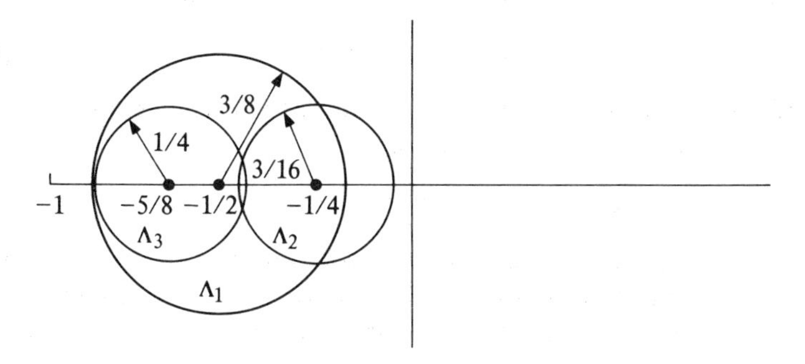

Figure 7.1: *Gerschgorin's Disks in the Complex Plane*

To illustrate the second part of Gerschgorin's Theorem, suppose that the second row of the matrix of (7.1.17) is changed to $\frac{1}{16}(-1, 6, 2)$. Then the disk Λ_2 is centered at $+\frac{3}{8}$, again with radius $\frac{3}{16}$. Since Λ_2 is now disjoint from the other two disks, it contains exactly one eigenvalue of A. Moreover, since any complex eigenvalues of A must occur in conjugate pairs, this eigenvalue must be real and therefore lies in the interval $[\frac{3}{16}, \frac{9}{16}]$.

The proof of the first part of Gerschgorin's Theorem is very easy. Let λ be any eigenvalue of A, and $\mathbf{x}$ a corresponding eigenvector. Then, by (7.1.1),

$$(\lambda - a_{ii})x_i = \sum_{\substack{j=1 \\ j \neq i}}^{n} a_{ij}x_j, \qquad i = 1, \ldots, n.$$

If we let x_k be the component of largest absolute value in the vector $\mathbf{x}$, then

$$|\lambda - a_{kk}| \leq \sum_{\substack{j=1 \\ j \neq k}}^{n} |a_{kj}|\frac{|x_j|}{|x_k|} \leq \sum_{\substack{j=1 \\ j \neq k}}^{n} |a_{kj}|.$$

Thus λ is in the disk centered at a_{kk} and therefore in the union of all the disks. The proof of the second part of the theorem is more complicated and relies on the fact that the eigenvalues of a matrix are continuous functions of the elements of the matrix.

By a simple similarity transformation, it is sometimes possible to use Gerschgorin's Theorem to extract additional information about the eigenvalues. For example, consider the matrix

$$A = \begin{bmatrix} 8 & 1 & 0 \\ 1 & 12 & 1 \\ 0 & 1 & 10 \end{bmatrix}.$$

Since A is symmetric its eigenvalues are real, and by Gerschgorin's Theorem we conclude that they lie in the union of the intervals $[7, 9]$, $[10, 14]$, $[9, 11]$. Since these intervals are not disjoint we cannot yet conclude that any of them contains an eigenvalue. However, if we do a similarity transformation with the matrix $D = \text{diag}(d, 1, 1)$ we obtain

$$DAD^{-1} = \begin{bmatrix} 8 & d & \\ d^{-1} & 12 & 1 \\ & 1 & 10 \end{bmatrix}.$$

By Gerschgorin's Theorem, the eigenvalues of this matrix (which are the same as those of A) lie in the union of the intervals $[8-d, 8+d]$, $[11-d^{-1}, 13+d^{-1}]$, $[9, 11]$. As long as $1 > d > \frac{1}{2}[3 - \sqrt{5}]$, the first interval is disjoint from the others and thus contains exactly one eigenvalue. In particular, the interval $[7.6, 8.4]$ contains one eigenvalue.

Another important use of Gerschgorin's Theorem is in ascertaining the change in the eigenvalues of a matrix due to changes in the coefficients. Let A be a given $n \times n$ matrix with eigenvalues $\lambda_1, \ldots, \lambda_n$ and suppose that E is a matrix whose elements are small compared to those of A; for example, E may be the rounding errors committed in entering the matrix A into a computer. Suppose that $\mu_1, \ldots, \mu_n$ are the eigenvalues of $A + E$. Then, what can one say about the changes $|\lambda_i - \mu_i|$? We next give a relatively simple result in the case that A has n linearly independent eigenvectors. Recall, from Appendix 2, that the infinity norm of a matrix is the maximum value of the sums of the absolute values of the elements in each row.

THEOREM 7.1.4 *Assume that $A = PDP^{-1}$, where D is the diagonal matrix of eigenvalues of A, and let $d = ||P^{-1}EP||_\infty$. Then every eigenvalue of $A + E$ is within d of an eigenvalue of A.*

The proof of this theorem is a simple consequence of Gerschgorin's Theorem. Set $C = P^{-1}(A + E)P$. Then C has the same eigenvalues $\mu_1, \ldots, \mu_n$ as

$A+E$. Let $B = P^{-1}EP$. Then $C = D + B$, and the diagonal elements of C are $\lambda_i + b_{ii}$, $i = 1, \ldots, n$. Hence, by Gerschgorin's Theorem, the eigenvalues $\mu_1, \ldots, \mu_n$ are in the union of the disks

$$\{z : |z - \lambda_i - b_{ii}| \leq \sum_{\substack{j=1 \\ j \neq i}}^{n} |b_{ij}|\}.$$

Therefore, given any μ_k, there is an i such that

$$|\mu_k - \lambda_i - b_{ii}| \leq \sum_{\substack{j=1 \\ j \neq i}}^{n} |b_{ij}|,$$

or

$$|\mu_k - \lambda_i| \leq \sum_{j=1}^{n} |b_{ij}| \leq d,$$

which was to be shown.

Ill-conditioned Eigenvalues

Note that the quantity d need not be small even though $||E||_\infty$ is small; this will depend on P. In general, the more ill-conditioned the matrix P (in the sense of Chapter 4), the more ill-conditioned will be the eigenvalues of A, and the more the eigenvalues may change because of small changes in the coefficients of A. We give a simple example of this. Let

$$A = \begin{bmatrix} 1 & 1 \\ 0 & 1 + 10^{-10} \end{bmatrix}, \qquad A + E = \begin{bmatrix} 1 & 1 \\ 10^{-10} & 1 + 10^{-10} \end{bmatrix}.$$

Then the eigenvalues of A are 1 and $1 + 10^{-10}$, and those of $A + E$ are approximately 1 ± 10^{-5}. Thus a change of 10^{-10} in one element of A has caused a change 10^5 times as large in the eigenvalues. The reason for this is that the matrix P of eigenvectors of A is very ill-conditioned. It is easy to verify that

$$P = \begin{bmatrix} 1 & 1 \\ 0 & 10^{-10} \end{bmatrix}, \qquad P^{-1} = \begin{bmatrix} 1 & -10^{10} \\ 0 & 10^{10} \end{bmatrix}.$$

Therefore the matrix $P^{-1}EP$ of Theorem 7.1.4 is

$$P^{-1}EP = \begin{bmatrix} 1 & -10^{10} \\ 0 & 10^{10} \end{bmatrix} \begin{bmatrix} 0 & 0 \\ 10^{-10} & 0 \end{bmatrix} \begin{bmatrix} 1 & 1 \\ 0 & 10^{-10} \end{bmatrix} = \begin{bmatrix} -1 & -1 \\ 1 & 1 \end{bmatrix},$$

and thus $d = ||P^{-1}EP||_\infty = 2$. Note that the actual change in the eigenvalues is far smaller than this bound.

It is an interesting and important fact that the eigenvalues of a symmetric matrix are always well-conditioned; this is the interpretation of the following theorem, stated without proof.

THEOREM 7.1.5 *Let A and B be real symmetric $n \times n$ matrices with eigenvalues $\lambda_1, \ldots, \lambda_n$ and $\mu_1, \ldots, \mu_n$, respectively. Then given any μ_j, there is a λ_i such that*

$$|\lambda_i - \mu_j| \leq \|A - B\|_2.$$

Note that in this theorem it is the 2-norm (see Appendix 2) that is used, and hence the result does not follow directly from Theorem 7.1.4.

In this section we have given various examples of eigenvalue problems and some of the basic mathematical theory. In the remainder of this chapter we will discuss the foundation of various methods for computing eigenvalues and eigenvectors.

Supplementary Discussion and References: 7.1

Further discussion of the use of eigenvalues for solving linear ordinary differential equations can be found in most elementary differential equation textbooks. See also Ortega [1987]. Discussions of the theory of matrix eigenvalue problems in a form most suitable for scientific computing are given in Golub and Van Loan [1989], Ortega [1987], and Ortega [1990]. See also Wilkinson [1965], Householder [1964], Stewart [1973], and Parlett [1980].

EXERCISES 7.1

7.1.1. Compute the characteristic equations $\det(A - \lambda I)$ for the matrices

$$A = \begin{bmatrix} 2 & -1 \\ -1 & 2 \end{bmatrix}, \qquad A = \begin{bmatrix} 2 & -1 \\ 1 & 2 \end{bmatrix}.$$

Next compute the eigenvalues of A by obtaining the roots of these polynomials, and then compute the eigenvectors by solving the homogeneous equations (7.1.2).

7.1.2. Give the solution of the initial-value problem

$$\mathbf{y}'(t) = A\mathbf{y}(t), \qquad \mathbf{y}(0) = \begin{bmatrix} 1 \\ 1 \end{bmatrix}$$

in terms of the eigenvalues and eigenvectors that were computed in Exercise 7.1.1.

7.1.3. If A is the matrix

$$A = \frac{1}{4}\begin{bmatrix} 1 & 1 \\ -1 & 2 \end{bmatrix},$$

determine whether the zero solution of $\mathbf{y}' = A\mathbf{y}$ is asymptotically stable.

7.1.4. Compute an eigenvector of the matrix (7.1.8) and show that there are no other linearly independent eigenvectors.

7.1.5. Assume that a matrix A has two eigenvalues $\lambda_1 = \lambda_2$ and corresponding linearly independent eigenvectors $\mathbf{x}_1$, $\mathbf{x}_2$. Show that any linear combination $c_1\mathbf{x}_1 + c_2\mathbf{x}_2$ is also an eigenvector.

7.1.6. Suppose that A has eigenvalues $\lambda_1, \ldots, \lambda_n$ and eigenvectors $\mathbf{x}_1, \ldots, \mathbf{x}_n$. Show that for any constants α and β, $\alpha A + \beta I$ has eigenvalues $\alpha\lambda_i + \beta$ and corresponding eigenvectors $\mathbf{x}_i$. Use this result in combination with Exercise 4.4.5 to show that the matrix

$$A = \begin{bmatrix} a & b & & & & \\ b & a & b & & & \\ & b & a & b & & \\ & & \ddots & \ddots & \ddots & \\ & & & b & a & b \\ & & & & b & a \end{bmatrix}$$

has eigenvalues $\lambda_k = a + 2b\cos[k\pi/(n+1)]$, $k = 1, \ldots, n$. What are the eigenvectors?

7.1.7. A polynomial in a matrix A is $p(A) = \alpha_0 + \alpha_1 A + \cdots + \alpha_m A^m$. If A has an eigenvalue λ and corresponding eigenvector $\mathbf{x}$, show that $p(\lambda) = \alpha_0 + \alpha_1\lambda + \cdots + \alpha_m\lambda^m$ is an eigenvalue of $p(A)$ with corresponding eigenvector $\mathbf{x}$. Formulate and prove the corresponding result for a rational function of a matrix.

7.1.8. If A and B are $n \times n$ matrices at least one of which is nonsingular, show that AB and BA have the same eigenvalues.

7.1.9. Find the Gerschgorin disks for the matrix

$$A = \begin{bmatrix} 4 & 2 & 2 \\ 1 & 8 & 1 \\ 1 & 1 & 12 \end{bmatrix}.$$

Use the fact that A and A^T have the same eigenvalues to conclude that A has an eigenvalue that satisfies $|\lambda - 4| \leq 2$ by applying Gerschgorin's Theorem to A^T.

7.1.10. Use Gerschgorin's Theorem to prove that a symmetric strictly diagonally dominant matrix with positive diagonal elements is positive definite.

7.1.11. If $p(\lambda) = a_0 + a_1\lambda + \cdots + a_{n-1}\lambda^{n-1} + \lambda^n$, the matrix

$$A = \begin{bmatrix} & 1 & & & \\ & & 1 & & \\ & & & \ddots & \\ & & & & 1 \\ -a_0 & -a_1 & & \cdots & -a_{n-1} \end{bmatrix}$$

is called the *companion matrix* (or *Frobenius matrix*) of p. Show that $p(\lambda)$ is the characteristic polynomial of A. Then apply Gerschgorin's Theorem to both A and A^T to obtain bounds for the roots of p.

7.1.12. Let A be a real, symmetric matrix. Show that Schur's Theorem implies that there exists an orthogonal matrix Q such that $Q^T AQ = D$, where D is a diagonal matrix.

7.1.13. A matrix A is *skew-symmetric* if $A^T = -A$. Let A be a real, skew-symmetric matrix and $PAP^T = T$, where T is given by the Murnaghan-Wintner Theorem. Describe the structure of T in this case.

7.1.14. Show how to write the differential equation

$$y''(t) + ay'(t) + by(t) = 0$$

as a system of first-order equations in the form (7.1.3).Then give conditions on a and b so that $y(t) \to 0$ as $t \to \infty$ for any initial conditions.

7.1.15. Suppose that the matrix A has p zero eigenvalues and corresponding linearly independent eigenvectors. Show how to obtain the solution of the differential equation $A\mathbf{y}' = \mathbf{y}$ even though A^{-1} does not exist. What does this imply about initial or boundary conditions?

7.1.16. Compare the eigenvalues of (7.1.10) with those of the matrix

$$B = h^{-2}A, \quad h = \frac{\pi}{n+1},$$

where A is the $(2, -1)$ tridiagonal matrix of (3.1.10). Which eigenvalues of B are accurate approximations of those of the differential equation?

7.1.17. Consider the equation (7.1.14) where A and B are symmetric and B is positive definite. Show that we can construct matrices F and D such that

$$A = FDF^T \text{ and } B = FF^T,$$

where D is a diagonal matrix whose entries are the eigenvalue of (7.1.14).

7.1.18. Consider a matrix J_i of the form (7.1.9b). Show that there exists a diagonal matrix D so that DJ_iD^{-1} is the same as J_i except that the off-diagonal 1's are replaced by ϵ.

7.1.19. Assume that $A = A^T$. In Theorem 7.1.4, give an upper bound for $d = \|P^{-1}EP\|_\infty$.

7.2 The QR Method

We now begin the study of methods to compute the eigenvalues and eigenvectors of an $n \times n$ matrix A. We will assume that A is real but, in general, it may have complex eigenvalues and eigenvectors. We will first consider a method that applies to such matrices, and then specialize to the important special case in which A is symmetric and thus has real eigenvalues.

We begin with the QR factorization of Section 4.5:

$$A = QR, \tag{7.2.1}$$

where Q is orthogonal and R is upper triangular. Now form a new matrix by multiplying these factors in reverse order:

$$A_1 = RQ. \tag{7.2.2}$$

Since

$$A = QR = QRQQ^{-1} = QA_1Q^{-1} = QA_1Q^T,$$

A_1 is similar to A and has the same eigenvalues (see also Exercise 7.1.8). We then compute the QR factorization of A_1 and reverse the order of the factors to obtain another matrix A_2:

$$A_1 = Q_1R_1, \qquad A_2 = R_1Q_1.$$

Again, A_2 is similar to A_1, and hence to A. We continue this process, alternately doing a QR factorization and then reversing the order of the factors to generate a sequence of matrices

$$A_k = Q_kR_k, \qquad A_{k+1} = R_kQ_k, \qquad k = 0, 1, \ldots, \tag{7.2.3}$$

where $A_0 = A$. All of these matrices are similar and thus have the same eigenvalues as A. The generation of the matrices A_k of (7.2.3) is called the *QR algorithm.* For this algorithm we have the following basic convergence theorem, which we state without proof.

THEOREM 7.2.1. *(QR Convergence) If the eigenvalues of A satisfy*

$$|\lambda_1| > |\lambda_2| > \cdots > |\lambda_n|, \tag{7.2.4}$$

then the matrices A_k of (7.2.3) converge to an upper triangular matrix whose diagonal elements are the eigenvalues of A. Moreover, if $A = PDP^{-1}$, where $D = \text{diag}(\lambda_1, \ldots, \lambda_n)$, and if P^{-1} has an LU decomposition, then

$$A_k \to T = \begin{bmatrix} \lambda_1 & * & \cdots & * \\ & \ddots & & \vdots \\ & & & * \\ & & & \lambda_n \end{bmatrix}, \quad \textit{as} \quad k \to \infty, \tag{7.2.5}$$

and the rate of convergence to zero of the off-diagonal elements $a_{ij}^{(k)}$ of A_k is given by

$$a_{ij}^{(k)} = 0\left(\frac{|\lambda_i|^k}{|\lambda_j|^k}\right), \qquad k \to \infty, \qquad i > j. \tag{7.2.6}$$

The technical condition that P^{-1} have an LU decomposition ensures that the eigenvalues appear on the main diagonal of T in descending order of magnitude. This is the usual situation, although if the condition is not satisfied the order of the eigenvalues may be different. The more stringent condition is (7.2.4), which precludes not only multiple eigenvalues but also complex conjugate pairs of eigenvalues. If the matrix A is real, then all the factors Q_k and R_k are also real, and there is, of course, no possibility that the A_k could converge to a triangular matrix with complex eigenvalues. However, what does occur – which is the best that one could hope – is that the A_k will "converge" to an almost-triangular form illustrated by the matrix

$$\begin{bmatrix} \Box & * & & \cdots & & * \\ & \lambda_3 & & & & \vdots \\ & & \Box & & & \\ & & & \Box & & \\ & & & & \lambda_8 & * \\ & & & & & \lambda_9 \end{bmatrix}. \tag{7.2.7}$$

In this example we have assumed that there are three real eigenvalues λ_3, λ_8, λ_9 with distinct absolute values and three complex conjugate pairs of eigenvalues, again with distinct absolute values. The latter eigenvalues are determined by the three 2×2 matrices indicated by the blocks on the main diagonal. Actually, the elements of these 2×2 matrices do not converge, but their eigenvalues do converge to eigenvalues of A. Hence the computation of complex eigenvalues of real matrices does not present any problem. Note that (7.2.7) is the Murnaghan-Wintner form. Thus the QR algorithm attempts to obtain the Schur triangular form of the matrix when possible, and the Murnaghan-Wintner form otherwise.

Hessenberg Form

The QR algorithm as described so far is too inefficient to be effective, and two important modifications must be made. The first problem is that each step of (7.2.3) requires $0(n^3)$ operations, which makes the process very slow. We can circumvent this difficulty by making a preliminary reduction of the matrix A to a form for which the decomposition can be more rapidly computed. This is the *Hessenberg form*

$$\begin{bmatrix} * & * & & \cdots & * \\ * & & \ddots & & \\ 0 & \ddots & \ddots & & \vdots \\ \vdots & \ddots & & * & \\ 0 & \cdots & 0 & * & * \end{bmatrix}, \tag{7.2.8}$$

which has one non-zero diagonal below the main diagonal, while the elements above the main diagonal are, in general, non-zero . The reduction by similarity transformations of the original matrix A to Hessenberg form can be effected by the Householder transformations used in Section 4.5, as we now discuss.

Let $P_2 = I - 2\mathbf{w}_2\mathbf{w}_2^T$ be a Householder transformation such that P_2A has zeros in its first column below the second position:

$$P_2A = \begin{bmatrix} * & * & \cdots & * \\ * & * & & \\ 0 & * & & \vdots \\ \vdots & \vdots & & \\ 0 & * & \cdots & * \end{bmatrix}. \tag{7.2.9}$$

The vector $\mathbf{w}_2$ has a zero in the first component; otherwise, it is defined analogously to (4.5.13) by

$$\mathbf{w}_2 = \mu_2\mathbf{u}_2, \ \mathbf{u}_2^T = (0, a_{21} - s_2, a_{31}, \ldots, a_n), \tag{7.2.10}$$

where

$$s_2 = \pm(\sum_{j=2}^{n} a_{ji}^2)^{1/2}, \ \mu_2 = (2s_2^2 - 2a_{21}s_2)^{-1/2}.$$

Since we are performing similarity transformations, we must also multiply on the right by P_2^T:

$$A_2 = P_2AP_2^T = A - 2\mathbf{w}_2\mathbf{w}_2^TA - 2A\mathbf{w}_2\mathbf{w}_2^T + 4\mathbf{w}_2^TA\mathbf{w}_2\mathbf{w}_2\mathbf{w}_2^T. \tag{7.2.11}$$

Since the first component of $\mathbf{w}_2$ is zero, the multiplication on the right by P_2^T does not change the zeros introduced in the first column by P_2A. Thus A_2 is similar to A and has the form shown in (7.2.9). We now continue this process. P_3 is determined by a vector $\mathbf{w}_3$ whose first two components are zero and is otherwise chosen analogously to (7.2.10) to produce zeros in the third column below the third element. And so on. After $n-2$ Householder transformations, the matrix

$$PAP^T = H, \qquad P = P_{n-1} \cdots P_2, \tag{7.2.12}$$

will have the Hessenberg form (7.2.8). Since (7.2.12) is a similarity transformation, A and H will have the same eigenvalues. This reduction requires $0(n^3)$ operations; more precisely, it requires roughly twice as many operations as a QR factorization.

A particularly important special case of (7.2.12) is when A is symmetric. In this case H must also be symmetric (Exercise 7.2.1); thus it is tridiagonal. We summarize the discussion above as:

THEOREM 7.2.2. *An $n \times n$ real matrix can be reduced to Hessenberg form* (7.2.8) *by $n-2$ Householder similarity transformations. If A is symmetric, the Hessenberg form is tridiagonal.*

We now apply the QR method (7.2.3) to the Hessenberg matrix H. The QR factorization of H can be carried out by Householder transformations, as before, but since there is only one non-zero element below the main diagonal in each column it is slightly easier to use Givens transformations (see Section 4.5). Each Givens transformation will eliminate one zero below the main diagonal and thus $n-1$ Givens transformations produce the QR factorization. As discussed in Section 4.5, the first Givens transformation modifies the first two rows of H and requires $4n$ multiplications and $2n$ additions. At each stage the length of the rows decreases by one and hence the total number of operations for the row modifications is

$$4\sum_{k=2}^{n} k \text{ multiplications } + 2\sum_{k=2}^{n} k \text{ additions } = 0(n^2) \text{ operations.}$$

In addition to these row modifications it is necessary to obtain the multipliers but the overall operation count is still $0(n^2)$, as opposed to $0(n^3)$ for a full matrix. This is the advantage of using the Hessenberg form.

The initial reduction of A to Hessenberg form would not be effective if the QR method itself did not preserve the Hessenberg form. But it does. Let $Q^T = Q_{n-1} \cdots Q_1$ be the product of the Givens transformations so that $Q = Q_1^T \cdots Q_{n-1}^T$. Each Q_i has off-diagonal elements only in the $(i+1, i)$ and $(i, i+1)$ positions and hence Q itself is a Hessenberg matrix (Exercise 7.2.2). Then since R is upper triangular, the product RQ is a Hessenberg matrix and can be formed in $0(n^2)$ operations (Exercise 7.2.3). Thus all of the matrices generated by the QR method retain the Hessenberg form and each complete QR step requires $0(n^2)$ operations.

Shifting

Even with the initial reduction of the matrix to Hessenberg form, the QR method is still inefficient due to the possibly slow rate of convergence to zero of the subdiagonal elements. This rate of convergence is indicated by (7.2.6), which shows that if two eigenvalues, say λ_i and λ_{i+1}, are very close in absolute value, then the off-diagonal element in position $(i+1, i)$ will converge to zero very slowly.

We will attempt to mitigate this convergence problem by *shifting* the eigenvalues of H. Suppose that $\hat{\lambda}_n$ is a good approximation to the smallest eigenvalue, λ_n (assumed real), and consider the matrix $\hat{H} = H - \hat{\lambda}_n I$, which has eigenvalues $\lambda_1 - \hat{\lambda}_n, \ldots, \lambda_n - \hat{\lambda}_n$. If we apply the QR method to $\hat{H}$, then the off-diagonal element in the last row of the matrices H_k will converge to zero as powers of the quotient $(\lambda_n - \hat{\lambda}_n)/(\lambda_{n-1} - \hat{\lambda}_n)$, as opposed to the quotient λ_n/λ_{n-1}. For example, suppose that $\lambda_n = 0.99$, $\lambda_{n-1} = 1.1$, and $\hat{\lambda}_n = 1.0$. Then, $\lambda_n/\lambda_{n-1} = 0.9$ while $|\lambda_n - \hat{\lambda}_n|/|\lambda_{n-1} - \hat{\lambda}_n| = 0.1$, so that the convergence of the $(n, n-1)$ element is approximately 20 times faster for the matrix $\hat{H}$.

Of course, we usually will not know a good approximation $\hat{\lambda}_n$ to use as the shift parameter. However, as the QR process proceeds, if the (n, n) elements $h_{nn}^{(k)}$ of the matrices H_k are converging to the eigenvalue λ_n, we can use them as the shift parameters; that is, at the kth stage do the next QR step on the matrix $\hat{H}_k = H_k - h_{n,n}^{(k)} I$. Then we continue using the (n, n) element of the current matrix to make a shift at each stage. Each shift changes the eigenvalues of the original matrix by the amount of the shift, so we need to keep track of the accumulation of shifts that are made; indeed, it is this accumulation that converges to the eigenvalue λ_n. The convergence is signaled by the off-diagonal element in the last row becoming sufficiently small. When this occurs the last row and column of the matrix may be dropped, and to determine the eigenvalue λ_{n-1} we proceed with the resulting $(n-1) \times (n-1)$ submatrix. Note that the eigenvalues of this submatrix, and hence of the original matrix, have been changed by the total accumulation of shifts (which is the approximation to λ_n), and this must be added back to the other computed eigenvalues at the end of the computation. Alternatively, the shifts may be added back in at each stage, as illustrated by (7.2.13) in a different context, so that all of the matrices H_k retain the same eigenvalues.

The preceding discussion has been predicated on the assumption that the smallest eigenvalue, λ_n, is real. Now suppose that λ_n is complex. Then shifting by $h_{nn}^{(k)}$, which remains real, is not a particularly good strategy since the imaginary part of the eigenvalue cannot be approximated. Instead, as was discussed earlier, the eigenvalues of the lower right 2×2 submatrices of the matrices H_k produced by the unshifted QR algorithm will converge to the eigenvalue pair $\lambda_n, \lambda_{n-1} = \bar{\lambda}_n$. Hence we use the eigenvalues of these 2×2 submatrices as shift parameters. Consider the first step applied to the matrix H_1 and let $k_1, k_2 = \bar{k}_1$ be the eigenvalues of the lower right 2×2 submatrix. If we add back the shifts, we obtain

$$\begin{aligned} H_1 - k_1 I &= Q_1 R_1, \qquad & H_2 &= R_1 Q_1 + k_1 I, \\ H_2 - k_2 I &= Q_2 R_2, \qquad & H_3 &= R_2 Q_2 + k_2 I. \end{aligned} \tag{7.2.13}$$

If k_1 and k_2 are complex, the matrices H_1, H_2, Q_1, Q_2, R_1, and R_2 will generally be complex, and consequently the QR steps need to be carried out in complex arithmetic. However, an interesting fact is that H_3 is real (Exercise 7.2.5). Indeed, it is possible to carry out the transformation from H_1 to H_3 entirely in real arithmetic, although we will not go into the details of this here. This procedure is called the *double-shift QR method.* Even if the eigenvalues are real, it is a good strategy to shift twice using the eigenvalues of the lower right 2×2 submatrix. With this choice of shifts, as with shifting by the (n, n) element, the rate of convergence is usually at least quadratic.

There is another possibility that enhances the speed of the QR method. Suppose that the subdiagonal element $a_{i+1,i}$ of the Hessenberg matrix H is zero. Then H can be written in block form

$$H = \begin{bmatrix} H_1 & * \\ 0 & H_2 \end{bmatrix}, \tag{7.2.14}$$

and the eigenvalues of H are those of the matrices H_1 and H_2 (Exercise 7.2.6). Thus the QR method can be applied to these smaller matrices, which reduces the operation count. This observation can also be applied during the QR method: if it should happen that the elements in position $(i + 1, i)$ converge to zero more rapidly than other off-diagonal elements, then the problem can be decomposed into two smaller problems.

Householder's Method

We now return to the important special case in which A is symmetric. In this case the Hessenberg matrix is tridiagonal (Theorem 7.2.2) and the reduction of the original matrix by Householder transformations is known as *Householder's method.* The QR method can again be used to compute the eigenvalues of the tridiagonal matrix T (see the Supplementary Discussion for alternative methods). In this case there are two simplifications. Since T is symmetric, its eigenvalues are necessarily real and therefore there is no need to be concerned with complex shifts or convergence of 2×2 submatrices. Secondly, the QR steps are very rapid; each requires only $0(n)$ operations and the tridiagonal form is preserved (Exercise 7.2.4).

Computation of Eigenvectors

We next discuss the computation of eigenvectors, if they are desired. Assume that an approximate eigenvalue has been computed. Then there are two steps to obtain the corresponding approximate eigenvector. First, compute the approximate eigenvector of the Hessenberg (or tridiagonal) matrix. We postpone the discussion of this until the following section since it is a special case of methods to be given there. Second, transform this eigenvector back to an eigenvector of the original matrix A. We now consider this second step.

Let $\mathbf{y}$ be an eigenvector of H corresponding to the eigenvalue λ. Let $H = PAP^T$, where $P = P_{n-1} \cdots P_2$ is the product of the Householder transformations $P_i = I - 2\mathbf{w}_i\mathbf{w}_i^T$ used to obtain the Hessenberg form. Then $\mathbf{x} = P^T\mathbf{y}$ is the corresponding eigenvector of A since

$$A\mathbf{x} = AP^T\mathbf{y} = P^TPAP^T\mathbf{y} = P^TH\mathbf{y} = \lambda P^T\mathbf{y} = \lambda\mathbf{x}.$$

If $\mathbf{y}$ is only an approximate eigenvector of H, we still use the same transformation, $P^T\mathbf{y}$, to obtain an approximate eigenvector of A. Thus

$$\mathbf{x} = P^T\mathbf{y} = (P_{n-1} \cdots P_2)^T\mathbf{y} = P_2^T \cdots P_{n-1}^T\mathbf{y}.$$

This is very easy to carry out. The first step is

$$P_{n-1}^T\mathbf{y} = (I - 2\mathbf{w}_{n-1}\mathbf{w}_{n-1}^T)\mathbf{y} = \mathbf{y} - 2(\mathbf{w}_{n-1}^T\mathbf{y})\mathbf{w}_{n-1}.$$

Then $P_{n-2}^T = I - 2\mathbf{w}_{n-2}\mathbf{w}_{n-2}^T$ is applied to this vector, and so on. Note that we need to retain the vectors $\mathbf{w}_i$ that were used to produce the Hessenberg matrix H from A. The non-zero components of the $\mathbf{w}_i$ can be stored in the corresponding subdiagonal positions of A that are set to zero, if desired.

We now summarize briefly the main points of this section. For the QR method to be efficient we must first reduce the original matrix A to Hessenberg form (tridiagonal if A is symmetric), and then incorporate shifts into the basic QR algorithm applied to this Hessenberg matrix. As the iteration proceeds, the eigenvalues are obtained one by one (or two at a time in the case of a complex conjugate pair), the matrix is reduced in size, and the iteration proceeds toward the remaining eigenvalues. Properly implemented, the QR algorithm is the best general-purpose method for nonsymmetric matrices. We have not been able to give all of the details necessary for such an implementation and have tried only to present the basic flavor of the method. The Supplementary Discussion gives references for further reading.

Supplementary Discussion and References: 7.2

It is tempting to try to find an orthogonal matrix so that P^TAP is diagonal if A is symmetric. Unfortunately, this cannot be done with a finite number of operations, except in trivial cases, but there is a classical algorithm that attempts to find P as a limit of a sequence of products $P_1 \cdots P_k$. This is *Jacobi's method*, in which each P_i is a Givens matrix and, in the simplest case, the elements of A are zeroed in the order $(2,1), (3,1), \ldots, (n,1), (3,2), (4,2), \ldots$. Ideally, after all subdiagonal elements have been zeroed we would have a diagonal matrix but non-zero elements generally will appear in positions that had previously been zeroed. The process is then repeated and, under mild assumptions, the matrices $A_k = P_k^T \cdots P_1^T A P_1 \cdots P_k$ converge as $k \to \infty$ to a diagonal matrix containing the eigenvalues of A, and $P_1 \cdots P_k$ converges to a matrix P whose columns are the eigenvectors. Although Jacobi's method is slow relative to the methods discussed in this section, it has been enjoying a recent revival due to its good properties on parallel computers (see, e.g., Golub and Van Loan [1989]).

The idea of reducing the original symmetric matrix to tridiagonal form, rather than attempting to obtain a diagonal matrix as in Jacobi's method, is due to J. W. Givens in the early 1950's. He used the plane rotation matrices now associated with his name. Shortly thereafter A. Householder noted that the reduction could be done more efficiently using elementary reflection matrices, now called Householder matrices.

The QR method was introduced independently by Francis [1961, 1962] and Kublanovskaya [1961]. It was preceeded by the corresponding algorithm based on the LU decomposition and called the LR algorithm by H. Rutishauser in 1958. This method proceeds as in (7.2.3), but the QR factorizations are replaced by LR (i.e. LU) factorizations. Although the LU factorizations are faster (see Section 4.5), the QR method in general enjoys better numerical stability properties and has been the method of choice.

Excellent codes for the QR method, and the special case of Householder's method for symmetric matrices, are contained in EISPACK (Garbow et al. [1979]), which is now being transformed to the new LAPACK package (Dongarra and Anderson et al. [1990]).

J. Wilkinson contributed immensely to the understanding and extension of all of the methods of this section. A wealth of material, including the proof of Theorem 7.2.1, detailed rounding error analyses, and further discussions of the practicalities of different methods may be found in his classic book (Wilkinson [1965]). See also Householder [1964] for a more mathematical treatment of some of the topics of this chapter, Parlett [1980] for results pertaining primarily to symmetric matrices, and Stewart [1973]. In particular, this latter book gives an implicitly shifted version of the QR method as well as relationships between the QR method and the power, inverse power, and Rayleigh quotient methods to be discussed in the next section. For a more recent review of all of the methods in this chapter, see Golub and Van Loan [1989].

In Section 7.1 it was mentioned that the generalized eigenvalue problem $A\mathbf{x} = \lambda B\mathbf{x}$ can be converted to a standard problem $B^{-1}A\mathbf{x} = \lambda\mathbf{x}$ if B is nonsingular. An alternative is the QZ algorithm (Moler and Stewart [1973]; see also Golub and Van Loan [1989]), which is an extension of the QR algorithm to the generalized eigenvalue problem.

Although the QR algorithm is probably the best method, in general, for computing the eigenvalues of a symmetric tridiagonal matrix T, there are two attractive alternatives, each of which is sometimes very useful. The first is based on a *Sturm sequence*, which for a tridiagonal matrix with diagonal elements a_i and off-diagonal elements b_i (assumed to be non-zero), is a sequence of polynomials defined by

$$p_i(\lambda) = (a_i - \lambda)p_{i-1}(\lambda) - b_{i-1}^2 p_{i-2}(\lambda), \qquad i = 2, \ldots, n, \tag{7.2.15}$$

with $p_0(\lambda) \equiv 1$ and $p_1(\lambda) = a_1 - \lambda$. The polynomial p_k is the characteristic polynomial of the $k \times k$ leading principal submatrix of T. In particular, p_n is the characteristic polynomial of T and its roots are the eigenvalues of T. These polynomials have the remarkable property that the number of agreements in sign between consecutive terms in the sequence $1, p_1(\hat{\lambda}), \ldots, p_n(\hat{\lambda})$ is equal to the number of roots of p_n greater than or equal to $\hat{\lambda}$. (See Exercise 7.2.9 for a related result.) This property then allows a bisection type algorithm. In particular, by using two test points $\hat{\lambda}_1$ and $\hat{\lambda}_2$ it is possible to know the number of the roots in the interval $[\hat{\lambda}_1, \hat{\lambda}_2]$, which can be very useful. For further information see Parlett [1980], which also discusses the *spectrum splicing* method (which is essentially equivalent for tridiagonal matrices to Sturm sequences). This is based on an LDL^T decomposition of $T - \hat{\lambda}I$ and application of the Inertia Theorem to ascertain the number of eigenvalues of T greater than $\hat{\lambda}$ (for the Inertia Theorem see, e.g., Ortega [1987]).

The second alternative for symmetric tridiagonal matrices is an iterative method for the characteristic polynomial $p_n(\lambda)$ of T. Consider Newton's method (Section 5.2). We can differentiate the sequence (7.2.15) to obtain the corresponding sequence for the derivatives

$$p_i'(\lambda) = -p_{i-1}(\lambda) + (a_i - \lambda)p_{i-1}'(\lambda) - b_{i-1}^2 p_{i-2}'(\lambda), \qquad i = 2, \ldots, n, \tag{7.2.16}$$

where $p_0' = 0$ and $p_1' = -1$. The two sequences (7.2.15) and (7.2.16) can be evaluated together to yield $p_n(\lambda)$ and $p_n'(\lambda)$ to use in Newton's method. This can be combined with the Sturm sequence property to ascertain an interval in which a root is known to lie before applying Newton's method to achieve rapid convergence to the root. See, for example, Wilkinson [1965] for further details. Other root-finding techniques could also be used in place of Newton's method.

EXERCISES 7.2

7.2.1. If A is symmetric, show that PAP^T is also symmetric for any matrix P. Apply this, in particular, to (7.2.12) to conclude that the Hessenberg matrix H is tridiagonal if A is symmetric.

7.2.2. Show that if Q_i is a diagonal matrix except for non-zero elements in the $(i+1, i)$ and $(i, i+1)$ positions, then the product $Q_1 \cdots Q_{n-1}$ is a Hessenberg matrix.

7.2.3. Show that if Q is a Hessenberg matrix and R is upper-triangular, then the product RQ is a Hessenberg matrix. Show that this multiplication requires $0(n^2)$ operations.

7.2.4. Let A be a symmetric banded matrix. Show that the QR method (7.2.3) preserves the bandwidth of A. What is the operation count for one step? Specialize this to show that if the QR method is applied to a symmetric tridiagonal matrix, the tridiagonal form is preserved and each QR step requires $0(n)$ operations. What happens for nonsymmetric banded matrices?

7.2.5. Let H_1 be real and $k_1, k_2 = \bar{k}_1$ be complex scalars. Show that the matrix H_3 defined by (7.2.13) is real.

7.2.6. Suppose that the matrix A has the block form

$$A = \begin{bmatrix} A_1 & A_3 \\ 0 & A_2 \end{bmatrix}.$$

Show that $\det(A - \lambda I) = \det(A_1 - \lambda I)\det(A_2 - \lambda I)$ so that the eigenvalues of A are those of A_1 and A_2.

7.2.7. Let A be skew-symmetric ($A^T = -A$). Show that if H is the Hessenberg matrix obtained by Householder reduction, then H is tridiagonal and skew-symmetric. Note that a skew-symmetric matrix has zero main diagonal elements. Can you use this to simplify the QR algorithm?

7.2.8. Let

$$A = \begin{pmatrix} a_1 & & & b_1 \\ & \ddots & & \vdots \\ & & & b_{n-1} \\ c_1 & \cdots & c_{n-1} & a_n \end{pmatrix}.$$

Show that if $b_i c_i > 0$, then there exists a diagonal matrix D so that DAD^{-1} is symmetric.

7.2.9. Let $p_k(\lambda)$ be the characteristic polynomial of the leading principal $k \times k$ submatrix of A. Suppose that A is symmetric with eigenvalues $\lambda_1 \leq \cdots \leq \lambda_n$. Show that $p_k(\lambda) > 0$, $k = 1, \ldots, n$ if $\lambda < \lambda_1$, and that the $p_k(\lambda)$ alternate in sign if $\lambda > \lambda_n$.

7.2.10. Let A and B be symmetric $n \times n$ matrices with B positive definite. Show how to find a matrix S so that $A = STS^T$ and $B = SS^T$, where T is a tridiagonal matrix whose eigenvalues are the same as those of $B^{-1}A$.

7.2.11. Let A be a real $n \times n$ matrix. Suppose that we wish to solve the linear systems $(A + \mu I)\mathbf{x} = \mathbf{b}$ for several values of the real parameter μ. How may we use the decomposition $A = PHP^T$, where P is orthogonal and H is upper Hessenberg? How many operations does your algorithm take?

7.3 Other Iterative Methods

The Householder and QR methods of the previous section are of primary value when the matrix A is not particularly sparse, and all or a large number of the eigenvalues are desired. Conversely, they are not very useful for very large sparse matrices for which only a few eigenvalues are desired. Problems such as this arise in partial differential equations, discussed in Chapters 8 and 9, as well as in other areas. A typical problem of this type might involve a $5,000 \times 5,000$ matrix with only ten or fewer nonzero elements in each row, and for which only a few eigenvalues, perhaps four or five, are desired. For such a problem the QR method is unsatisfactory because the QR factorization may change zero elements of A into non-zero elements as the factorization proceeds. (See Section 9.2 for further discussion of "fill-in.") The purpose of the present section is to describe some alternative methods.

The Power Method

A classical method that has a certain usefulness – but also serious defects – for large sparse problems is the *power method.* Let A have eigenvalues $\lambda_1, \ldots, \lambda_n$, which we assume for the moment are real and satisfy

$$|\lambda_1| > |\lambda_2| \geq \cdots \geq |\lambda_n|. \tag{7.3.1}$$

For a given vector $\mathbf{x}^0$, consider the sequence of vectors generated by

$$\mathbf{x}^{k+1} = A\mathbf{x}^k, \qquad k = 0, 1, \ldots . \tag{7.3.2}$$

To analyze this sequence, assume that A has n linearly independent eigenvectors $\mathbf{v}_1, \ldots, \mathbf{v}_n$ corresponding to the eigenvalues $\lambda_1, \ldots, \lambda_n$, and expand $\mathbf{x}^0$ in terms of these eigenvectors:

$$\mathbf{x}^0 = c_1\mathbf{v}_1 + \cdots + c_n\mathbf{v}_n. \tag{7.3.3}$$

Then, since $\mathbf{x}^k = A^k\mathbf{x}^0$ and $A^k\mathbf{v}_i = \lambda_i^k\mathbf{v}_i$,

$$\begin{aligned} \mathbf{x}^k &= c_1\lambda_1^k\mathbf{v}_1 + c_2\lambda_2^k\mathbf{v}_2 + \cdots + c_n\lambda_n^k\mathbf{v}_n \qquad (7.3.4) \\ &= \lambda_1^k\left[c_1\mathbf{v}_1 + c_2\left(\frac{\lambda_2}{\lambda_1}\right)^k\mathbf{v}_2 + \cdots + c_n\left(\frac{\lambda_n}{\lambda_1}\right)^k\mathbf{v}_n\right]. \end{aligned}$$

Because of (7.3.1) the terms $(\lambda_i/\lambda_1)^k$, $i = 2, \ldots, n$ all tend to zero as k goes to infinity. Therefore, if $c_1 \neq 0$,

$$\lambda_1^{-k}\mathbf{x}^k \to c_1\mathbf{v}_1, \qquad \text{as } k \to \infty, \tag{7.3.5}$$

which shows that the vectors $\mathbf{x}^k$ tend to the direction of the eigenvector $\mathbf{v}_1$. The magnitude of the vectors $\mathbf{x}^k$, however, will tend to zero if $|\lambda_1| < 1$, or become unbounded if $|\lambda_1| > 1$. Therefore scaling of the vectors $\mathbf{x}^k$ is required, and the scaling process will also give approximations to the eigenvalue λ_1.

One way to choose the scaling factors is based on the observation that as $\mathbf{x}^k$ approaches the direction $\mathbf{v}_1$, then $A\mathbf{x}^k \doteq \lambda_1\mathbf{x}^k$. Hence ratios of the components of $\mathbf{x}^k$ and $A\mathbf{x}^k$ are approximations to λ_1. To avoid choosing components that are too small, let x_i^k be a component of maximum absolute value of $\mathbf{x}^k$ and define

$$\hat{\mathbf{x}}^{k+1} = A\mathbf{x}^k, \qquad \gamma_k = \frac{\hat{x}_i^{k+1}}{x_i^k}, \qquad \mathbf{x}^{k+1} = \frac{\hat{\mathbf{x}}^{k+1}}{\gamma_k}. \tag{7.3.6}$$

Then γ_k is an approximation to λ_1 and scaling $\hat{\mathbf{x}}^{k+1}$ by γ_k prevents the $\mathbf{x}^k$ from going to zero or infinity. In fact, it can be shown that

$$\gamma_k \to \lambda_1 \text{ and } \mathbf{x}^k \to c\mathbf{v}_1 \text{ as } k \to \infty, \tag{7.3.7}$$

where the last relation says that $\mathbf{x}^k$ tends to some multiple of the eigenvector $\mathbf{v}_1$.

There is a relationship between the power method and the QR method of the previous section (see Exercise 7.3.1). However, an advantage of the power method is that the vectors $\mathbf{x}^k$ can be generated by only matrix-vector multiplications (plus the work needed to compute the scaling factors); operations on the matrix A itself are unnecessary. The main disadvantage is the possibly slow rate of convergence, which, as shown by (7.3.4), is determined primarily by the ratio λ_2/λ_1. If this ratio is close to 1, as is typical for many problems, the convergence will be slow. One way to attempt to mitigate this problem is to use shifts as was done with the QR algorithm. If the power method is applied to the matrix $A - \sigma I$, whose eigenvalues are $\lambda_1 - \sigma, \ldots, \lambda_n - \sigma$ (Exercise 7.1.6), then the rate of convergence will be determined by the ratio $|\lambda_2 - \sigma|/|\lambda_1 - \sigma|$, provided that $\lambda_1 - \sigma$ remains the dominant eigenvalue. But even with this shift the convergence may still be painfully slow. For example, suppose that a $1,000 \times 1,000$ matrix has the eigenvalues 1,000, 999, $\ldots, 1$. Then, after a shift by $\sigma = 500$ the ratio is 0.998, which is barely better than the unshifted ratio of 0.999.

The power method also has other disadvantages. If there is more than one dominant eigenvalue, for example, $|\lambda_1| = |\lambda_2| > |\lambda_3|$, which would be the case for a real matrix with a dominant complex conjugate pair of roots, the sequence (7.3.6) may not converge. There are ways to circumvent this difficulty, but in the case of complex roots acceleration of the convergence by shifts is even less satisfactory. Another problem concerns computing the subdominant eigenvalues. Once we have approximated λ_1, we need to remove it in some fashion from the matrix or subsequent iterations will again converge to λ_1 rather than λ_2. We next show how to accomplish this by a process known as *deflation*. Deflation will also be useful in other methods to be discussed shortly.

Deflation

Assume that A is symmetric so that its eigenvalues λ_i are real and, by Theorem 7.1.3, the associated eigenvectors $\mathbf{v}_i$ can be assumed to be orthonormal. Again, let $\mathbf{x}^0$ be given by (7.3.3). Then by the orthonormality of the $\mathbf{v}_i$, we have $\mathbf{v}_1^T\mathbf{x}^0 = c_1$. Thus the vector

$$\hat{\mathbf{x}}^0 = \mathbf{x}^0 - (\mathbf{v}_1^T\mathbf{x}^0)\mathbf{v}_1 = c_2\mathbf{v}_2 + \ldots + c_n\mathbf{v}_n \tag{7.3.8}$$

is a linear combination of only $\mathbf{v}_2, \ldots, \mathbf{v}_n$, and the same will be true for the sequence (7.3.2) starting from $\hat{\mathbf{x}}^0$. If $|\lambda_2| > |\lambda_i|$, $i \geq 3$, then the power method will produce iterates converging to λ_2 and a multiple of $\mathbf{v}_2$. This idea extends to any number of eigenvectors (Exercise 7.3.2).

The above deflation procedure allows us, in principle, to remove the effect of λ_1 and $\mathbf{v}_1$ from the subsequent calculation of the remaining eigenvalues and eigenvectors. In practice, however, we will not know $\mathbf{v}_1$ exactly so that the vector $\hat{\mathbf{x}}^0$, formed with an approximation to $\mathbf{v}_1$, will still have a component in the direction $\mathbf{v}_1$, and the power method will still give convergence to λ_1 rather than λ_2. Even if $\mathbf{v}_1$ were known exactly, rounding error in the formation of $\hat{\mathbf{x}}^0$ and the power method computations would have the same effect. Therefore it is necessary to apply (7.3.8) periodically to the current iterates in order to keep the effect of $\mathbf{v}_1$ suppressed. That is, if $\mathbf{x}^k$ is the current power method iterate and $\hat{\mathbf{v}}_1$ our approximation to $\mathbf{v}_1$, we would form

$$\hat{\mathbf{x}}^k = \mathbf{x}^k - (\hat{\mathbf{v}}_1^T\mathbf{x}_k)\hat{\mathbf{v}}_1,$$

and then continue the iteration with $\hat{\mathbf{x}}^k$. This would be done only occasionally. Another way to carry out a deflation process is given in Exercise 7.3.3.

Inverse Iteration and Computation of Eigenvectors

We next consider a variation of the power method, called *inverse iteration* or the *inverse power method*, whose rate of convergence is potentially much faster than that of the power method. Consider the sequence $\{\mathbf{x}^k\}$ defined by

$$(A - \sigma I)\mathbf{x}^k = \mathbf{x}^{k-1}, \qquad k = 1, 2, \ldots, \tag{7.3.9}$$

for some parameter σ; that is, $\mathbf{x}^k$ is the solution of the linear system (7.3.9). This is the power method for the matrix $(A-\sigma I)^{-1}$. If A again has eigenvalues $\lambda_1, \ldots, \lambda_n$ and corresponding eigenvectors $\mathbf{v}_1, \ldots, \mathbf{v}_n$, then $(A - \sigma I)^{-1}$ has eigenvalues $(\lambda_i - \sigma)^{-1}$ and eigenvectors $\mathbf{v}_i$, and the sequence $\{\mathbf{x}^k\}$ of (7.3.9) obeys the relationship (7.3.4) with the λ_i replaced by $(\lambda_i - \sigma)^{-1}$:

$$\mathbf{x}^k = \frac{c_1}{(\lambda_1 - \sigma)^k}\mathbf{v}_1 + \ldots + \frac{c_n}{(\lambda_n - \sigma)^k}\mathbf{v}_n. \tag{7.3.10}$$

We will return to (7.3.9) shortly as the basis for a method of computing both the eigenvalues and eigenvectors of A, but we first note that inverse iteration is the standard way to compute the eigenvectors of a matrix once the eigenvalues have already been computed by, for example, the QR method. Suppose that σ is an approximation to λ_j and rewrite (7.3.10) for $k = 1$ as

$$\mathbf{x}^1 = \frac{c_j}{\lambda_j - \sigma}\mathbf{v}_j + \sum_{i \neq j} \frac{c_i}{\lambda_i - \sigma}\mathbf{v}_i. \tag{7.3.11}$$

Now suppose that $|\lambda_j - \sigma|$ is small (say, $0(10^{-6})$), λ_j is not particularly close to another eigenvalue λ_i, and c_j is not small. Then the dominant term in (7.3.11) will be $c_j(\lambda_j - \sigma)^{-1}\mathbf{v}_j$. Only the direction of $\mathbf{v}_j$ needs to be computed since we can scale this to any desired length. Thus the effect of solving the system $(A - \sigma I)\mathbf{x}^1 = \mathbf{x}^0$ is to approximate the direction of the desired eigenvector. Note that the better σ approximates λ_j, the closer to singular is the matrix $A - \sigma I$. This ill-conditioning of $A - \sigma I$ is not deleterious in this case, however, since any error in solving the system will be primarily in the direction $\mathbf{v}_j$ that we are approximating.

Two factors will affect the accuracy of this approximation to the eigenvector. First, if λ_j is very close to another eigenvalue, say λ_{j+1}, then $\lambda_{j+1} - \sigma$ will also be small, and the first term of (7.3.11) will no longer be dominant; we will then approximate some linear combination of $\mathbf{v}_j$ and $\mathbf{v}_{j+1}$. Closeness of the eigenvalues is an intrinsic property of the matrix and hampers any numerical method in the calculation of the eigenvectors. The second factor is the possibility that c_j is very small, and if this is the case, then again the first term of (7.3.11) may not be sufficiently dominant to give a good approximation to the desired eigenvector. In principle, we can insure that this will not happen by choosing the vector $\mathbf{x}^0$ so that c_j is not small. However, we can do that with certainty only if the eigenvectors are known, and of course that is not the case. It has been found that choosing $\mathbf{x}^0$ to be the vector with components all equal to 1 usually works very well. A similar strategy that sometimes works even better is to do Gaussian elimination on $A - \sigma I$ to produce the upper-triangular matrix U, and then solve the system $U\mathbf{y} = \mathbf{z}$ where $\mathbf{z}$ is a vector all of whose components are equal to 1. In this case the vector $\mathbf{x}^0$ is not specified explicitly: it is the vector that would give rise to a vector of all 1's under the Gaussian elimination calculation. Obviously, there is a great deal of flexibility in choosing the vector $\mathbf{x}^0$. Indeed, any "randomly" chosen vector would be very unlikely to yield a particularly small c_j. We note that one of the worst possible strategies would be to attempt to solve the homogeneous system $(A - \sigma I)\mathbf{x}^1 = 0$, which would be the mathematical definition of the eigenvector if $\sigma = \lambda_j$.

It is usually worthwhile to do another iteration using the approximate eigenvector $\mathbf{x}^1$ just computed. Even if the original choice of $\mathbf{x}^0$ is such that c_j is very small, $\mathbf{x}^1$ will have a $\hat{c}_j$ that is larger, and another iteration may then give a suitable approximation. This could be repeated as many times as desired, but one extra iteration is generally sufficient.

In the context of the QR method of the previous section, the above inverse iteration procedure would be applied to the Hessenberg matrix H (or the tridiagonal matrix T in case A is symmetric). Once the desired eigenvectors of H have been computed, they are transformed back to eigenvectors of the original matrix as discussed in Section 7.2. Note that the eigenvalues, and consequently the eigenvectors, may be complex. This does not affect the inverse iteration procedure except that complex arithmetic must be performed. However, a side benefit of complex eigenvalues of a real matrix is that the eigenvectors occur in complex conjugate pairs, as do the eigenvalues, so that if $\mathbf{u} + i\mathbf{v}$ is an eigenvector for $a + ib$, then $\mathbf{u} - i\mathbf{v}$ is an eigenvector for $a - ib$, and no further computation is needed for this second eigenvector (Exercise 7.3.9).

Computation of Eigenvalues

As we have seen, each step of the inverse iteration (7.3.4) can greatly improve an approximation to an eigenvector if σ is a good approximation to a corresponding eigenvalue. However, there remains the problem of approximating the eigenvalue itself for matrices that are not suitable for the QR method. Since (7.3.9) is the power method for $(A - \sigma I)^{-1}$, we can proceed as in (7.3.6):

$$(A - \sigma I)\hat{\mathbf{x}}^{k+1} = \mathbf{x}^k, \qquad k = 0, 1, \ldots \tag{7.3.12a}$$

$$\mathbf{x}^{k+1} = \frac{\hat{\mathbf{x}}^{k+1}}{\gamma_k}, \qquad k = 0, 1, \ldots \tag{7.3.12b}$$

where γ_k is defined as in (7.3.6). Then

$$\gamma_k \to \gamma = (\sigma - \lambda_j)^{-1}, \qquad \mathbf{x}^k \to c_j \mathbf{v}_j, \qquad \text{as } k \to \infty, \tag{7.3.13}$$

provided that

$$|(\sigma - \lambda_j)^{-1}| > |(\sigma - \lambda_i)^{-1}|, \qquad i \neq j. \tag{7.3.14}$$

The eigenvalue λ_j is then given by

$$\lambda_j = \sigma - \frac{1}{\gamma}. \tag{7.3.15}$$

If $|(\sigma - \lambda_i)^{-1}| = \max\{|(\sigma - \lambda_m)^{-1}|;\ m \neq j\}$, then the rate of convergence is governed by the ratio

$$\frac{|(\sigma - \lambda_i)^{-1}|}{|(\sigma - \lambda_j)^{-1}|} = \frac{|\sigma - \lambda_j|}{|\sigma - \lambda_i|}. \tag{7.3.16}$$

The closer σ is to λ_j, the smaller this ratio. Therefore it is reasonable to replace a fixed σ by estimates of λ_j depending on γ_k. From (7.3.13),

$$\sigma_k = \sigma - \frac{1}{\gamma_k} \to \lambda_j, \qquad \text{as } k \to \infty, \tag{7.3.17}$$

and thus we modify (7.3.12a) to

$$(A - \sigma_k I)\hat{\mathbf{x}}^{k+1} = \mathbf{x}^k, \qquad k = 0, 1, \ldots, \tag{7.3.18}$$

where σ_k is given by (7.3.17). The value of σ in (7.3.17) would be our best estimate of the eigenvalue we wish to approximate. For example, if we want the smallest eigenvalue in absolute value we might choose $\sigma = 0$, whereas if we want the largest we could choose $\sigma = ||A||$ for some norm. Once an eigenvalue has been approximated, we could use the same deflation procedure discussed for the power method to minimize the effect of that eigenvalue on further computations. However, the use of the shifts σ_k allow us to circumvent the need for deflation if we have good estimates for the eigenvalues to be computed. Of course, if we are solving (7.3.18) by LU factorization, each time we change σ_k we must refactor $A - \sigma_k I$.

The Rayleigh Quotient Method

We next describe another way to choose the shift parameters σ_k in the case that A is a symmetric matrix. For a given vector $\mathbf{v} \neq 0$, the *Rayleigh quotient* is the quantity

$$\sigma(\mathbf{v}) = \frac{\mathbf{v}^T A \mathbf{v}}{\mathbf{v}^T \mathbf{v}}. \tag{7.3.19}$$

The Rayleigh quotient has two basic properties (Exercise 7.4.5): If $\lambda_1 \leq \ldots \leq \lambda_n$ and $\mathbf{v}_1, \ldots, \mathbf{v}_n$ are the eigenvalues and corresponding orthonormal eigenvectors of A, then for any $\mathbf{v}$,

$$\lambda_1 \leq \sigma(\mathbf{v}) \leq \lambda_n \tag{7.3.20}$$

and if $\mathbf{v} = \mathbf{v}_i$, then

$$\sigma(\mathbf{v}) = \lambda_i. \tag{7.3.21}$$

Another basic property of the Rayleigh quotient is that if $\mathbf{v}$ is a good approximation to an eigenvector, then the Rayleigh quotient is a much better approximation to the corresponding eigenvalue. We make this precise in the following theorem.

THEOREM 7.3.1 *Let* $\mathbf{v} = \gamma\mathbf{v}_j + \mathbf{w}$, *where*

$$\mathbf{w} = \sum_{i \neq j} c_i \mathbf{v}_i.$$

If $\gamma = 1 + 0(\varepsilon)$ *and* $||\mathbf{w}||_2 = 0(\varepsilon)$, *then*

$$|\sigma(\mathbf{v}) - \lambda_j| = 0(\varepsilon^2). \tag{7.3.22}$$

PROOF: By assumption, $\mathbf{v}_j$ is orthogonal to $\mathbf{w}$ so that

$$\mathbf{v}^T\mathbf{v} = (\gamma\mathbf{v}_j + \mathbf{w})^T(\gamma\mathbf{v}_j + \mathbf{w}) = \gamma^2 + \mathbf{w}^T\mathbf{w}.$$

Since

$$A(\gamma\mathbf{v}_j + \mathbf{w}) = \gamma\lambda_j\mathbf{v}_j + \sum_{i\neq j}\lambda_i c_i\mathbf{v}_i = \gamma\lambda_j\mathbf{v}_j + \hat{\mathbf{w}},$$

$\hat{\mathbf{w}}$ is also orthogonal to $\mathbf{v}_j$. Thus

$$\mathbf{v}^TA\mathbf{v} = (\gamma\mathbf{v}_j + \mathbf{w})^T(\gamma\lambda_j\mathbf{v}_j + \hat{\mathbf{w}}) = \gamma^2\lambda_j + \mathbf{w}^T\hat{\mathbf{w}},$$

so that

$$\sigma(\mathbf{v}) = \frac{\gamma^2\lambda_j + \mathbf{w}^T\hat{\mathbf{w}}}{\gamma^2 + \mathbf{w}^T\mathbf{w}} = \frac{\lambda_j + \gamma^{-2}\mathbf{w}^T\hat{\mathbf{w}}}{1 + \gamma^{-2}\mathbf{w}^T\mathbf{w}}. \tag{7.3.23}$$

If $\lambda = \max\{|\lambda_i| : i \neq j\}$, then

$$\mathbf{w}^T\hat{\mathbf{w}} = \sum_{i\neq j}\lambda_i c_i^2 \leq \lambda^2\sum_{i\neq j}c_i^2 = \lambda^2 0(\varepsilon^2) = 0(\varepsilon^2),$$

since λ is fixed. By the estimates of Exercise 7.3.6, we then conclude from (7.3.23) that

$$\sigma(\mathbf{v}) = \frac{\lambda_j + [1 + 0(\varepsilon)]^2 0(\varepsilon^2)}{1 + [1 + 0(\varepsilon)]^2 0(\varepsilon^2)} = \lambda_j + 0(\varepsilon^2), \tag{7.3.24}$$

which was to be proved.

As a consequence of Theorem 7.3.1, if an approximate eigenvector $\mathbf{v}$ is known to m digits of accuracy, then $\sigma(\mathbf{v})$ is an eigenvalue to approximately $2m$ digits of accuracy. This forms the basis of a combined Rayleigh quotient/inverse iteration method of the form:

$$\sigma_k = \sigma(\mathbf{x}^k), \qquad (A - \sigma_k I)\mathbf{x}^{k+1} = \mathbf{x}^k, \qquad k = 0, 1, \ldots . \tag{7.3.25}$$

Thus starting from an initial $\mathbf{x}^0$, at each stage we compute a new Rayleigh quotient approximation σ_k to the eigenvalue, and then a new approximation to the eigenvector by an inverse iteration step. This procedure can be very rapidly convergent when it is successful; indeed, it can be shown that there is a cubic rate of convergence of the sequence $\{\sigma_k\}$ to a simple eigenvalue. However, there are several drawbacks. The first is that of obtaining a satisfactory starting vector $\mathbf{x}^0$, since if $\mathbf{x}^0$ is not a reasonable approximation to the direction $\mathbf{v}_j$ of interest, the sequences (7.3.25) will not necessarily converge to the eigenvalue and eigenvector pair $\lambda_j, \mathbf{v}_j$. One way to obtain a suitable $\mathbf{x}^0$ is to use the power method for several steps to obtain an approximation to the eigenvector corresponding to the largest eigenvalue in absolute value. One then could switch to (7.3.25). For the other eigenvalue/eigenvector pairs, one would need to use deflation to obtain approximate eigenvectors by the power method to use in (7.3.25).

Another drawback of (7.3.25), or of inverse iteration in general, is the necessity of solving the linear system of equations at each stage. For large sparse matrices such as arise in the solution of partial differential equations, this is a major problem and is the subject of Chapter 9.

Lanczos' Method

We end this chapter by considering one more method for large sparse symmetric matrices. As with the other methods of this section, Lanczos' method is usually used to approximate only a few eigenvalues. In contrast to the other methods, it approximates both the largest and smallest eigenvalues simultaneously, although the rate of convergence to the smallest eigenvalues may be slower.

Recall from the previous section that a symmetric matrix can be reduced to a tridiagonal matrix T by an orthogonal similarity transformation:

$$A = QTQ^T. \tag{7.3.26}$$

The orthogonal matrix Q was constructed by means of Householder transformations, but we will now obtain T by an entirely different approach which does not destroy the sparsity of A. Suppose that (7.3.26) holds and

$$T = \begin{bmatrix} \alpha_1 & \beta_1 & & \\ \beta_1 & \alpha_2 & \ddots & \\ & \ddots & \ddots & \beta_{n-1} \\ & & \beta_{n-1} & \alpha_n \end{bmatrix}. \tag{7.3.27}$$

If we multiply (7.3.26) on the right by Q,

$$AQ = QT, \tag{7.3.28}$$

and equate the columns of the two sides of (7.3.28), we have

$$A\mathbf{q}_1 = \alpha_1\mathbf{q}_1 + \beta_1\mathbf{q}_2, \tag{7.3.29a}$$

$$A\mathbf{q}_i = \beta_{i-1}\mathbf{q}_{i-1} + \alpha_i\mathbf{q}_i + \beta_i q_{i+1}, \qquad i = 2, \ldots, n-1, \tag{7.3.29b}$$

$$A\mathbf{q}_n = \beta_{n-1}\mathbf{q}_{n-1} + \alpha_n\mathbf{q}_n, \tag{7.3.29c}$$

where $\mathbf{q}_i$ is the ith column of Q. Since the orthogonality of Q implies that $\mathbf{q}_i^T\mathbf{q}_j = 0$, $i \neq j$ and $\mathbf{q}_i^T\mathbf{q}_i = 1$, if we multiply (7.3.29) by $\mathbf{q}_i^T$ we see that

$$\alpha_i = \mathbf{q}_i^T A\mathbf{q}_i, \qquad i = 1, \ldots, n. \tag{7.3.30}$$

To characterize the β_i we write (7.3.29a) as

$$\beta_1 \mathbf{q}_2 = A\mathbf{q}_1 - \alpha_1 \mathbf{q}_1,$$

and take norms of both sides to obtain

$$\beta_1 = \pm ||A\mathbf{q}_1 - \alpha_1 \mathbf{q}_1||_2. \tag{7.3.31}$$

The sign of β_1 is immaterial and we shall elect to take the β_i as positive. Then in a similar way we obtain from (7.3.29b)

$$\beta_i = ||A\mathbf{q}_i - \alpha_i \mathbf{q}_i - \beta_{i-1}\mathbf{q}_{i-1}||_2, \qquad i = 2, \ldots, n-1. \tag{7.3.32}$$

We can now use the above characterizations as the basis for an algorithm to obtain Q and T. Let $\mathbf{q}_1$ be an arbitrary vector such that $\mathbf{q}_1^T \mathbf{q}_1 = 1$. Form $A\mathbf{q}_1$ and then α_1 and β_1 from (7.3.30) and (7.3.31). Next, from (7.3.29a),

$$\mathbf{q}_2 = \frac{1}{\beta_1}(A\mathbf{q}_1 - \alpha_1 \mathbf{q}_1).$$

Now we can obtain α_2 and β_2 from (7.3.30) and (7.3.32) and then $\mathbf{q}_3$ from (7.3.29b). Continuing in this way, we may compute all of the α_i, β_i, and $\mathbf{q}_i$. The algorithm is summarized in Figure 7.2. One of the main strengths of the Lanczos algorithm is that the matrix A is never modified, as it is in the Householder reduction to tridiagonal form. As in the power method, A need not even be known explicitly as long as the matrix-vector product $A\mathbf{q}$ can be formed.

Choose $\mathbf{q}_1$ with $\mathbf{q}_1^T \mathbf{q}_1 = 1$. Set $\beta_0 = 0$
For $i = 1$ *to* $n-1$
$\quad \mathbf{p}_i = A\mathbf{q}_i$
$\quad \alpha_i = \mathbf{q}_i^T \mathbf{p}_i$
$\quad \mathbf{w}_i = \mathbf{p}_i - \alpha_i \mathbf{q}_i - \beta_{i-1}\mathbf{q}_{i-1}$
$\quad \beta_i = ||\mathbf{w}_i||_2$
$\quad \mathbf{q}_{i+1} = \beta_i^{-1}\mathbf{w}_i$
$\alpha_n = \mathbf{q}_n^T A \mathbf{q}_n$

Figure 7.2: *Lanczos Algorithm*

We have to verify that the $\mathbf{q}_i$ generated by the Lanczos algorithm of Figure 7.2 are indeed orthonormal. Clearly, the choice of β_i guarantees that $||\mathbf{q}_{i+1}||_2 = 1$, provided that $\beta_i \neq 0$; we shall return to this point shortly. For now we assume that all $\beta_i \neq 0$ and show that the $\mathbf{q}_i$ are orthogonal. First,

$$\mathbf{q}_1^T \mathbf{q}_2 = \beta_1^{-1}\mathbf{q}_1^T(A\mathbf{q}_1 - \alpha_1 \mathbf{q}_1) = \beta_1^{-1}(\alpha_1 - \alpha_1) = 0, \tag{7.3.33}$$

and, by induction,

$$\mathbf{q}_i^T \mathbf{q}_{i+1} = \beta_i^{-1} \mathbf{q}_i^T (A\mathbf{q}_i - \alpha_i \mathbf{q}_i - \beta_{i-1}\mathbf{q}_{i-1}) = \beta_i^{-1}(\alpha_i - \alpha_i) = 0, \qquad (7.3.34)$$

for $i = 2, \ldots, n-1$. We now show by induction that all of the $\mathbf{q}_i$ are orthogonal. We make the induction hypothesis that

$$\mathbf{q}_j^T \mathbf{q}_{i+1} = 0, \qquad j = 1, \ldots, i, \qquad (7.3.35)$$

which we have shown to be true for $i = 1$. We then wish to prove that

$$\mathbf{q}_j^T \mathbf{q}_{i+2} = 0, \qquad j = 1, \ldots, i+1. \qquad (7.3.36)$$

By (7.3.34) we have shown this for $j = i + 1$, so we assume that $j \leq i$. Then, by (7.3.35),

$$\begin{aligned} \mathbf{q}_j^T \mathbf{q}_{i+2} &= \beta_{i+1}^{-1} \mathbf{q}_j^T (A\mathbf{q}_{i+1} - \alpha_{i+1}\mathbf{q}_{i+1} - \beta_i \mathbf{q}_i) \\ &= \beta_{i+1}^{-1} (\mathbf{q}_j^T A\mathbf{q}_{i+1} - \beta_i \mathbf{q}_j^T \mathbf{q}_i). \end{aligned} \qquad (7.3.37)$$

Now by (7.3.29b) and the symmetry of A,

$$\mathbf{q}_j^T A\mathbf{q}_{i+1} = \mathbf{q}_{i+1}^T A\mathbf{q}_j = \mathbf{q}_{i+1}^T (\beta_{j-1}\mathbf{q}_{j-1} + \alpha_j \mathbf{q}_j + \beta_j \mathbf{q}_{j+1}). \qquad (7.3.38)$$

If $j < i$, then all of the inner products in (7.3.38) are zero by (7.3.35), as is $\mathbf{q}_j^T \mathbf{q}_i$ in (7.3.37); hence $\mathbf{q}_j^T \mathbf{q}_{i+2} = 0$. If $j = i$, then (7.3.38) shows that $\mathbf{q}_i^T A\mathbf{q}_{i+1} = \beta_i$, which cancels the β_i in (7.3.37) so that, again, $\mathbf{q}_j^T \mathbf{q}_{i+2} = 0$ and the induction is complete.

We now return to the assumption that $\beta_i \neq 0$. Suppose that $\beta_1 = ||\mathbf{w}_1||_2 = 0$. Then $A\mathbf{q}_1 = \alpha_1 \mathbf{q}_1$, so that $\mathbf{q}_1$ and α_1 are an eigenvector and a corresponding eigenvalue. More generally, if $\beta_i = ||\mathbf{w}_i||_2 = 0$, then the Lanczos process stops and we have at this point

$$AQ_i = Q_i T_i, \qquad (7.3.39)$$

where $Q_i = (\mathbf{q}_1, \ldots, \mathbf{q}_i)$ and T_i is the $i \times i$ leading principal submatrix of T. If $\hat{\lambda}_1, \ldots, \hat{\lambda}_i$ are the eigenvalues of T_i, then we can write

$$T_i = \hat{Q}_i D_i \hat{Q}_i^T, \qquad (7.3.40)$$

where $D = \mathrm{diag}(\hat{\lambda}_1, \ldots, \hat{\lambda}_i)$ and the columns of $\hat{Q}_i$ are the corresponding orthonormal eigenvectors of T_i. Putting (7.3.40) into (7.3.39) we obtain

$$AQ_i \hat{Q}_i = Q_i \hat{Q}_i D_i. \qquad (7.3.41)$$

The columns of $Q_i \hat{Q}_i$ are orthonormal (Exercise 7.3.9), and thus (7.3.41) shows that the $\hat{\lambda}_i$ are eigenvalues of A with corresponding eigenvectors which are the columns of $Q_i \hat{Q}_i$. Thus the emergence of a $\beta_i = 0$ signals that we can find i eigenvalues of A by computing the eigenvalues of T_i. The process can then be restarted by choosing a $\mathbf{q}_{i+1}$ that is orthogonal to $\mathbf{q}_1, \ldots, \mathbf{q}_i$. It is easy to show (Exercise 7.3.10) that $\beta_i = 0$ if and only if $\mathbf{q}_1$ is a linear combination of i eigenvectors of A; this is highly unlikely to happen in practice.

Assuming that no $\beta_i = 0$ we could carry the Lanczos process to completion and obtain the tridiagonal matrix T of (7.3.27). In practice this is rarely done for the large sparse matrices for which the Lanczos method is most useful. It turns out that if we stop the process at the kth step for some moderate size of k, the largest and smallest eigenvalues of the corresponding tridiagonal matrix T_k may be surprisingly good approximations to the corresponding eigenvalues of A . Moreover, other eigenvalues of T_k will approximate the corresponding small or large eigenvalues of A, although generally not as well as the extremal eigenvalue approximations.

Summary

In this section we have discussed several approaches to obtaining at least a few eigenvalues and corresponding eigenvectors of large sparse matrices. None of these methods is completely reliable and their usefulness depends in large measure on the location of the eigenvalues of A. In general, the symmetric eigenvalue problem is much easier than the nonsymmetric, but even for symmetric matrices there are no foolproof methods yet known.

Supplementary Discussion and References: 7.3

As with Section 7.2, the books by Golub and Van Loan [1989], Stewart [1973], and Wilkinson [1965] provide excellent information on more advanced aspects of the methods considered. See also Parlett [1980] for symmetric matrices. The analysis of the algorithms of this section was restricted to matrices with n linearly independent eigenvectors, but it can be extended to general matrices; see Golub and Van Loan [1989] and Wilkinson [1965], in particular.

The Rayleigh quotient and Lanczos methods were presented in the text only for symmetric matrices, but extensions to nonsymmetric matrices are possible; see Golub and Van Loan [1989].

Many of the methods of this section have extensions to "block" methods in which several eigenvalues and eigenvectors are approximated simultaneously. For example, "subspace" iteration is an extension of the power or inverse power iterations. For further discussion of block methods, see Golub and Van Loan [1989] and Parlett [1980].

As we have shown, the vectors $\mathbf{q}_i$ generated by the Lanczos algorithm are orthogonal. One serious problem with the algorithm is that due to rounding error this orthogonality is lost and with relatively few steps can be lost so significantly that the algorithm is no longer satisfactory. For ways to circumvent this problem by "reorthogonalization;" see Golub and Van Loan [1989].

EXERCISES 7.3

7.3.1. (Stewart [1973]). Let A_k be the sequence of matrices generated by the QR algorithm and let $\hat{Q}_k = Q_0 \cdots Q_{k-1}$, $\hat{R}_k = R_{k-1} \ldots R_0$. Show by induction that $A_k = \hat{Q}_k \hat{R}_k$. Conclude from this that the first column of $\hat{Q}_k$ is a multiple of $A^k \mathbf{e}_1$.

7.3.2. Let A be symmetric with eigenvalues λ_i and corresponding orthonormal eigenvectors $\mathbf{v}_i$. If $\mathbf{x} = c_1\mathbf{v}_1 + \cdots + c_n\mathbf{v}_n$, show that

$$\mathbf{x} - (\mathbf{v}_1^T\mathbf{x})\mathbf{v}_1 - \cdots - (\mathbf{v}_m^T\mathbf{x})\mathbf{v}_m = c_{m+1}\mathbf{v}_{m+1} + \cdots + c_n\mathbf{v}_n.$$

7.3.3. Let A be symmetric with eigenvalues λ_i and corresponding orthonormal eigenvectors $\mathbf{v}_i$. Show that the matrix $A_2 = A - \lambda_1\mathbf{v}_1\mathbf{v}_1^T$ has eigenvalues $0, \lambda_2, \ldots, \lambda_n$ and corresponding eigenvectors $\mathbf{v}_1, \ldots, \mathbf{v}_n$. Discuss how to compute $A_2\mathbf{x}$ without forming A_2 explicitly.

7.3.4. Let A be a real matrix with complex eigenvalue $a + ib$ and corresponding eigenvector $\mathbf{u} + i\mathbf{v}$. Show that $\mathbf{u} - i\mathbf{v}$ is the eigenvector corresponding to $a - ib$.

7.3.5. Let A be a symmetric matrix with eigenvalues $\lambda_1 \leq \cdots \leq \lambda_n$ and eigenvectors $\mathbf{v}_1, \ldots, \mathbf{v}_n$. Show that if $\mathbf{v} = \mathbf{v}_i$, the Rayleigh quotient of (7.3.19) is $\sigma = \lambda_i$. Use the fact that any vector $\mathbf{v}$ can be written as $\mathbf{v} = c_1\mathbf{v}_1 + \cdots + c_n\mathbf{v}_n$ to show that (7.3.20) holds.

7.3.6. From the geometric series

$$\frac{1}{1-\alpha} = 1 + \alpha + \alpha^2 + \cdots$$

conclude that

$$\frac{1}{1 + 0(\varepsilon)} = 1 + 0(\varepsilon), \qquad \frac{1}{1 + 0(\varepsilon^2)} = 1 + 0(\varepsilon^2).$$

Show also that $[1 + 0(\varepsilon)]^2 = 1 + 0(\varepsilon)$.

7.3.7. The matrix

$$A = \begin{bmatrix} 2 & 1 \\ 1 & 2 \end{bmatrix}$$

has eigenvalues 1 and 3 with corresponding eigenvectors $(1, -1)$ and $(1, 1)$. Apply several steps of the power method (7.3.6) to this matrix, starting with the vector $\mathbf{x}_0 = (1, 0)^T$. Carry the iteration far enough for the rate of convergence to become apparent.

7.3.8. Apply the power method to the shifted matrix $A - \frac{1}{2}I$, where A is given in Exercise 7.3.7. Discuss the improvement in the rate of convergence.

7.3.9. Let Q_i be an $n \times i$ matrix whose columns are orthonormal and P_i an $i \times i$ orthogonal matrix. Show that the columns of Q_iP_i are orthonormal.

7.3.10. Use (7.3.41) to show that if $\beta_i = 0$ in the Lanczos algorithm (and $\beta_i \neq 0$, $j < i$), then $\mathbf{q}_1$ is a linear combination of i eigenvectors of A.

7.3.11. Let $\mathbf{q}$ be an arbitrary vector with $\|\mathbf{q}\|_2 = 1$ and assume that $A = A^T$. Set $\sigma = \mathbf{q}^T A\mathbf{q}$ and $\mathbf{z} = A\mathbf{q} - \sigma\mathbf{q}$. Show that the interval $|\lambda - \sigma| \leq \|\mathbf{z}\|_2$ must contain at least one eigenvalue of A.

7.3.12. Develop the Lanczos algorithm for skew-symmetric matrices.

7.3.13. Let A be an arbitrary real matrix with eigenvalues λ_i and corresponding eigenvectors $\mathbf{v}_i$. Assume $\lambda_2 = \bar{\lambda}_1$, so that $\mathbf{v}_2 = \bar{\mathbf{v}}_1$ (Exercise 7.3.4). Given a vector $\mathbf{r}$ such that $\mathbf{r} = c\mathbf{v}_1 + \bar{c}\bar{\mathbf{v}}_1$, show how $\lambda_1, \bar{\lambda}_1, \mathbf{v}_1$ and $\bar{\mathbf{v}}_1$ can be calculated from $\mathbf{r}, \mathbf{s} = A\mathbf{r}$, and $\mathbf{t} = A\mathbf{s}$. Does this suggest an acceleration scheme for computing eigenvalues of non-symmetric matrices?

7.3.14. (Steepest Descent) Let A be a real symmetric matrix, $\mathbf{q}$ be an arbitrary vector with $\|\mathbf{q}\|_2 = 1$, and

$$\gamma^{-1}\mathbf{z} = A\mathbf{q} - \sigma\mathbf{q}, \quad \sigma = \mathbf{q}^T A\mathbf{q},$$

where γ is choosen so that $\|\mathbf{z}\|_2 = 1$. Show that $\mathbf{q}^T\mathbf{z} = 0$. In order to compute the smallest eigenvalue of A, let $\mathbf{w} = a\mathbf{q} + b\mathbf{z}$.

a. Show how to compute a and b so that $\mathbf{w}^T A\mathbf{w}/\mathbf{w}^T\mathbf{w}$ is a minimum.

b. Consider the sequence $\mathbf{w}_k = a_k\mathbf{q}_k + b_k\mathbf{z}_k$ and

$$\mu_k = \min\{\mathbf{w}_k^T A\mathbf{w}_k/\mathbf{w}_k^T\mathbf{w}_k : \mathbf{w}_k \neq 0\}.$$

Show that $\mu_{k+1} \leq \mu_k$.

7.3.15. Let $A = D+B$ and $A(\alpha) = D+\alpha B$, where D contains the diagonal elements of A and B the off-diagonal. Use the continuation method described in the Supplementary Discussion of Section 5.3 to give an algorithm for computing the smallest eigenvalue in magnitude of $A(1) = A$.

7.3.16. Let T be a symmetric tridiagonal matrix and $K = T + E$, where $E = (e_{ij})$ is a symmetric tridiagonal matrix with $|e_{ij}| \leq \varepsilon$. Use Exercise 7.3.11 to give bounds on how far each eigenvalue of K is from an eigenvalue of T.

Chapter 8

Space and Time

8.1 Partial Differential Equations

In previous chapters we have considered differential equations in a single independent variable. This independent variable was either time, as in the case of the trajectory or predator-prey problems of Chapter 2, or a space variable as in the problems of Chapter 3. We now begin the study of the numerical solution of differential equations in two or more independent variables – partial differential equations. In the present chapter the two independent variables will be time and a single space variable; in the next chapter we shall treat problems in more than one space variable.

Example Equations

As two examples, we will concentrate on

$$u_t = cu_{xx} \tag{8.1.1}$$

and

$$u_{tt} = cu_{xx}, \tag{8.1.2}$$

which are known, respectively, as the *heat* (or *diffusion*) *equation* and the *wave equation*. In both (8.1.1) and (8.1.2) c is a given constant, and subscripts denote partial derivatives. The heat equation (8.1.1) is the prototype example of a *parabolic equation*, and (8.1.2) is an example of a *hyperbolic equation* (see the Supplementary Discussion.) The third standard type of partial differential equation is called *elliptic*, and the simplest example of this type is *Poisson's equation*

$$u_{xx} + u_{yy} = f, \tag{8.1.3}$$

where f is a given function of x and y. If $f \equiv 0$, then (8.1.3) is *Laplace's equation.* Elliptic equations will be treated in Chapter 9.

Initial and Boundary Conditions for the Heat Equation

As we saw with ordinary differential equations, a solution of a differential equation is not determined without appropriate initial and/or boundary conditions, and we expect the same to be true for partial differential equations. For (8.1.1) we need to prescribe an initial condition, and if x ranges over only a finite interval, we also need boundary conditions at the endpoints of this interval. Thus if the domain of x is $0 \leq x \leq 1$, the initial condition would be

$$u(0,x) = g(x), \qquad 0 \leq x \leq 1, \tag{8.1.4}$$

where g is a given function. We will also prescribe the boundary conditions

$$u(t,0) = \alpha, \qquad u(t,1) = \beta, \qquad t \geq 0, \tag{8.1.5}$$

for given constants α and β. The equation (8.1.1) together with the conditions (8.1.4) and (8.1.5) is the mathematical model for the temperature distribution in a thin rod whose ends are held at fixed temperatures α and β, and whose initial temperature distribution is $g(x)$. The solution $u(x,t)$ then gives the temperature within the rod as a function of time. For this model the constant c in (8.1.1) is $c = k/(s\rho)$, where $k > 0$ is the thermal conductivity, s is the specific heat of the material, and ρ is the mass density. Thus $c > 0$.

There are several variations on this problem that can be modeled by changing the boundary conditions or the equation itself. For example, suppose that we assume that the right end of the rod is, like the sides, perfectly insulated; by definition, then, we expect no heat loss or change in temperature across this end, so the boundary conditions (8.1.5) are changed to

$$u(t,0) = \alpha, \qquad u_x(t,1) = 0. \tag{8.1.6}$$

Another variation is to suppose that the rod is not homogeneous – as has been tacitly assumed – but is made of an alloy whose components vary as a function of x. Then the density as well as the thermal conductivity and specific heat will generally also vary with x, so that $c = c(x)$. Thus the differential equation is now one with a variable rather than constant coefficient. Going one step further, the thermal conductivity will, in general, depend not only on the material but also on the temperature itself. For many problems this dependence is so slight that it can be ignored, but for others it cannot. Thus we may have $c = c(u)$ so that the equation (8.1.1) is now nonlinear.

The heat equation is also a mathematical model of various other physical phenomena such as the diffusion of a gas.

The Wave Equation

Consider next the wave equation (8.1.2). This equation, or more general forms of it, models various types of wave propagation phenomena, as, for example, in acoustics. One classical problem is a vibrating string. Consider, for example, a taut string along the x-axis that is fastened at $x = 0$ and $x = 1$. If the string is plucked, it will vibrate. We assume that the string is "ideal" – that is, it is perfectly flexible, and the tension T is constant as a function of both x and t and is large compared to the weight of the string. We denote the deflection of the string at a point x and time t by $u(t, x)$, and we assume that the deflections u are small compared to the length of the string. We assume, moreover, that the slope of the deflected string at any point is small compared to unity and that the horizontal displacement of the string is negligible compared to the vertical displacement (this is sometimes called *transverse motion*). We are also tacitly assuming that the motion of the string is only in a plane. The constant c in (8.1.2) is then equal to gT/w, where g is the gravitational constant and w is the weight of the string per unit length. In more general situations, T, and hence c, may not be constant.

In addition to the differential equation, we again need suitable initial and boundary conditions. Since the ends of the string are fixed at $x = 0$ and $x = 1$, we have the boundary conditions

$$u(t, 0) = 0 \qquad u(t, 1) = 0. \tag{8.1.7}$$

For the initial conditions we must specify the initial deflection as well as the initial velocity of the string; thus for a given function f, we will use

$$u(0, x) = f(x), \qquad u_t(0, x) = 0, \tag{8.1.8}$$

where the latter condition implies zero initial velocity of the string. With the differential equation (8.1.2), the boundary conditions (8.1.7), and the initial conditions (8.1.8), the problem is now fully specified. For other problems different boundary or initial conditions may be given, but since the equation is second order in t two initial conditions must generally be given, just as for ordinary differential equations.

The purpose of this book is, of course, to study techniques for the numerical solution of problems. It is worth recalling here, however, that there is a classical analytical technique for representing the solution of both (8.1.1) and (8.1.2) by means of Fourier series. This technique is valid only under very restrictive conditions, but it does apply to the heat and wave equations together with the types of initial and boundary conditions we have considered and for c constant. It is the *method of separation of variables*, and we shall review it rather briefly.

Separation of Variables

Assume that the solution of (8.1.1) can be written as a product of a function that depends only on t and a function that depends only on x:

$$u(t, x) = v(t)w(x). \tag{8.1.9}$$

If we substitute (8.1.9) into (8.1.1), we obtain

$$v'(t)w(x) = cv(t)w''(x), \tag{8.1.10}$$

or, assuming that neither v nor w is zero,

$$\frac{v'(t)}{v(t)} = c\frac{w''(x)}{w(x)}. \tag{8.1.11}$$

Since the left side of (8.1.11) is a function only of t and the right side is a function only of x, it follows that both sides must be equal to some constant, say μ; thus,

$$v'(t) = \mu v(t), \qquad cw''(x) = \mu w(x). \tag{8.1.12}$$

The general solution of the first equation is

$$v(t) = c_1 e^{\mu t}, \qquad c_1 = \text{ constant.} \tag{8.1.13}$$

The second equation of (8.1.12) is the eigenvalue problem discussed in Section 7.1. As we saw there, the eigenfunctions are

$$w(x) = \sin k\pi x, \qquad 0 \leq x \leq 1, \quad k = 1, 2, \ldots, \tag{8.1.14}$$

and the eigenvalues are $-k^2\pi^2$. Because of the constant c, the corresponding values of μ are

$$\mu = -ck^2\pi^2, \qquad k = 1, 2, \ldots\ . \tag{8.1.15}$$

Hence any function w of the form (8.1.14) is a solution of $cw'' = \mu w$ provided that μ is given by (8.1.15). Note that such a solution would vanish at certain points in the interval $[0, 1]$, and (8.1.11) would not be valid there. However, the use of (8.1.11) is meant only to be suggestive; the final criterion is whether v and w satisfy (8.1.10), and this will be the case provided that (8.1.12) holds. Thus any function of the form

$$u(t, x) = e^{-ck^2\pi^2 t} \sin k\pi x, \tag{8.1.16}$$

where k is any positive integer, satisfies the heat equation (8.1.1), as may be verified directly (Exercise 8.1.1).

Even though (8.1.16) satisfies the differential equation, it need not satisfy the initial or boundary conditions. Consider the special case of the boundary conditions (8.1.5) for which

$$u(t, 0) = 0, \qquad u(t, 1) = 0, \tag{8.1.17}$$

and suppose that the initial condition can be written as a finite trigonometric sum

$$u(0, x) = \sum_{k=1}^{n} a_k \sin k\pi x. \tag{8.1.18}$$

It is then easy to verify (Exercise 8.1.1) that

$$u(t,x) = \sum_{k=1}^{n} a_k e^{-ck^2\pi^2 t} \sin k\pi x \tag{8.1.19}$$

is a solution of (8.1.1) and, moreover, satisfies the boundary and initial conditions (8.1.17) and (8.1.18).

The solution (8.1.19) is predicated on the finite expansion (8.1.18), but by the theory of Fourier series a very large class of functions, and hence initial conditions, can be represented by the infinite series

$$g(x) = \sum_{k=1}^{\infty} a_k \sin k\pi x, \qquad a_k = 2\int_0^1 g(z)\sin(k\pi z)dz. \tag{8.1.20}$$

In this case the solution can be given, analogously to (8.1.19), by

$$u(t,x) = \sum_{k=1}^{\infty} a_k e^{-ck^2\pi^2 t} \sin k\pi x, \tag{8.1.21}$$

although it is no longer as simple as in the case of (8.1.19) to verify rigorously that this is a solution.

The method of separation of variables can also be applied to the wave equation, and we will just indicate the result corresponding to (8.1.21) for equation (8.1.2) together with the boundary and initial conditions (8.1.7) and (8.1.8). Again, if we assume that the first initial condition of (8.1.8) can be represented by

$$f(x) = \sum_{k=1}^{\infty} a_k \sin(k\pi x), \tag{8.1.22}$$

then the solution is

$$u(t,x) = \sum_{k=1}^{\infty} a_k \sin(k\pi x)\cos(k\pi\sqrt{c}t). \tag{8.1.23}$$

In the case that (8.1.22) is a finite sum, analogous to (8.1.18), it is easy to verify this result directly (Exercise 8.1.2) without any technical difficulties.

We have not meant to imply that these series expansions are to be the basis for numerical methods, although in certain special cases they can be. Rather, such representations are sometimes useful in ascertaining qualitative information about the solution of the differential equation. For example, (8.1.19) clearly shows, since $c > 0$, that $u(t,x) \to 0$ as $t \to \infty$, and this same conclusion can be reached from the infinite series (8.1.21). We shall use this information in the following section on finite difference methods. We shall also use the technique of separation of variables, applied to difference equations, to study stability properties of the numerical methods in the following sections.

Supplementary Discussion and References: 8.1

Consider a partial differential equation of the form

$$au_{xx} + bu_{xt} + cu_{tt} + du_x + eu_t + fu = g,$$

where the coefficients $a, b, \ldots$ are functions of x and t. Then the equation is *elliptic* if

$$[b(x,t)]^2 < a(x,t)c(x,t) \tag{8.1.24}$$

for all x, t in the region of interest. Laplace's equation is the special case in which $a = c = 1$ and all other coefficients are zero. The equation is *hyperbolic* if $b^2 > ac$; this is the case for the wave equation. Finally, the equation is *parabolic* if $b = ac$, which holds for the heat equation since $b = c = 0$.

Very interesting, important, and difficult problems occur for equations of *mixed type* in which the condition (8.1.24) holds in part of the domain while the opposite inequality holds in another part; that is, the equation is elliptic in part of the domain and hyperbolic in another part. Such problems arise, for example, in transonic airflow in which the flow is subsonic in part of the region (the elliptic part) and supersonic in another part (the hyperbolic part). In these problems the variable t is a second space variable y.

Many, if not most, partial differential equation models of physical phenomena involve systems of equations rather than a single equation. The classification system of elliptic, hyperbolic, and parabolic can be extended to systems of equations, although relatively few systems that model realistic physical situations fit into this nice classification.

The method of separation of variables together with the use of Fourier series is a classical technique for solving certain simple equations and is discussed in most beginning textbooks on partial differential equations. Most such books will contain additional examples, derivations of equations, and classification theory. See, for example, Haberman [1983] and Keener [1988], and, for more advanced treatments, Courant and Hilbert [1953, 1962], and Garabedian [1986].

EXERCISES 8.1

8.1.1. Show that the function of (8.1.16) satisfies (8.1.1) for any integer k. Then verify that (8.1.19) is a solution of (8.1.1) that satisfies the initial and boundary conditions (8.1.17) and (8.1.18).

8.1.2. **a.** Show that

$$u(x,t) = \sum_{k=1}^{n} a_k \sin(k\pi x)\cos(k\pi\sqrt{c}t) \tag{8.1.25}$$

satisfies the wave equation (8.1.2) as well as the boundary and initial conditions $u(t,0) = u(t,1) = u_t(0,x) = 0$, $u(0,x) = \sum_{k=1}^{n} a_k \sin(k\pi x)$.

b. Write a program to display the special solution $u(x,t) = \sin(\pi x)\cos(\pi t)$ on a graphics terminal in such a way that the motion of the string is clear. Do the same for more complicated solutions consisting of two and three terms of (8.1.25), for $c = 1$.

8.2 Explicit Methods and Stability

We begin in this section the study of finite difference methods for partial differential equations and, in particular, for the equations discussed in the previous section. We will treat first the heat equation

$$u_t = cu_{xx}, \qquad 0 \le x \le 1, \quad t \ge 0, \tag{8.2.1}$$

with the initial and boundary conditions

$$u(0,x) = g(x), \qquad 0 \le x \le 1, \tag{8.2.2}$$

$$u(t,0) = \alpha, \qquad u(t,1) = \beta, \qquad t \ge 0. \tag{8.2.3}$$

Difference Equations for the Heat Equation

We set up a grid in the x,t plane with grid spacings Δx and Δt as illustrated in Figure 8.1. The idea of the simplest finite difference method for (8.2.1) is to replace the second derivative on the right-hand side of (8.2.1) with a central difference quotient in x, and replace u_t with a forward difference in time. Then one advances the approximate solution forward in time one time level after another. More precisely, if we let u_j^m denote the approximate solution at $x_j = j\Delta x$ and $t_m = m\Delta t$, then the finite difference analog of (8.2.1) is

$$\frac{u_j^{m+1} - u_j^m}{\Delta t} = \frac{c}{(\Delta x)^2}(u_{j+1}^m - 2u_j^m + u_{j-1}^m), \tag{8.2.4}$$

or

$$u_j^{m+1} = u_j^m + \mu(u_{j+1}^m - 2u_j^m + u_{j-1}^m), \qquad j = 1,\ldots,n, \tag{8.2.5}$$

where

$$\mu = \frac{c\Delta t}{(\Delta x)^2}. \tag{8.2.6}$$

The boundary conditions (8.2.3) give the values

$$u_0^m = \alpha, \qquad u_{n+1}^m = \beta, \quad m = 0,1,\ldots,$$

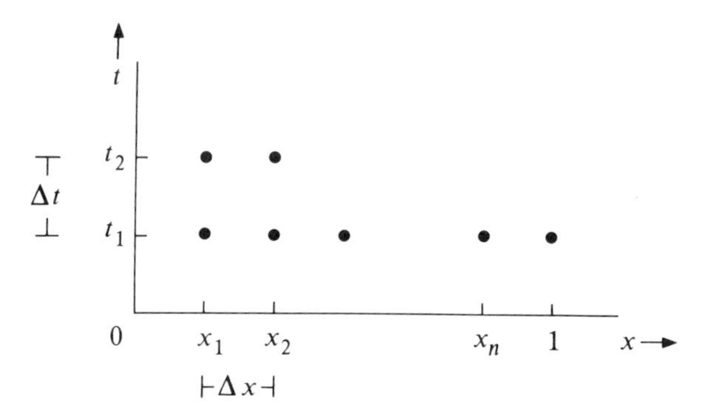

Figure 8.1: *Grid Spacings*

and the initial condition (8.2.2) furnishes

$$u_j^0 = g(x_j), \qquad j = 1, \ldots, n.$$

Therefore (8.2.5) provides a prescription for marching the approximate solution forward one time step after another: the values u_j^1, $j = 1, \ldots, n$, are first all obtained, and knowing these we can obtain u_j^2, $j = 1, \ldots, n$, and so on.

How accurate will be the approximate solution obtained by (8.2.5)? A rigorous answer to this question is a difficult problem that is beyond the scope of this book, but we will attempt to obtain some insight by considering two aspects of the error analysis.

Discretization Error

Let $u(t, x)$ be the exact solution of (8.2.1) together with the initial and boundary conditions (8.2.2) and (8.2.3). If we put this exact solution into the difference formula (8.2.4), the amount by which the formula fails to be satisfied is called the *local discretization error* (or *local truncation error*); at the point (t, x) the local discretization error, e, is

$$\frac{u(t + \Delta t, x) - u(t, x)}{\Delta t} - \frac{c}{(\Delta x)^2}[u(t, x + \Delta x) - 2u(t, x) + u(t, x - \Delta x)]. \quad (8.2.7)$$

This is entirely analogous to the previous definitions of local discretization error for ordinary differential equations and enjoys similar properties. For example, suppose that we know the exact solution $u(t, x)$ for some t and all $0 \leq x \leq 1$, and we use (8.2.4) to estimate the solution at $t + \Delta t$. Call this estimate $\hat{u}(t + \Delta t, x)$. Then, by definition

$$\frac{\hat{u}(t + \Delta t, x) - u(t, x)}{\Delta t} = \frac{c}{(\Delta x)^2}[u(t, x + \Delta x) - 2u(t, x) + u(t, x - \Delta x)],$$

so if we subtract this from (8.2.7), we obtain

$$\hat{u}(t+\Delta t, x) - u(t+\Delta t, x) = \Delta te. \tag{8.2.8}$$

Thus the error caused by one time step using the difference scheme (8.2.4) is Δt times the local discretization error.

It is easy to estimate the quantity e of (8.2.7) in terms of Δt and Δx. If we consider u as only a function of t for fixed x, we can apply the Taylor expansion

$$u(t+\Delta t, x) = u(t,x) + u_t(t,x)\Delta t + O[(\Delta t)^2]$$

to conclude that

$$\frac{u(t+\Delta t, x) - u(t,x)}{\Delta t} = u_t(t,x) + O(\Delta t).$$

Similarly, by Taylor expansions in x, we have

$$\frac{u(t, x+\Delta x) - 2u(t,x) + u(t, x-\Delta x)}{(\Delta x)^2} = u_{xx}(t,x) + O[(\Delta x)^2].$$

If we put these expressions into (8.2.7) and use $u_t = cu_{xx}$ (since u is the exact solution of the differential equation), we obtain

$$e = 0(\Delta t) + 0[(\Delta x)^2]. \tag{8.2.9}$$

The fact that Δt appears to the first power and Δx to the second power in this expression for the local discretization error is usually described by the statement that the finite difference method (8.2.4) is *first-order accurate in time* and *second-order accurate in space.*

Stability

It is tempting to conclude from (8.2.8) and (8.2.9) that the discretization error in u_j^m, as obtained from (8.2.5), converges to zero as Δt and Δx tend to zero. Unfortunately, this conclusion is not warranted since (8.2.8) gives the error in the approximate solution only for a single time step. To show that the discretization error tends to zero on a whole time interval $[0, T]$ is difficult and, in general, requires additional conditions on how Δt and Δx tend to zero. A relationship of the type (8.2.9), or more generally a statement that the local discretization error tends to zero with Δt and Δx, is essentially a necessary condition for the global discretization error itself to tend to zero, and is called *consistency* of the difference scheme. The reason that consistency of the difference method does not necessarily imply convergence of the discretization error is connected with *stability* of the difference scheme, and we now discuss certain aspects of this for the difference scheme (8.2.5).

In a way exactly analogous to the method of separation of variables and the use of Fourier series that were applied to the differential equation $u_t = cu_{xx}$ in the previous section, we can obtain the exact solution of (8.2.5) together with the boundary and initial conditions

$$\begin{aligned} u_0^m &= u_{n+1}^m = 0, \qquad m = 1, 2, \ldots, \\ u_j^0 &= g(x_j), \qquad j = 1, \ldots, n. \end{aligned} \tag{8.2.10}$$

Assume that the solution u_j^m can be written as

$$u_j^m = v_m w_j, \qquad j = 1, \ldots, n, \qquad m = 0, 1, \ldots . \tag{8.2.11}$$

This is the paradigm of separation of variables for difference equations. Putting (8.2.11) into (8.2.5) and collecting terms yields

$$\frac{v_{m+1} - v_m}{\mu v_m} = \frac{w_{j+1} - 2w_j + w_{j-1}}{w_j}, \qquad j = 1, \ldots, n, \quad m = 0, 1, \ldots .$$

Since the left side is independent of j and the right side is independent of m, both sides must be equal to some constant, say $-\lambda$; thus

$$v_{m+1} - v_m = -\lambda \mu v_m, \qquad m = 0, 1, \ldots, \tag{8.2.12}$$

$$w_{j+1} - 2w_j + w_{j-1} = -\lambda w_j, \qquad j = 1, \ldots, n, \tag{8.2.13}$$

where $w_0 = w_{n+1} = 0$ from the boundary conditions (8.2.10). Equation (8.2.13) represents the eigenvalue problem for the $(2, -1)$ tridiagonal matrix of (3.1.10). The eigenvalues of this matrix are (see Exercise 4.4.5)

$$\lambda_k = 2 - 2\cos k\pi\Delta x, \qquad k = 1, \ldots, n, \tag{8.2.14}$$

with corresponding eigenvectors (Exercise 4.4.5)

$$\mathbf{w} = [\sin(k\pi\Delta x), \sin(2k\pi\Delta x), \ldots, \sin(nk\pi\Delta x)]^T, \qquad k = 1, \ldots, n,$$

where $\Delta x = 1/(n+1)$. Thus for each $\lambda = \lambda_k$

$$w_j = \sin(jk\pi\Delta x), \qquad j = 0, 1, \ldots, n+1, \tag{8.2.15}$$

is a solution of (8.2.13). Clearly

$$v_m = (1 - \lambda\mu)^m v_0, \qquad m = 0, 1, \ldots,$$

is a solution of (8.2.12) for any λ, so that

$$u_j^m = v_m w_j = (1 - \lambda_k \mu)^m \sin(jk\pi\Delta x), \qquad m = 0, 1, \ldots, \ j = 0, 1, \ldots, n+1,$$

is a solution of (8.2.5) for each k. As with the differential equation, any linear combination of these solutions is also a solution; thus

$$u_j^m = \sum_{k=1}^{n} a_k(1 - \lambda_k \mu)^m \sin(jk\pi\Delta x) \tag{8.2.16}$$

is a solution of (8.2.5) for any constants a_k. If the a_k are chosen so that

$$a_k = \sum_{l=1}^{n} g(x_l) \sin k\pi l\Delta x, \tag{8.2.17}$$

then (Exercise 8.2.2) u_j^m also satisfies the initial condition

$$u_j^0 = g(x_j), \qquad j = 1, \ldots, n. \tag{8.2.18}$$

We now use the representation (8.2.16) in the following way. From our discussion in the previous section, the equation $u_t = cu_{xx}$ together with the boundary conditions $u(t, 0) = u(t, 1) = 0$ is a model of the temperature distribution in a thin insulated rod whose ends are held at zero temperature. Since there is no source of heat, we expect that the temperature of the rod will decrease to zero so that $u(t, x) \to 0$ as $t \to \infty$. This conclusion can also be obtained mathematically from the series representations (8.1.19) or (8.1.21) of the solution since the exponential terms all tend to zero. Therefore it is reasonable to demand that the finite difference approximations u_j^m also tend to zero as m tends to infinity, for any initial conditions; by (8.2.16), this will be the case if and only if

$$|1 - \mu\lambda_k| < 1, \qquad k = 1, \ldots, n. \tag{8.2.19}$$

Since μ and all the λ_k are positive, (8.2.19) will hold if and only if

$$-(1 - \mu\lambda_k) < 1, \qquad k = 1, \ldots, n,$$

or

$$\mu < \min_k \frac{2}{\lambda_k} = \frac{1}{1 - \cos \pi n\Delta x} = \frac{1}{1 + \cos \pi\Delta x}, \tag{8.2.20}$$

since the largest λ_k is λ_n. Thus, with $\mu = c\Delta t/(\Delta x)^2$ from (8.2.6), (8.2.20) becomes

$$\Delta t < \frac{(\Delta x)^2}{c(1 + \cos \pi\Delta x)}. \tag{8.2.21}$$

This gives a restriction on the relative sizes of Δt and Δx, which if not satisfied will, in general, mean that the approximate solution u_j^m of the difference scheme will ultimately diverge as m tends to infinity and, obviously, will become an increasingly poor approximation to the solution of the differential

Table 8.1: *Unstable Behavior*

t/x	0.2	0.4	0.6	0.8
0	0.59	0.95	0.95	0.59
0.16	0.08	0.13	0.13	0.08
0.24	0.03	0.05	0.05	0.03
0.28	0.02	0.03	0.02	0.004
0.32	0.01	0.02	0.09	0.20
0.36	0.005	−0.14	−1.22	−2.43
0.40	0.12	3.20	19.0	32.2
0.44	−3.97	−62.0	−286.9	−428.

equation, which tends to zero. In Table 8.1 we give an example of this instability for (8.2.5) with $c = 1$, $g(x) = \sin \pi x$, $\alpha = \beta = 0$ and with $\Delta x = 0.1$ and $\Delta t = 0.04$ so that (8.2.21) is not satisfied. Note that the instability has begun to develop noticeably by $t = 0.32$ and then rapidly worsens.

We can replace (8.2.21) by the slightly stronger condition

$$\Delta t \le \frac{(\Delta x)^2}{2c}, \tag{8.2.22}$$

which always implies (8.2.21). This relation is called the *stability condition* for the difference method (8.2.5). Our derivation of it has been in the context of the behavior of u_j^m as $m \to \infty$ for fixed Δt and Δx. But it is also relevant to the problem of the convergence of the discretization error to zero as Δt and Δx tend to zero. In fact, although it is beyond our scope to prove this, the approximate solutions will converge to the exact solution as Δt and Δx tend to zero if (8.2.22) holds as Δt and Δx tend to zero. This is a special case of a more general principle known as the *Lax Equivalence Theorem*, which states that for quite general differential equations and consistent difference schemes, the global discretization error will tend to zero if and only if the method is stable.

The condition (8.2.22) imposes an increasingly stringent limitation on the time step Δt as the space increment Δx becomes small, as Table 8.2 shows for the case $c = 1$. Thus we may require a time step far smaller than otherwise necessary to resolve the time-dependent nature of the solution of the differential equation itself. Although the analysis that we have done has been restricted to the simplest differential equation and simplest difference scheme, the requirement of small time steps for explicit finite difference methods for parabolic and similar equations is a general problem and is a primary motivation for the implicit methods to be discussed in the next section.

Table 8.2: *Maximum Time Steps for Given Δx and $c = 1$*

Δx	Δt
0.1	$0.5{\cdot}10^{-2}$
0.01	$0.5{\cdot}10^{-4}$
0.001	$0.5{\cdot}10^{-6}$

The Wave Equation

We turn now to hyperbolic equations, and in particular to the wave equation

$$u_{tt} = cu_{xx}, \qquad 0 \le x \le 1, \qquad t \ge 0, \tag{8.2.23}$$

together with the initial and boundary conditions

$$u(0,x) = f(x), \quad u_t(0,x) = g(x), \quad u(t,0) = \alpha, \quad u(t,1) = \beta. \tag{8.2.24}$$

As we discussed in the previous section, the problem (8.2.23), (8.2.24) is a mathematical model for a vibrating string.

The simplest finite difference scheme for (8.2.23) is

$$\frac{u_j^{m+1} - 2u_j^m + u_j^{m-1}}{(\Delta t)^2} = \frac{c}{(\Delta x)^2}(u_{j+1}^m - 2u_j^m + u_{j-1}^m). \tag{8.2.25}$$

To obtain (8.2.25) we have used the usual centered difference formula for u_{xx}, just as in (8.2.4), as well as for u_{tt}. Note that (8.2.25) now involves three time levels and requires that both u_j^m and u_j^{m-1} be known in order to advance to the $(m+1)$st time level. This requires additional storage for u_j^{m-1}, as opposed to the method (8.2.5) for $u_t = cu_{xx}$, but is a natural consequence of the fact that the differential equation contains a second derivative in time. We also require both u_j^0 and u_j^1 in order to start, and these can be obtained from the initial conditions (8.2.24):

$$u_j^0 = f(x_j) \qquad u_j^1 = f(x_j) + \Delta t g(x_j), \qquad j = 1, \ldots, n, \tag{8.2.26}$$

where the second condition is obtained by approximating $u_t(0,x) = g(x)$ by $[u(\Delta t, x) - u(0,x)]/\Delta t = g(x)$. From the boundary conditions, we have

$$u_0^m = \alpha, \qquad u_{n+1}^m = \beta, \qquad m = 0, 1, \ldots\,. \tag{8.2.27}$$

Thus the values at the $(m+1)$st time level are obtained by

$$u_j^{m+1} = 2u_j^m - u_j^{m-1} + \mu(u_{j+1}^m - 2u_j^m + u_{j-1}^m), \tag{8.2.28}$$

where

$$\mu = \frac{c(\Delta t)^2}{(\Delta x)^2}.$$

It is easy to show that the local discretization error for (8.2.25) is $0(\Delta t)^2 + 0(\Delta x)^2$, so that the method is second-order accurate in both space and time (Exercise 8.2.4). For the stability analysis we can again proceed by the method of separation of variables and assume that the boundary conditions α and β are zero. Let $u_j^m = v_m w_j$. Putting this into (8.2.25) leads to the two conditions

$$v_{m+1} - 2v_m + v_{m-1} = -\lambda\mu v_m, \qquad m = 1, \ldots, \tag{8.2.29}$$

$$w_{j+1} - 2w_j + w_{j-1} = -\lambda w_j, \qquad j = 1, \ldots, n. \tag{8.2.30}$$

The second set of equations is the same as (8.2.13), and thus its solutions are given by (8.2.14),(8.2.15). The equations (8.2.29), although ostensibly of the same form as (8.2.30), are for an initial-value problem in which v_0 and v_1 are known. From Section 2.5 [see (2.5.11) - (2.5.14)], the solution of (8.2.29) is given by

$$v_m = \gamma_1 \eta_+^m + \gamma_2 \eta_-^m, \qquad m = 0, 1, \ldots,$$

where $\eta_\pm$ are the roots of the characteristic equation $\eta^2 + (\lambda\mu - 2)\eta + 1 = 0$ and are given by

$$\eta_\pm = \tfrac{1}{2}(2 - \lambda\mu \pm \sqrt{\lambda^2\mu^2 - 4\lambda\mu}).$$

The γ_i can be obtained from the initial conditions v_0 and v_1. Thus the solution of (8.2.25) can be written as

$$u_j^m = \sum_{k=1}^{n} a_k(\gamma_{k,1}\eta_{k,+}^m + \gamma_{k,2}\eta_{k,-}^m)\sin(jk\pi\Delta x), \tag{8.2.31}$$

where the subscript k indicates that the corresponding γ and η have been computed for $\lambda = \lambda_k$. For u_j^m to remain bounded for arbitrary a_k and initial conditions, it is necessary and sufficient that $|\eta_{k,\pm}| \le 1$. If

$$\lambda_k\eta - 4 \le 0, \tag{8.2.32}$$

then it is easy to verify that $|\eta_{k,\pm}| = 1$, and if $\lambda_k\eta - 4 > 0$, then $\eta_{k,-} < -1$. Hence (8.2.32) is the necessary and sufficient condition for stability. Since the eigenvalues λ_k satisfy $0 < \lambda_k < 4$, a sufficient condition for (8.2.32) is that $\mu \le 1$, or

$$\Delta t \le \frac{\Delta x}{\sqrt{c}}. \tag{8.2.33}$$

This is also essentially a necessary condition in the sense that $\lambda_n \to 4$ as $n \to \infty$, so any condition weaker that (8.2.33) would allow (8.2.32) to be violated for sufficiently large n. The stability condition (8.2.33) is much less stringent on Δt than was (8.2.22) for the heat equation. Indeed, (8.2.33) shows that Δt need decrease only proportionately to Δx, rather than as the square of Δx, as was the case with (8.2.22).

Although this section has dealt only with the heat and wave equations, the same principles of obtaining finite difference methods apply to more general initial-boundary value problems for either single equations or systems of equations. In all such cases the user must be alert to the possibility of instability, although for most equations a simple analysis of the form given in this section will not be possible.

Supplementary Discussion and References: 8.2

We have only touched the surface of finite difference methods for parabolic and hyperbolic equations. In particular, we have considered only first- and second-order methods, although a variety of higher-order methods have been developed. More importantly, most of the useful methods for parabolic equations are implicit, and these will be dealt with in the next section. The books by Ames [1977] and Hall and Porsching [1990] give many additional methods. See also Isaacson and Keller [1966] and Richtmyer and Morton [1967] for a discussion of methods as well as a rigorous analysis of discretization error and stability criteria.

The separation of variables analysis leading to (8.2.16) can also be viewed in matrix terms. The matrix-vector formulation of (8.2.5) is

$$\mathbf{u}^{m+1} = \mathbf{u}^m - \mu A \mathbf{u}^m, \qquad m = 0, 1, \ldots,$$

where A is the $(2, -1)$ matrix of (8.2.13), and $\mathbf{u}^m$ is the vector with components $u_1^m, \ldots, u_n^m$. The matrix $H = I - \mu A$ has eigenvalues $1 - \mu\lambda_k$ and eigenvectors $\mathbf{w}_k$, where the λ_k and $\mathbf{w}_k$ are the eigenvalues and eigenvectors of A. Therefore if

$$\mathbf{u}^0 = \sum_{k=1}^{n} a_k \mathbf{w}_k$$

is the expansion of $\mathbf{u}^0$ in terms of the eigenvectors, then

$$\mathbf{u}^m = H\mathbf{u}^{m-1} = \cdots = H^m \mathbf{u}^0 = \sum_{k=1}^{n} a_k (1 - \mu\lambda_k)^m \mathbf{w}_k,$$

which is (8.2.16).

EXERCISES 8.2

8.2.1. Use (8.2.5) to approximate a solution to $u_t = u_{xx}$ for the boundary and initial conditions $u(t, 0) = 0$, $u(t, 1) = 1$, and $u(0, x) = \sin \pi x + x$. Use different values of Δt and Δx and discuss your approximate solutions. For what ratios of Δt and Δx do you conclude that your approximate solution is stable?

8.2.2. Substitute (8.2.16) into (8.2.5) and verify that it is indeed a solution. Verify also that (8.2.18) holds if the a_k are given by (8.2.17).

8.2.3. Repeat and verify the calculation of Table 8.1.

8.2.4. Use the approach discussed in the text for the heat equation to conclude that the method (8.2.25) for the wave equation is second-order accurate in both time and space.

8.2.5. Write a computer program to solve the wave equation (8.2.23) with the initial and boundary conditions (8.2.24) by the difference method (8.2.25). Apply your program for various values of Δt and Δx, and conclude that the calculation is stable if (8.2.33) is satisfied.

8.3 Implicit Methods

The finite difference method (8.2.5) discussed in the previous section is called explicit because the values of u_j^{m+1} at the next time level are obtained by an explicit formula in terms of the values at the previous time level. In contrast, consider again the heat equation

$$u_t = cu_{xx}, \qquad 0 \le x \le 1, \qquad t \ge 0, \tag{8.3.1}$$

and the difference approximation

$$\frac{u_j^{m+1} - u_j^m}{\Delta t} = \frac{c}{(\Delta x)^2}(u_{j+1}^{m+1} - 2u_j^{m+1} + u_{j-1}^{m+1}), \qquad j = 1, \ldots, n. \tag{8.3.2}$$

This is similar in form to (8.2.4) but has the important difference that the values of u_j on the right side are now evaluated at the $(m+1)$st time level rather than the mth. Consequently, if we know u_j^m, $j = 1, \ldots, n$, and are ready to compute u_j^{m+1}, $j = 1, \ldots, n$, we see that the variables u_j on the right-hand side of (8.3.2) are all unknown. Thus we must view (8.3.2) as a system of equations that implicitly defines the values u_j^{m+1}, $j = 1, \ldots, n$. This is one of the basic differences between implicit and explicit methods: in an explicit method we have a formula for u_j^{m+1}, such as (8.2.5), in terms of known values of u_j at previous time levels, whereas with an implicit method we must solve a system of equations to advance to the next time level.

If, as in the previous section, we set $\mu = c\Delta t/(\Delta x)^2$, then we can rewrite (8.3.2) as

$$(1 + 2\mu)u_j^{m+1} - \mu(u_{j+1}^{m+1} + u_{j-1}^{m+1}) = u_j^m, \qquad j = 1, \ldots, n, \tag{8.3.3}$$

or in matrix-vector form,

$$(I + \mu A)\mathbf{u}^{m+1} = \mathbf{u}^m + \mathbf{b}, \qquad m = 0, 1, \ldots . \tag{8.3.4}$$

Here A is the $(2, -1)$ tridiagonal matrix of (3.1.10), and $\mathbf{u}^{m+1}$ and $\mathbf{u}^m$ are vectors with components u_i^{m+1} and u_i^m, $i = 1, \ldots, n$, respectively. If we use the same boundary conditions

$$u(t,0) = \alpha, \qquad u(t,1) = \beta \tag{8.3.5}$$

as in the previous section, then $u_0^k = \alpha$ and $u_{n+1}^k = \beta$ for $k = 0, 1, \ldots$; thus the vector $\mathbf{b}$ in (8.3.4) is zero except for $\mu\alpha$ and $\mu\beta$ in the first and last components. We also assume the initial condition

$$u(0,x) = f(x), \qquad 0 \le x \le 1, \tag{8.3.6}$$

so, as before, $u_j^0 = f(x_j)$, $j = 1, \ldots, n$.

The implicit method (8.3.4) is now carried out by solving the linear system of equations (8.3.4) at each time step to obtain the u_j^{m+1} from u_j^m. The matrix in (8.3.4) is tridiagonal and also diagonally dominant (Section 3.1) since $c > 0$, and thus $\mu > 0$. Therefore, as we saw in Section 4.3, the system of equations can be efficiently solved by Gaussian elimination without pivoting. In the particular case of (8.3.4), we could compute the L and U factors once and for all, although we may not be able to do this for more general problems.

Even though each time step of (8.3.4) can be carried out relatively efficiently, this method is more costly per time step than the explicit method (8.2.5). However, in return for this additional cost we obtain a substantial benefit in the stability properties of the method, which in many cases will allow us to use a much larger time step than does the explicit method and thus will greatly cut the overall computing costs. We will now indicate the stability analysis following the lines of the previous section.

Stability

We assume, as before, that $\alpha = \beta = 0$. Then, corresponding to (8.2.16),

$$u_j^m = \sum_{k=1}^{n} a_k \gamma_k^m \sin(k\pi j \Delta x) \tag{8.3.7}$$

identically satisfies the difference scheme (8.3.3) for any constants a_k, provided that

$$\gamma_k = \frac{1}{1 + 2\mu(1 + \cos k\pi\Delta x)}, \qquad k = 1, \ldots, n. \tag{8.3.8}$$

Moreover, u_j^0 satisfies the initial condition if

$$a_k = \frac{2}{n+1} \sum_{l=1}^{n} f(x_l) \sin(k\pi l \Delta x). \tag{8.3.9}$$

The verification of these results is left to Exercise 8.3.2.

Now recall from our discussion in the previous section that we require that the approximate solution $u_j^m \to 0$ as $m \to \infty$ if it is to mirror the solution of the differential equation itself. From (8.3.7) we see that this will be the case, in general, if and only if

$$|\gamma_k| < 1, \qquad k = 1, \ldots, n. \tag{8.3.10}$$

But from (8.3.8), since $\mu > 0$,

$$0 < \gamma_k < 1, \qquad k = 1, \ldots, n, \tag{8.3.11}$$

so that (8.3.10) indeed holds. Most importantly, we see that (8.3.11) is true for *any* $\mu > 0$; thus, since $\mu = c\Delta t/(\Delta x)^2$, (8.3.11) is true for *any* ratio of Δt and Δx. We say in this case that the method is *unconditionally stable*, meaning that it is stable without restrictions on the relative size of Δt and Δx.

Discretization Error

The fact that the method (8.3.3) is unconditionally stable does *not* mean that we can expect to obtain a good approximate solution for any Δt and Δx. As usual, these must be chosen sufficiently small to control discretization error. Now it is the case (see Exercise 8.3.3) that (8.3.2), like the corresponding explicit method (8.2.5), is first-order accurate in time and second-order accurate in space; that is, the local discretization error will be

$$e = 0(\Delta t) + 0(\Delta x)^2.$$

Suppose that

$$e = c_1 \Delta t + c_2 (\Delta x)^2.$$

Then for the contributions to the total error from the discretization in time and the discretization in space to be commensurate, we require that

$$\Delta t \doteq c_3 (\Delta x)^2,$$

which is reminiscent of the stability condition (8.2.22) for the explicit method. Thus we see that although the stability requirements for our implicit method do not impose any restrictions on the relative sizes of Δt and Δx, the accuracy requirements may.

The Crank-Nicolson Method

A potentially better implicit method in this regard is the famous *Crank-Nicolson method*, which is an average of the explicit method (8.2.4) and the implicit method (8.3.2):

$$u_j^{m+1} - u_j^m = \frac{c\Delta t}{2(\Delta x)^2}(u_{j+1}^{m+1} - 2u_j^{m+1} + u_{j-1}^{m+1} + u_{j+1}^m - 2u_j^m + u_{j-1}^m). \tag{8.3.12}$$

This can be written in matrix-vector form as

$$(I + \frac{\mu}{2}A)\mathbf{u}^{m+1} = (I - \frac{\mu}{2}A)\mathbf{u}^m + \mathbf{b}, \qquad m = 0, 1, \ldots, \tag{8.3.13}$$

where A is again the $(2, -1)$ matrix. Hence (8.3.12) is carried out by solving a tridiagonal system of equations at each time step. The advantage of (8.3.12) is that it is not only unconditionally stable, as is (8.3.4), but it is second-order accurate in time as well as in space. (The verification of these assertions is left to Exercise 8.3.5). These properties have made it one of the most often used methods for parabolic equations.

One easy way to recall the three different methods (8.2.4), (8.3.2), and (8.3.12) is by their "stencils" of grid points as illustrated in Figure 8.2. These show which grid points enter into the difference method.

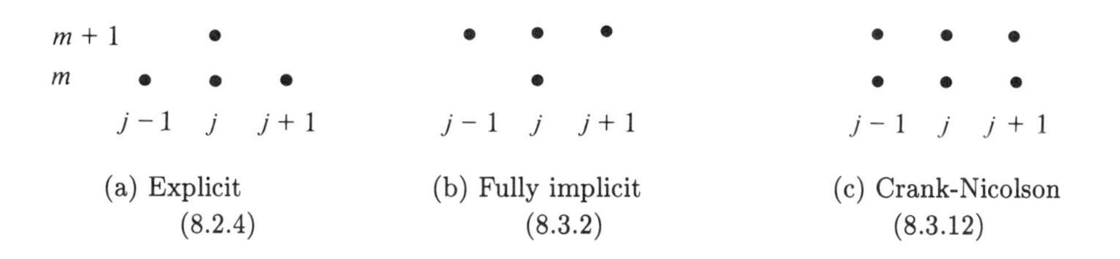

Figure 8.2: *Stencils for the Methods*

It has become common practice in the numerical solution of parabolic-type partial differential equations to use implicit methods since their good stability properties outweigh the additional work required per time step. Most of the methods in actual use are more complicated than the Crank-Nicolson method, but the principles are the same. However, for problems involving more than one space dimension, straightforward extensions of the implicit methods of this section are not satisfactory, and additional techniques are required. One such technique will be discussed in Section 9.1.

It is possible to formulate implicit methods for hyperbolic equations, such as the wave equation, in much the same way. However, as was seen with the wave equation, the stability requirements of explicit methods typically do not impose a stringent restriction on the time step. Consequently, implicit methods for hyperbolic equations are rather infrequently used in practice, and we shall not discuss them further.

Supplementary Discussion and References: 8.3

The references given in Section 8.2 are also relevant for implicit methods. In particular, Richtmyer and Morton [1967, pp. 189-91] summarize graphically a number of implicit finite difference methods for parabolic equations in terms of their stencils.

EXERCISES 8.3

8.3.1. Write a program to carry out (8.3.4) and apply it to the problem of Exercise 8.2.1. Use various values of Δt and Δx and verify numerically the stability of the method. Discuss your results compared to those for Exercise 8.2.1, including the relative ease and efficiency of carrying out the two methods.

8.3.2. Proceed along the lines of the analysis of the previous section to verify that (8.3.7) satisfies (8.3.3).

8.3.3. Proceed along the lines of the analysis of the previous section to show that the local discretization error for the method (8.3.3) satisfies (8.2.9).

8.3.4. Modify your program of Exercise 8.3.1 to carry out the Crank-Nicolson method (8.3.12). Discuss your results and compare this method to (8.3.4).

8.3.5. For the Crank-Nicolson method (8.3.12) with the boundary conditions $\alpha = \beta = 0$, verify that the solution of (8.3.12) is of the form (8.3.7), where now $\gamma_k = (1 - \frac{1}{2}\mu\lambda_k)/(1 + \frac{1}{2}\mu\lambda_k)$, with $\lambda_k = 2 - 2\cos(k\pi\Delta x)$. Conclude that the method is unconditionally stable. Show also that the method is second-order accurate in both space and time.

8.3.6. The Dufort-Frankel method for the heat equation is

$$u_j^{m+1} - u_j^{m-1} = [\Delta tc/(\Delta x)^2](u_{j+1}^m - u_j^{m+1} - u_j^{m-1} + u_{j-1}^m).$$

Show that this method is unconditionally stable. Give an explicit formulation of it.

8.3.7. Consider the nonlinear parabolic equation

$$u_t = u_{xx} - u - x^2 - u^3,$$

with boundary and initial conditions $u(t, 0) = u(t, 1) = 0$, $u(0, x) = sin\pi x$.

a. Formulate an explicit method and do a stability analysis for the linear equation $u_t = u_{xx} - u$. Test numerically your stability criterion against the nonlinear equation.

b. Formulate a completely implicit method and write a program to solve the resulting nonlinear system at each time step by Newton's method. Verify numerically that your method is unconditionally stable.

c. The corresponding steady-state equation is the two-point boundary-value problem $v'' = v + x^2 + v^3$, $v(0) = v(1) = 0$. If you are only interested in the steady-state solution, would it be better to attack this equation directly by the methods of Chapter 5 or to integrate the partial differential equation to steady-state by the methods of parts **a** and **b**?

8.4 Semidiscrete Methods

We now consider another approach to initial boundary value problems which utilizes the projection principles of Chapter 6 and reduces the partial differential equation to an approximating system of ordinary differential equations. This approach can be applied in principle to both parabolic- and hyperbolic-type equations. We will first illustrate it for the heat equation

$$u_t = cu_{xx}, \qquad 0 \le x \le 1, \qquad t \le 0, \tag{8.4.1}$$

with the initial and boundary conditions

$$u(0, x) = f(x), \qquad u(t, 0) = 0, \qquad u(t, 1) = 0. \tag{8.4.2}$$

As in Section 6.1, let $\phi_1(x), \ldots, \phi_n(x)$ be a set of basis functions that satisfy the boundary conditions:

$$\phi_k(0) = 0 \qquad \phi_k(1) = 0, \qquad k = 1, \ldots, n. \tag{8.4.3}$$

We attempt to find an approximate solution $\hat{u}$ of (8.4.1) of the form

$$\hat{u}(t, x) = \sum_{i=1}^{n} \alpha_i(t)\phi_i(x), \tag{8.4.4}$$

where the α_i are to be determined. Note that this is the same approach taken in Chapter 6, with the exception that now we allow the coefficients α_i of the linear combination of the basis functions to be functions of t to reflect the time-dependent nature of the problem.

To determine the unknown coefficients α_i, we can apply any of the criteria of Section 6.1, and we will consider first collocation in the following way. We again let $0 \le x_1 < \cdots < x_n \le 1$ be (not necessarily equally spaced) grid points in the x variable, and we require that the approximate solution (8.4.4) satisfy the differential equation at these points; thus, since

$$\hat{u}_t(t, x) = \sum_{i=1}^{n} \alpha_i'(t)\phi_i(x), \qquad \hat{u}_{xx}(t, x) = \sum_{i=1}^{n} \alpha_i(t)\phi_i''(x),$$

we require that

$$\sum_{i=1}^{n} \alpha_i'(t)\phi_i(x_j) = c\sum_{i=1}^{n} \alpha_i(t)\phi_i''(x_j), \qquad j = 1, \ldots, n. \tag{8.4.5}$$

If we introduce the $n \times n$ matrices

$$A = (\phi_j(x_i)), \qquad B = c(\phi_j''(x_i)), \tag{8.4.6}$$

and the n -vectors

$$\boldsymbol{\alpha}(t) = (\alpha_1(t), \ldots, \alpha_n(t))^T, \qquad \boldsymbol{\alpha}'(t) = (\alpha_1'(t), \ldots, \alpha_n'(t))^T,$$

we can write (8.4.5) as

$$A\boldsymbol{\alpha}'(t) = B\boldsymbol{\alpha}(t). \tag{8.4.7}$$

Thus if we assume that A is nonsingular and set $C = A^{-1}B$,

$$\boldsymbol{\alpha}'(t) = C\boldsymbol{\alpha}(t). \tag{8.4.8}$$

Equation (8.4.8) is a system of n ordinary differential equations for the α_i. To solve this system we will need an initial condition. If we require that the approximate solution $\hat{u}$ satisfy the initial condition (8.4.2) at the points $x_1, \ldots, x_n$, we have

$$\sum_{i=1}^{n} \alpha_i(0)\phi_i(x_j) = f(x_j), \qquad j = 1, \ldots, n,$$

or

$$A\boldsymbol{\alpha}(0) = \mathbf{f}, \tag{8.4.9}$$

where $\mathbf{f} = (f(x_1), \ldots, f(x_n))^T$. By our assumption that A is nonsingular, we then have

$$\boldsymbol{\alpha}(0) = A^{-1}\mathbf{f}. \tag{8.4.10}$$

The conceptual problem is then to solve the system of ordinary differential equations (8.4.8) with the initial condition (8.4.10). If we could do this exactly, then $\hat{u}$ as given by (8.4.4) would be the approximate solution. Such a method is called *semidiscrete* because we have discretized only in space by means of the basis functions ϕ_i and grid points x_i while leaving time as a continuous variable. In practice, however, the system of differential equations (8.4.8) must be solved numerically, so a discretization of time is introduced by that process and the term semidiscrete is somewhat of a misnomer. Nevertheless, the conceptual viewpoint of considering discretization of only the space variables and thus reducing the problem to a system of ordinary differential equations is useful.

Let us consider the Euler method discussed in Chapter 2 for the numerical integration of the system (8.4.8). If $t_0, t_1, \ldots$ are equally spaced points in time with the spacing Δt, then Euler's method is

$$\boldsymbol{\alpha}^{k+1} = \boldsymbol{\alpha}^k + \Delta t C\boldsymbol{\alpha}^k, \qquad k = 0, 1, \ldots, \tag{8.4.11}$$

where $\boldsymbol{\alpha}^k$ denotes the approximate solution at the kth time step. In practice we will not carry out (8.4.11) in the manner indicated since we will not actually form $C = A^{-1}B$. Rather, we will work directly with the differential equation (8.4.7) and the corresponding Euler method

$$A\boldsymbol{\alpha}^{k+1} = A\boldsymbol{\alpha}^k + \Delta t B\boldsymbol{\alpha}^k, \qquad k = 0, 1, \ldots . \tag{8.4.12}$$

Thus at each time step we will solve the linear system of equations (8.4.12) with coefficient matrix A. The LU decomposition of A can, in this case, be done once and for all and the factors saved to be used at each time step.

In principle one could apply to (8.4.8) any of the higher-order methods discussed in Chapter 2; the formulation of some of these methods is left to Exercise 8.4.2. However, the equations (8.4.8) are typically rather "stiff," in the sense discussed in Chapter 2, and the use of explicit methods will require a rather small time step. Hence it may be advantageous to use a method such as the trapezoid rule. Applied to (8.4.8), the trapezoid rule (2.5.38) becomes

$$\boldsymbol{\alpha}^{k+1} = \boldsymbol{\alpha}^k + \frac{\Delta t}{2}(A^{-1}B\boldsymbol{\alpha}^{k+1} + A^{-1}B\boldsymbol{\alpha}^k),$$

or, multiplying through by A, collecting coefficients of $\boldsymbol{\alpha}^{k+1}$ and $\boldsymbol{\alpha}^k$, and setting $\mu = \Delta t/2$,

$$(A - \mu B)\boldsymbol{\alpha}^{k+1} = (A + \mu B)\boldsymbol{\alpha}^k. \qquad (8.4.13)$$

To carry out this method requires the solution at each time step of a linear system similar to (8.4.12) but with the coefficient matrix $A - \mu B$. Assuming that this matrix is nonsingular and that Δt is held constant, we may proceed as before to compute the LU factors once and for all and use these in subsequent solutions of (8.4.13) at the different time steps. If the subtraction of the matrix μB from A does not materially affect the difficulty of computing the LU decomposition, then the work in carrying out (8.4.13) will not be much more than that for Euler's method. However, the trapezoid method is second-order accurate in time, as discussed in Chapter 2, and allows a larger time step for stiff equations; it should, therefore, be more suitable for this problem.

Hyperbolic Equations

The same approach can be applied to hyperbolic equations, and we illustrate this for the wave equation

$$u_{tt} = cu_{xx}, \qquad 0 \le x \le 1, \qquad t \ge 0, \qquad (8.4.14)$$

with the initial and boundary conditions

$$u(0,x) = f(x), \qquad u_t(0,x) = g(x), \qquad u(t,0) = 0, \qquad u(t,1) = 0. \quad (8.4.15)$$

It is common practice to reduce a hyperbolic equation like (8.4.14) to a system of equations in which only the first derivative with respect to time appears. This is analogous to the situation for ordinary differential equations: recall that in Chapter 2 we reduced higher-order equations to first-order systems of ordinary differential equations (see Appendix 1). In the case of (8.4.14) we will use a reduction made by introducing a function $v(t,x)$ such that

$$u_t = av_x, \qquad v_t = au_x, \qquad (8.4.16)$$

where $a = \sqrt{c}$. If u and v are solutions of the system (8.4.16) and are sufficiently differentiable, then by differentiating the first equation of (8.4.16) with respect to t and the second with respect to x, we obtain

$$u_{tt} = av_{xt} = av_{tx} = a^2 u_{xx} = cu_{xx},$$

so that u is a solution of (8.4.14). The initial and boundary conditions (8.4.15) for (8.4.14) become

$$\begin{aligned} &u(0,x) = f(x), \qquad v(0,x) = \frac{1}{a}\int_0^x g(s)ds, \\ &u(t,0) = 0, \qquad u(t,1) = 0. \end{aligned} \tag{8.4.17}$$

Now let $\phi_1(x), \ldots, \phi_n(x)$ and $\psi_1(x), \ldots, \psi_n(x)$ be two sets of basis functions that satisfy

$$\phi_i(0) = \phi_i(1) = \psi_i(0) = \psi_i(1) = 0, \qquad i = 1, \ldots, n. \tag{8.4.18}$$

We shall seek approximate solutions $\hat{u}$ and $\hat{v}$ of (8.4.16) of the form

$$\hat{u}(x,t) = \sum_{i=1}^{n} \alpha_i(t)\phi_i(x), \qquad \hat{v}(x,t) = \sum_{i=1}^{n} \beta_i(t)\psi_i(x). \tag{8.4.19}$$

If we require that these approximate solutions satisfy the equations (8.4.16) at the grid points $x_1, \ldots, x_n$, we obtain

$$\sum_{i=1}^{n} \alpha_i'(t)\phi_i(x_j) = a\sum_{j=1}^{n} \beta_i(t)\psi_i'(x_j), \qquad j = 1, \ldots, n, \tag{8.4.20a}$$

$$\sum_{i=1}^{n} \beta_i'(t)\psi_i(x_j) = a\sum_{i=1}^{n} \alpha_i(t)\phi_i'(x_j), \qquad j = 1, \ldots, n, \tag{8.4.20b}$$

which is a coupled system of $2n$ ordinary differential equations for the unknown functions $\alpha_1, \ldots, \alpha_n$ and $\beta_1, \ldots, \beta_n$. As before, the initial conditions are obtained from (8.4.17) by

$$\sum_{i=1}^{n} \alpha_i(0)\phi_i(x_j) = f(x_j), \quad \sum_{i=1}^{n} \beta_i(0)\psi_i(x_j) = \frac{1}{a}\int_0^x g(s)ds, \quad j = 1, \ldots, n,$$

where we will assume that the $n \times n$ matrices $(\phi_i(x_j))$ and $(\psi_i(x_j))$ are nonsingular. Thus the semidiscrete method for the wave equation (8.4.14) is entirely analogous to that for the heat equation, with the exception that there are now twice as many unknown functions in the system of ordinary differential equations.

The Method of Lines

We have used the collocation principle for the discretization of the space variable in both of the preceding examples, but the Galerkin principle discussed in Chapter 6 could have been used. Finite difference discretizations can also be used in the same fashion, as we now discuss briefly for the heat equation (8.4.1). If u is the exact solution of (8.4.1), then the approximate relation

$$u_t(t, x_i) \doteq \frac{c}{(\Delta x)^2}[u(t, x_{i+1}) - 2u(t, x_i) + u(t, x_{i-1})] \tag{8.4.21}$$

holds at the grid points $x_1, \ldots, x_n$. This leads to the following procedure. We seek n functions $v_1(t), \ldots, v_n(t)$ such that

$$v_i(t) \doteq u(t, x_i), \qquad i = 1, \ldots, n.$$

The approximate relationship (8.4.21) suggests attempting to find these functions as the solution of the system of ordinary differential equations

$$v_i'(t) = \frac{c}{(\Delta x)^2}[v_{i+1}(t) - 2v_i(t) + v_{i-1}(t)], \qquad i = 1, \ldots, n, \tag{8.4.22}$$

in which the functions v_0 and v_{n+1} are taken to be identically zero from the boundary conditions (8.4.2). Moreover, from the initial condition (8.4.2) we will take

$$v_i(0) = f(x_i), \qquad i = 1, \ldots, n. \tag{8.4.23}$$

The system (8.4.22) can be written in matrix form as

$$\mathbf{v}'(t) = \frac{-c}{(\Delta x)^2} A\mathbf{v}(t), \tag{8.4.24}$$

where A is the $(2, -1)$ tridiagonal matrix (3.1.10). If we apply Euler's method to this system, we have

$$\mathbf{v}^{m+1} = \mathbf{v}^m - \frac{c\Delta t}{(\Delta x)^2} A\mathbf{v}^m, \qquad m = 0, 1, \ldots .$$

Written out in component form, this is

$$v_i^{m+1} = v_i^m + \frac{c\Delta t}{(\Delta x)^2}(v_{i+1}^m - 2v_i^m + v_{i-1}^m), \quad i = 1, \ldots, n, \quad m = 0, 1, \ldots,$$

which is the explicit method (8.2.5). Similarly, the implicit method (8.3.2) is obtained by applying the backward Euler method (2.5.33) to (8.4.24), and the Crank-Nicolson method (8.3.12) arises by applying the trapezoid rule (2.5.38) to (8.4.24).

The above procedure leading to the system of ordinary differential equations (8.4.24) is called the *method of lines*, and some authors use this term for any semi-discrete method, whether or not it arises from finite differences.

Supplementary Discussion and References: 8.4

In the text we have described only the use of very simple methods for solving the ordinary differential equations arising from discretization only in the space variable. However, one of the advantages of this approach is the possibility of using high quality packages for solving the ordinary differential equations. For further reading, see Schiesser [1991].

EXERCISES 8.4

8.4.1. **a.** Write out the system of equations (8.4.7) explicitly for the basis functions $\phi_k(x) = \sin k\pi x$, $k = 1, \ldots, n$, assuming that the grid points $x_1, \ldots, x_n$ are equally spaced. Write a program to carry out Euler's method (8.4.12) with $n = 10$ and $\Delta t = 0.1$. Run the program for 20 time steps and for different initial conditions.

b. Do the same if the ϕ_k are quadratic splines as discussed in Section 6.2.

8.4.2. Write out explicitly the second- and fourth-order Runge-Kutta methods and the second- and fourth-order Adams-Bashforth methods for (8.4.8).

8.4.3. Write a program to carry out the trapezoid method (8.4.13) for the problems of Exercise 8.4.1.

8.4.4. Repeat Exercise 8.4.1 for the equations (8.4.20) for the wave equation, assuming that the ϕ_i and ψ_i are both trigonometric functions in part **a**, and quadratic splines in part **b**.

Chapter 9

The Curse of Dimensionality

9.1 Two and Three Space Dimensions

In the previous chapter we considered partial differential equations in two independent variables: time and one space variable. Since physical phenomena occur in a three-dimensional world, mathematical models in only one space dimension are usually considerable simplifications of the actual physical situation although in many cases they are sufficient for phenomena that exhibit various symmetries or in which events are happening in two of the three space dimensions at such a slow rate that those directions can be ignored. However, large-scale scientific computing is now increasingly concerned with more detailed analyses of problems in which all three space directions, or at least two, are of concern. This chapter, then, will be concerned with problems in more than one space dimension, although for simplicity of exposition we will mainly discuss only two-dimensional problems.

In the previous chapter we considered the heat equation

$$u_t = cu_{xx} \tag{9.1.1}$$

as a mathematical model of the temperature in a long, thin rod. If the body of interest is a three-dimensional cube, as shown in Figure 9.1, (9.1.1) extends to three dimensions with partial derivatives in all three variables x, y, and z. Thus

$$u_t = c(u_{xx} + u_{yy} + u_{zz}), \tag{9.1.2}$$

where the constant c is again the ratio of the thermal conductivity and the product of the specific heat and mass density under the assumption that these quantities are not functions of space or time.

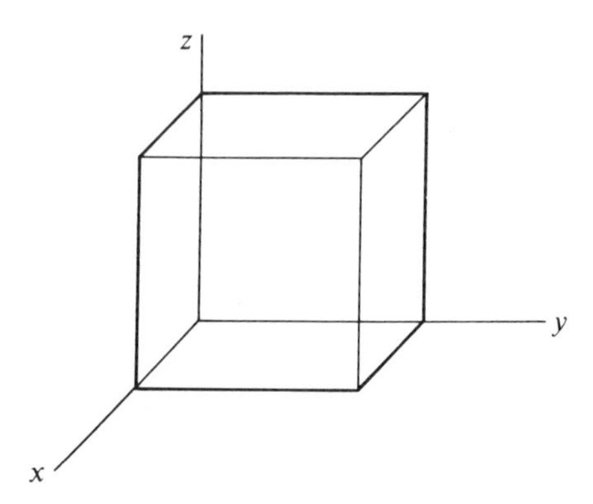

Figure 9.1: *Three-Dimensional Cube*

Equation (9.1.2) is a model of the temperature u as a function of time and at points within the interior of the body. As usual, to complete the model we need to specify boundary conditions, and for this purpose it is simplest for exposition to treat the corresponding problem in two space dimensions:

$$u_t = c(u_{xx} + u_{yy}). \tag{9.1.3}$$

We can consider (9.1.3) to be the mathematical model of the temperature in a flat, thin plate as shown in Figure 9.2, where we have taken the plate to be the unit square.

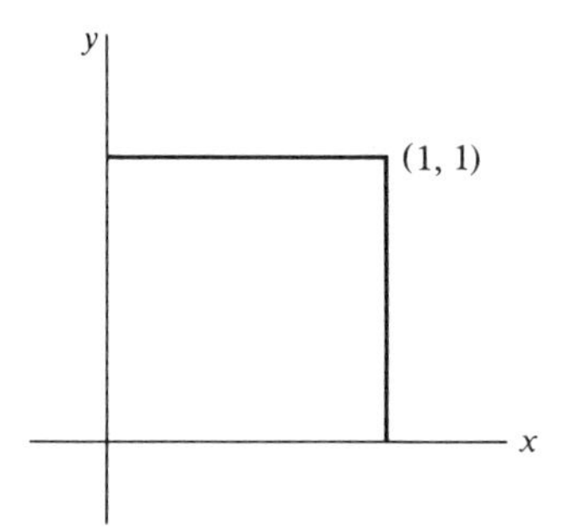

Figure 9.2: *Flat, Thin Plate*

The simplest boundary conditions occur when the temperature is prescribed on the four sides of the plate:

$$u(t, x, y) = g(x, y), \qquad (x, y) \text{ on boundary,} \tag{9.1.4}$$

where g is a given function. Another possibility is to assume that one of the sides, say $x = 0$, is perfectly insulated; thus, there is no heat loss across that side and no change in temperature, so the boundary condition is

$$u_x(t, 0, y) = 0, \qquad 0 \leq y \leq 1, \tag{9.1.5}$$

combined with the specification (9.1.4) on the other sides. A boundary condition of the form (9.1.5) is usually called a *Neumann condition*, and that of the form (9.1.4) is a *Dirichlet condition*. Clearly, various other such combinations are possible, including a specified temperature change (other than zero) across a boundary. Boundary conditions for the three-dimensional problem can be given in a similar fashion. We also must specify a temperature distribution at some time which we take to be $t = 0$; such an initial condition for (9.1.3) is of the form

$$u(0, x, y) = f(x, y). \tag{9.1.6}$$

Given the initial condition (9.1.6) and boundary conditions of the form (9.1.4) and/or (9.1.5), it is intuitively clear that the temperature distribution should evolve in time to a final steady state that is determined only by the boundary conditions. In many situations it is this steady-state solution that is of primary interest, and since it no longer depends on time it should satisfy the equation (9.1.3) with $u_t = 0$:

$$u_{xx} + u_{yy} = 0. \tag{9.1.7}$$

This is Laplace's equation and, as mentioned in the previous chapter, is the prototype of an elliptic equation. If we wish only the steady-state solution of the temperature distribution problem that we have been discussing, we can proceed, in principle, in two ways: solve equation (9.1.3) for u as a function of time until convergence to a steady state is reached, or solve (9.1.7) only for the steady-state solution.

Finite Differences for Poisson's Equation

We will return to the time-dependent problem shortly, after considering the finite difference method for (9.1.7) and, more generally, Poisson's equation

$$u_{xx} + u_{yy} = f, \tag{9.1.8}$$

where f is a given function of x and y. We assume that the domain of the problem is the unit square $0 \leq x, y \leq 1$, and that Dirichlet boundary conditions

$$u(x, y) = g(x, y), \qquad (x, y) \text{ on boundary} \tag{9.1.9}$$

are given, where g is a known function. We impose a mesh of grid points on the unit square with spacing h between the points in both the horizontal and vertical directions; this is illustrated in Figure 9.3.

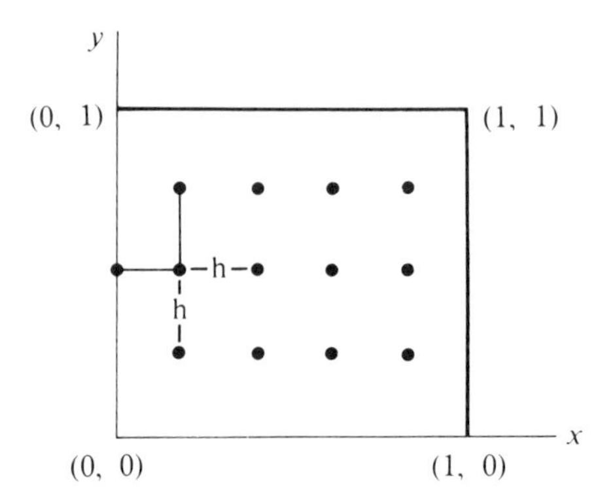

Figure 9.3: *Mesh Points on the Unit Square and Discretization Stencil*

The interior grid points are given by

$$(x_i, y_j) = (ih, jh), \qquad i, j = 1, \dots N, \tag{9.1.10}$$

where $(N+1)h = 1$. Now consider a typical grid point (x_i, y_j). We approximate u_{xx} and u_{yy} at this point by the centered difference approximations

$$u_{xx}(x_i, y_j) \doteq \frac{1}{h^2}[u(x_{i-1}, y_j) - 2u(x_i, y_j) + u(x_{i+1}, y_j)], \tag{9.1.11a}$$

$$u_{yy}(x_i, y_j) \doteq \frac{1}{h^2}[u(x_i, y_{j-1}) - 2u(x_i, y_j) + u(x_i, y_{j+1})]. \tag{9.1.11b}$$

If we put these approximations into the differential equation (9.1.8), we obtain

$$\begin{aligned} u(x_{i-1}, y_j) + u(x_{i+1}, y_j) + u(x_i, y_{j-1}) + u(x_i, y_{j+1}) - 4u(x_i, y_j) \\ \doteq h^2 f(x_i, y_j), \end{aligned} \tag{9.1.12}$$

which is an approximate relationship that the exact solution u of (9.1.8) satisfies at any grid point in the interior of the domain.

We now define approximations u_{ij} to the exact solution $u(x_i, y_j)$ at the N^2 interior grid points by requiring that they satisfy exactly the relationship (9.1.12); that is,

$$-u_{i+1,j} - u_{i-1,j} - u_{i,j+1} - u_{i,j-1} + 4u_{ij} = -h^2 f_{ij}, \qquad i, j = 1, \dots, N, \tag{9.1.13}$$

where we have multiplied (9.1.12) by -1. This is a linear system of equations in the $(N+2)^2$ variables u_{ij}. Note, however, that the variables $u_{0,j}$, $u_{N+1,j}$, $j = 0, \dots, N+1$, and $u_{i,0}$, $u_{i,N+1}$, $i = 0, \dots, N+1$, correspond to the grid points on the boundary and thus are given by the boundary condition (9.1.9):

$$\begin{aligned} u_{0,j} &= g(0, y_j), \qquad & u_{N+1,j} &= g(1, y_j), \qquad & j &= 0, 1, \dots, N+1, \\ u_{i,0} &= g(x_i, 0), \qquad & u_{i,N+1} &= g(x_i, 1), \qquad & i &= 0, 1, \dots, N+1. \end{aligned} \tag{9.1.14}$$

Therefore (9.1.13) is a linear system of N^2 equations in the N^2 unknowns u_{ij}, $i,j = 1,\ldots,N$, corresponding to the interior grid points. The stencil in Figure 9.3 shows how u_{ij} is coupled to its north, south, east, and west neighbors in (9.1.13). It is easy to show (Exercise 9.1.1) that the local discretization error in the u_{ij} is $0(h^2)$. Note that (9.1.13) is the natural extension to two space variables of the discrete equations

$$-u_{i+1} + 2u_i - u_{i-1} = -h^2 f_i, \qquad i = 1,\ldots,N,$$

obtained in Chapter 3 for the "one-dimensional Poisson equation" $u'' = f$.

We now wish to write the system (9.1.13) in matrix-vector form, and for this purpose we will number the interior grid points in the manner shown in Figure 9.4, which is called the *natural* or *row-wise ordering.* Corresponding to this ordering of the grid points, we order the unknowns $\{u_{ij}\}$ into the vector

$$(u_{11},\ldots,u_{N1},u_{12},\ldots,u_{N2},\ldots,u_{1N},\ldots,u_{NN}), \tag{9.1.15}$$

and write the system of equations in the same order. We illustrate this for $N = 2$ (Exercise 9.1.2):

$$\begin{bmatrix} 4 & -1 & -1 & 0 \\ -1 & 4 & 0 & -1 \\ -1 & 0 & 4 & -1 \\ 0 & -1 & -1 & 4 \end{bmatrix} \begin{bmatrix} u_{11} \\ u_{21} \\ u_{12} \\ u_{22} \end{bmatrix} = -h^2 \begin{bmatrix} f_{11} \\ f_{21} \\ f_{12} \\ f_{22} \end{bmatrix} + \begin{bmatrix} u_{01} + u_{10} \\ u_{20} + u_{31} \\ u_{02} + u_{13} \\ u_{32} + u_{23} \end{bmatrix}, \tag{9.1.16}$$

in which we have put the known boundary values on the right-hand side of the equation.

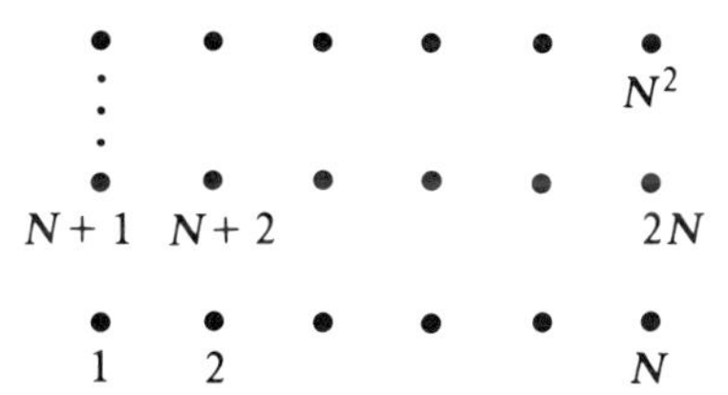

Figure 9.4: *Natural Ordering of the Interior Grid Points*

The equations (9.1.16) begin to illustrate the structure of the linear system. For general N, a typical row of the matrix will be

$$-1 \quad 0 \quad \cdots \quad 0 \quad -1 \quad 4 \quad -1 \quad 0 \quad \cdots \quad 0 \quad -1$$

where $N-2$ zeros separate the -1's in both directions. The equations corresponding to an interior grid point adjacent to a boundary point will contain a known boundary value, and this value will be moved to the right side of the

$$
\begin{array}{cccccccccccccccc}
4 & -1 & & & -1 & & & & & & & & & & & \\
-1 & 4 & -1 & & & -1 & & & & & & & & & & \\
 & -1 & 4 & -1 & & & -1 & & & & & & & & & \\
 & & -1 & 4 & & & & -1 & & & & & & & & \\
-1 & & & & 4 & -1 & & & -1 & & & & & & & \\
 & -1 & & & -1 & 4 & -1 & & & -1 & & & & & & \\
 & & -1 & & & -1 & 4 & -1 & & & -1 & & & & & \\
 & & & -1 & & & -1 & 4 & & & & -1 & & & & \\
 & & & & -1 & & & & 4 & -1 & & & -1 & & & \\
 & & & & & -1 & & & -1 & 4 & -1 & & & -1 & & \\
 & & & & & & -1 & & & -1 & 4 & -1 & & & -1 & \\
 & & & & & & & -1 & & & -1 & 4 & & & & -1 \\
 & & & & & & & & -1 & & & & 4 & -1 & & \\
 & & & & & & & & & -1 & & & -1 & 4 & -1 & \\
 & & & & & & & & & & -1 & & & -1 & 4 & -1 \\
 & & & & & & & & & & & -1 & & & -1 & 4
\end{array}
$$

Figure 9.5: *Coefficient Matrix of* (9.1.13) *for* $N = 4$

equation, eliminating the corresponding -1 from the matrix. This happened in each equation in (9.1.16) because of the size of N. We show in Figure 9.5 the coefficient matrix for $N = 4$ (see Exercise 9.1.2).

Although Figure 9.5 illustrates the structure of the coefficient matrix, this form is cumbersome for large N; it is much easier to write it in a block matrix form. To do this we define the $N \times N$ tridiagonal matrix

$$
T_N = \begin{bmatrix} 4 & -1 & & & \\ -1 & & \ddots & & \\ & & \ddots & & \\ & & \ddots & & -1 \\ & & & -1 & 4 \end{bmatrix} \tag{9.1.17}
$$

and let I_N denote the $N \times N$ identity matrix. Then the $N^2 \times N^2$ coefficient matrix of (9.1.13) is the *block tridiagonal matrix*

$$
A = \begin{bmatrix} T_N & -I_N & & & \\ -I_N & T_N & \ddots & & \\ & -I_N & \ddots & & \\ & & \ddots & & -I_N \\ & & & -I_N & T_N \end{bmatrix}. \tag{9.1.18}
$$

The matrix of (9.1.16) is the special case of (9.1.18) for $N = 2$, and Figure 9.5 shows the matrix for $N = 4$.

If we also define the vectors

$$\mathbf{u}_i = (u_{1i}, \ldots, u_{Ni})^T, \qquad \mathbf{f}_i = (f_{1i}, \ldots, f_{Ni})^T, \qquad i = 1, \ldots, N,$$
$$\mathbf{b}_1 = (u_{01} + u_{10}, u_{20}, \ldots, u_{N-1,0}, u_{N,0} + u_{N+1,1})^T,$$
$$\mathbf{b}_i = (u_{0i}, 0, \ldots, 0, u_{N+1,i})^T, \qquad i = 2, \ldots, N-1,$$
$$\mathbf{b}_N = (u_{0,N} + u_{1,N+1}, u_{2,N+1}, \ldots, u_{N-1,N}, u_{N,N+1} + u_{N+1,N})^T,$$

then we can write the system (9.1.13) in the compact form

$$\begin{bmatrix} T_N & -I_N & & \\ -I_N & \ddots & \ddots & \\ & \ddots & & -I_N \\ & & -I_N & T_N \end{bmatrix} \begin{bmatrix} \mathbf{u}_1 \\ \mathbf{u}_2 \\ \vdots \\ \mathbf{u}_N \end{bmatrix} = \begin{bmatrix} \mathbf{b}_1 - h^2\mathbf{f}_1 \\ \mathbf{b}_2 - h^2\mathbf{f}_2 \\ \vdots \\ \mathbf{b}_N - h^2\mathbf{f}_N \end{bmatrix}. \qquad (9.1.19)$$

We now make several comments about this system of equations. If N is of moderate size, say $N = 100$, then there are $N^2 = 10^4$ unknowns, and the matrix in (9.1.19) is $10,000 \times 10,000$. In each row of the matrix there are at most five nonzero elements, regardless of the size of N, so the distribution of nonzero to zero elements is very "sparse" if N is at all large. Such matrices are called *large sparse matrices* and arise in a variety of ways besides the numerical solution of partial differential equations.

It is the property of being sparse that allows such large systems of equations to be solved on today's computers with relative ease. Recall that in Chapter 4, we saw that Gaussian elimination requires on the order of n^3 arithmetic operations to solve an $n \times n$ linear system. Hence if a $10^4 \times 10^4$ linear system were "dense," that is, few of its elements were zero, and Gaussian elimination were used to solve the system, then on the order of 10^{12} operations would be required. At a rate of 10^6 operations per second, it would require several hours to solve such a system. Moreover, for the corresponding three dimensional problem the size of the system would be 10^6, requiring 10^{18} operations, which is completely beyond the capacity of the fastest computers. However, by utilizing the special structure and sparsity of systems such as (9.1.19), we shall see in the next two sections that they can be accurately solved relatively quickly and accurately, despite their large size.

The Heat Equation

We end this section by applying the discretization of Poisson's equation to the heat equation (9.1.3) in two space variables where, again for simplicity in exposition, we will assume that the x, y domain is the unit square of Figure 9.3

and that the Dirichlet boundary conditions (9.1.4) are given on the sides of the square. We also assume the initial condition (9.1.6).

Corresponding to the method (8.2.5) in the case of a single space variable, we can consider the following explicit method for (9.1.3):

$$u_{ij}^{m+1} = u_{ij}^m + \frac{c\Delta t}{h^2}(u_{i,j+1}^m + u_{i,j-1}^m + u_{i+1,j}^m + u_{i-1,j}^m - 4u_{ij}^m), \tag{9.1.20}$$

for $m = 0, 1, \ldots,$ and $i, j = 1, \ldots, N$. Here u_{ij}^m denotes the approximate solution at the i, j gridpoint and at the mth time level $m\Delta t$, and u_{ij}^{m+1} is the approximate solution at the next time level. The terms in parentheses on the right-hand side of (9.1.20) correspond exactly to the discretization (9.1.13) with $f_{ij} = 0$. The prescription (9.1.20) has the same properties as its one-dimensional counterpart (8.2.5): it is first-order accurate in time and second-order accurate in space, and it is easy to carry out. It is also subject to a similar stability condition

$$\Delta t \leq \frac{h^2}{4c}, \tag{9.1.21}$$

and thus has the problem that if h is small, very small time steps are required.

We can attempt to circumvent this restriction on the time step in the same way that we did in Section 8.3 by the use of implicit methods. For example, the implicit method (8.3.2) now becomes

$$u_{ij}^{m+1} = u_{ij}^m + \frac{c\Delta t}{h^2}(u_{ij+1}^{m+1} + u_{ij-1}^{m+1} + u_{i+1,j}^{m+1} + u_{i-1,j}^{m+1} - 4u_{ij}^{m+1}), \tag{9.1.22}$$

which is unconditionally stable. However, to carry out this method requires the solution at each time step of the system of linear equations

$$\left(4 + \frac{h^2}{c\Delta t}\right) u_{ij}^{m+1} - u_{ij+1}^{m+1} - u_{ij-1}^{m+1} - u_{i+1,j}^{m+1} - u_{i-1,j}^{m+1} = \frac{h^2}{c\Delta t} u_{ij}^m, \tag{9.1.23}$$

for $i, j = 1, \ldots, N$. This system has the same form as the system (9.1.13) for Poisson's equation, with the exception that the coefficent of u_{ij}^{m+1} is modified. Indeed, the left-hand sides of the equations (9.1.23) have exactly the same form as the finite difference equations for the *Helmholtz equation* $u_{xx} + u_{yy} - \sigma u = 0$, where σ is a given function of x and y. The term $-\sigma u$ in this differential equation becomes $-h^2\sigma_{ij}u_{ij}$ in the difference equations (9.1.13), with $\sigma_{ij} = \sigma(x_i, y_j)$. The particular constant function $\sigma = -1/(c\Delta t)$ then corresponds to the left-hand side of (9.1.23).

In the case of a single space variable, the use of an implicit method such as (8.3.2) does not cause much computational difficulty since the solution of tridiagonal systems of equations can be accomplished so rapidly. However, each time step of (9.1.23) requires the solution of a two-dimensional Poisson-type equation, which is a much more difficult computational problem. The

Crank-Nicolson method (8.3.12) can also be easily extended to equation (9.1.3) (Exercise 9.1.3) but suffers from the same difficulty that Poisson-type equations must be solved at each time step. We shall consider, instead, a different class of methods, in which the basic computational step is the solution of tridiagonal systems of equations. These are *time-splitting* methods, in which the time interval $(t, t + \Delta t)$ is further subdivided and, in essence, only one-dimensional problems are solved at each timestep.

An Alternating Direction Method

One of the most classical time-splitting methods is the *Peaceman-Rachford alternating direction implicit (ADI)* method, which has the form

$$u_{ij}^{m+1/2} = u_{ij}^m + \frac{1}{2}\frac{c\Delta t}{h^2}(u_{i+1,j}^{m+1/2} + u_{i-1,j}^{m+1/2} - 2u_{ij}^{m+1/2} + u_{i,j+1}^m + u_{i,j-1}^m - 2u_{ij}^m), \tag{9.1.24a}$$

$$u_{ij}^{m+1} = u_{ij}^{m+1/2} + \frac{1}{2}\frac{c\Delta t}{h^2}(u_{i,j+1}^{m+1} + u_{i,j-1}^{m+1} - 2u_{ij}^{m+1} + u_{i+1,j}^{m+1/2} + u_{i-1,j}^{m+1/2} - 2u_{ij}^{m+1/2}). \tag{9.1.24b}$$

This is a two-step method in which intermediate values $u_{ij}^{m+1/2}$, $i, j = 1, \ldots, N$, are computed at the first step (9.1.24a). These $u_{ij}^{m+1/2}$ are to be interpreted as approximate values of the solution at the intermediate time level $m + \frac{1}{2}$; thus the factor $\frac{1}{2}$ appears on the right-hand side of (9.1.24a) because the time step is $\frac{1}{2}\Delta t$. The computation in (9.1.24a) involves the solution of the N tridiagonal systems of equations

$$(2+\alpha)u_{ij}^{m+1/2} - u_{i+1,j}^{m+1/2} - u_{i-1,j}^{m+1/2} = (\alpha - 2)u_{ij}^m + u_{i,j+1}^m + u_{i,j-1}^m, \qquad i = 1, \ldots, N, \tag{9.1.25}$$

for $j = 1, \ldots, N$, where $\alpha = 2h^2/(c\Delta t)$; that is, for each fixed j, (9.1.25) is a tridiagonal system whose solution is $u_{ij}^{m+1/2}$, $i = 1, \ldots, N$. The coefficient matrix of each of these systems is $\alpha I + A$, where A is the $(2, -1)$ tridiagonal matrix (3.1.10). Since $\alpha I + A$ is diagonally dominant, these systems can be rapidly solved by Gaussian elimination without interchanges.

Once the intermediate values $u_{ij}^{m+1/2}$ have been computed, the final values u_{ij}^{m+1} are obtained from (9.1.24b) by solving the N tridiagonal systems

$$(2+\alpha)u_{ij}^{m+1} - u_{i,j+1}^{m+1} - u_{i,j-1}^{m+1} = (\alpha - 2)u_{ij}^{m+1/2} + u_{i+1,j}^{m+1/2} + u_{i-1,j}^{m+1/2}, \qquad j = 1, \ldots, N, \tag{9.1.26}$$

for $i = 1, \ldots, N$. Again, the coefficient matrices of these systems are $\alpha I + A$. Thus the computational process requires the solution of $2N$ tridiagonal systems of dimension N to move from the mth time level to the $(m+1)$st. It can be shown that this method is unconditionally stable.

The term *alternating direction* derives from the paradigm that in some sense we are approximating values of the solution in the x-direction by (9.1.24a), and then in the y-direction by (9.1.24b). There are many variants of ADI methods and, more generally, of time-splitting methods, and such methods are widely used for parabolic-type equations.

Supplementary Discussion and References: 9.1

The discussion of this section has been restricted to Poisson's equation in two variables on a square domain and the corresponding heat equation. However, problems that arise in practice will generally deviate considerably from these ideal conditions: the domain may not be square; the equation may have nonconstant coefficients or even be nonlinear; the boundary conditions may be a mixture of Dirichlet and Neumann conditions; there may be more than a single equation – that is, there may be a coupled system of partial differential equations; the equations may have derivatives of order higher than two; and there may be three or more independent variables. The general principles of finite difference discretization of this section still apply, but each of the preceding factors causes complications.

One of the classical references for the discretization of elliptic equations by finite difference methods is Forsythe and Wasow [1960]. See also Roache [1972] for problems that arise in fluid dynamics, and Hall and Porsching [1990]. Discussions and analyses of alternating direction methods and related methods such as the method of fractional steps are found in a number of books; see, for example, Varga [1962] and Richtmyer and Morton [1967].

In the last several years finite element and other projection-type methods have played an increasingly important role in the solution of elliptic- and parabolic-type equations. Although the mathematical basis of the finite element method goes back to the 1940s, its development into a viable procedure was carried out primarily by engineers in the 1950s and 1960s, especially for problems in structural analysis. Since then the mathematical basis has been extended and broadened and its applicability to general elliptic and parabolic equations well demonstrated. One of the method's main advantages is its ability to handle curved boundaries. For introductions to the finite element method, see Strang and Fix [1973], Becker, Carey and Oden [1981], Carey and Oden [1984], Axelsson and Barker [1984], and Hall and Porsching [1990].

EXERCISES 9.1

9.1.1. Assume that the function u is as many times continuously differentiable as needed. Expand u in Taylor series about (x_i, y_j) and show that the approximations (9.1.11) are second order accurate:

$$u_{xx}(x_i, y_j) - \frac{1}{h^2}[u(x_{i-1}, y_j) - 2u(x_i, y_j) + u(u_{i+1}, y_j)] = 0(h^2),$$

and similarly for the approximation to u_{yy}. Conclude that the local discretization error in the approximate solution of (9.1.13) is $0(h^2)$.

9.1.2. Verify that with the ordering (9.1.15), the system of equations (9.1.13) takes the form (9.1.16) for $N = 2$. Verify also that Figure 9.5 gives the coefficient matrix for $N = 4$. What is the right-hand side of the system in this case?

9.1.3. Formulate the Crank-Nicolson method for (9.1.3).

9.1.4. Consider the equation

$$u_{xx} + u_{yy} + \sigma u_x = f$$

with the boundary conditions (9.1.9) on the unit square and with σ a positive constant. Approximate u_x by central differences:

$$u_x(x, y) \doteq \frac{u(x+h, y) - u(x-h, y)}{2h}.$$

Write the difference equations corresponding to (9.1.13). Under what condition on σh will the coefficient matrix be diagonally dominant?

9.2 Direct Methods

In the previous section we obtained the system of linear equations (9.1.19) after discretizing Poisson's equation. We now consider ways to solve such systems. In the next section we study iterative methods; in the present section, we treat Gaussian elimination and other direct methods.

Fill in Gaussian Elimination

The coefficient matrix of (9.1.19) is banded with semi bandwidth N, but it is very sparse within the band: whatever the size of N, there are at most five non-zero elements in each row. However, if we apply Gaussian elimination or Cholesky factorization to this banded system, almost all elements that were zero within the band in the original matrix will become non-zero as the factorization proceeds. Such non-zero elements are called *filled-in elements* or simply *fill*. The way that fill occurs can be seen by examining the basic step in Gaussian elimination (see Section 4.2):

$$a_{ij}^{(k+1)} = [a_{ij}^{(k)} - a_{ik}^{(k)} a_{kj}^{(k)}]/a_{kk}^{(k)}. \tag{9.2.1}$$

At the kth stage of the elimination, the element $a_{ij}^{(k)}$ is modified to become $a_{ij}^{(k+1)}$, as shown in (9.2.1). If $a_{ij}^{(k)}$ is zero but $a_{ik}^{(k)}$ and $a_{kj}^{(k)}$ are both non-zero, $a_{ij}^{(k+1)}$ is non-zero and fill will occur in the i, j position at this stage. The more fill that occurs, the higher the operation count since elements that have become non-zero have to be eliminated later in the process. Ideally no fill would occur, so that the operation count would be based on the original number of non-zero elements in A.

In many, if not most, problems it is difficult to ascertain where fill will occur without carrying out the elimination process. In some cases, however, especially where A has a simple structure, it is possible to determine the fill pattern easily. We next do this for the matrix of (9.1.19). We will consider only the block 3×3 form of this matrix since this clearly exhibits the fill pattern. It is convenient to do the analysis in terms of the LU decomposition of A; if we partition L and U corresponding to A we have:

$$\begin{bmatrix} T & -I & \\ -I & T & -I \\ & -I & T \end{bmatrix} = \begin{bmatrix} L_{11} & & \\ L_{21} & L_{22} & \\ & L_{32} & L_{33} \end{bmatrix} \begin{bmatrix} U_{11} & U_{12} & \\ & U_{22} & U_{23} \\ & & U_{33} \end{bmatrix}. \qquad (9.2.2)$$

Equating the corresponding submatrices in (9.2.2), gives

$$L_{11}U_{11} = T, \qquad (9.2.3a)$$

$$L_{21}U_{11} = -I \text{ or } L_{21} = -U_{11}^{-1}; \quad U_{12} = -L_{11}^{-1}, \qquad (9.2.3b)$$

$$L_{22}U_{22} = T - L_{21}U_{12} = T - U_{11}^{-1}L_{11}^{-1} = T - T^{-1}, \qquad (9.2.3c)$$

$$L_{32} = -U_{22}^{-1}, \qquad U_{23} = -L_{22}^{-1}, \qquad (9.2.3d)$$

$$L_{33}U_{33} = T - L_{32}U_{23} = T - U_{22}^{-1}L_{22}^{-1}. \qquad (9.2.3e)$$

L_{11} and U_{11} are the LU factors of the tridiagonal matrix T. By (4.2.18), these factors have the form

$$L_{11} = \begin{bmatrix} 1 & & & \\ l_2 & 1 & & \\ & \ddots & \ddots & \\ & & l_N & 1 \end{bmatrix}, \quad U_{11} = \begin{bmatrix} u_1 & -1 & & \\ & u_2 & \ddots & \\ & & \ddots & -1 \\ & & & u_N \end{bmatrix}, \qquad (9.2.4)$$

where the -1's in U_{11} are the off-diagonal elements of T. Even though L_{11} has only two non-zero diagonals, the same is not true of L_{11}^{-1}; it is a full lower triangular matrix. To see why this is true, recall from (4.2.20) that the ith column of L_{11}^{-1} is the solution of the system

$$L_{11}\mathbf{x}_i = \mathbf{e}_i, \qquad (9.2.5)$$

where $\mathbf{e}_i$ is the vector with 1 in the ith position and zero elsewhere. The solution of (9.2.5) for $i = 1$ is

$$x_1 = 1, x_2 = -l_2 x_1 = -l_2, x_3 = -l_3 x_2 = l_2 l_3, \cdots, x_N = \pm l_2 \cdots l_N.$$

Thus provided that none of the l_i is zero, which is the case if L_{11} is the factor of T, all components of the first column of L_{11}^{-1} are non-zero. Doing the analogous computation for general i, one sees that the first non-zero component in the solution is the ith position and then each subsequent component of the solution is non-zero. It follows that each column of L_{11}^{-1} has all non-zero elements below the main diagonal. The same is true for U_{11}^T so that U_{11}^{-1} is full above the main diagonal. It is easy to verify (Exercise 9.2.1) that the product $U_{11}^{-1} L_{11}^{-1}$ is then a completely full matrix, and therefore the factors L_{22} and U_{22} in (9.2.3c) are full below and above the main diagonal, respectively. The same is true of the factors L_{32}, L_{33}, U_{23}, and U_{33}. Thus the non-zero structure of the factor L of (9.2.2) is as shown in Figure 9.6.

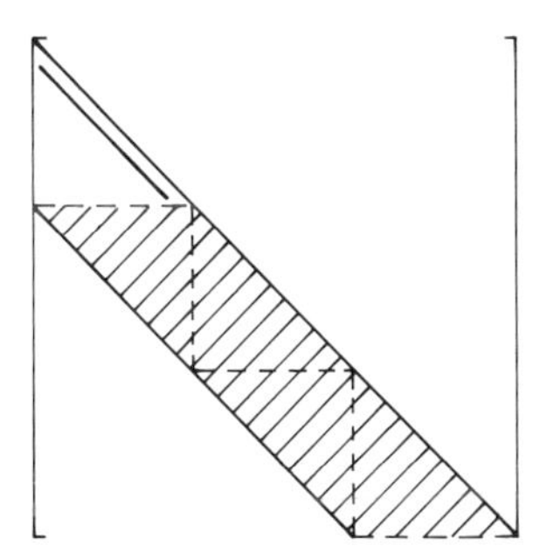

Figure 9.6: *Non-zero Structure of* L *and* U^T

U^T has the same structure and complete fill has occurred within the band, except for the first block. The same is true no matter how many blocks are in the matrix A; the 3×3 block structure of (9.2.2) was just used as an example. Thus, the amount of work to carry out the LU factorization (or Cholesky factorization, where the same fill pattern occurs) is almost as much as if the matrix A were a full banded matrix: the sparse structure of A within the band has essentially been lost because of the fill. This phenomenon is not dependent upon the particular $4, -1$ entries in A, and is true in much more general situations. Moreover, the QR factorization will also suffer from the same type of fill (Exercise 9.2.7).

One approach to circumventing this problem of fill is a reordering of the equations and unknowns. Consider the matrix with the non-zero structure of Figure 9.7(a). If Gaussian elimination is applied to this matrix, all elements will, in general, fill. On the other hand, for the matrix of Figure 9.7.(b) no elements will fill. The matrix of Figure 9.7(b) may be obtained from that of 9.7(a) by a reordering of the unknowns and equations (Exercise 9.2.2). In

Figure 9.7: *Arrowhead Matrices*

general, it will not be known in advance how to do a reordering that will minimize fill, but algorithms are known that can approximate this; see the Supplementary Discussion.

Domain Decomposition Reordering

We now consider a way to order the systems of equations (9.1.13) for the discrete Poisson problem so that Gaussian elimination can be carried out with less fill, and therefore fewer arithmetic operations, than if we used the natural ordering. We consider for illustration a rectangular grid of 22 interior points, as shown in Figure 9.8. We partition this grid into three subdomains, as well as two vertical lines of grid points called the *separator set*, labeled S. Such a partitioning is an example of a *domain decomposition.* We next number the grid points in the first subdomain using the natural ordering, followed by the points in the second, and third subdomains, and then finally those in the separator set. This is illustrated by the grid point numbers in the example of Figure 9.8.

2	6	4	20	8	12	10	22	14	18	16
1	5	3	19	7	11	9	21	13	17	15
	S_1		S		S_1		S		S_1	

Figure 9.8: *Domain Decomposition*

We now order the equations and unknowns according to the grid point numbering of Figure 9.8. The resulting coefficient matrix is shown in Figure 9.9. Also shown in Figure 9.9 is the fill pattern that results from Gaussian elimination (or Cholesky factorization). The original elements of the matrix are 4 and -1. The other integers i indicate elements that were zero in the original matrix but have become non-zero when elements in the ith column are zeroed in the Gaussian elimination process. It is left to Exercise 9.2.3 to verify the details of Figure 9.9. We note that the matrix (before fill) of Figure 9.9 is related by a permutation matrix to the matrix that would result from using the natural ordering (Exercise 9.2.4).

$$
\begin{array}{cccccccccccccccccccccc}
4 & -1 & & -1 & & & & & & & & & & & & & & & & & & \\
-1 & 4 & -1 & 1 & -1 & & & & & & & & & & & & & & & & & \\
& -1 & 4 & 2 & 2 & -1 & & & & & & & & & & & & & -1 & & & \\
-1 & 1 & 2 & 4 & -1 & & & & & & & & & & & & & & 3 & & & \\
& -1 & 2 & -1 & 4 & -1 & & & & & & & & & & & & & 3 & & & \\
& & -1 & 3 & -1 & 4 & & & & & & & & & & & & & 3 & -1 & & \\
& & & & & & 4 & -1 & & -1 & & & & & & & & & -1 & & & \\
& & & & & & -1 & 4 & -1 & 7 & -1 & & & & & & & & 7 & & & \\
& & & & & & & -1 & 4 & 8 & 8 & -1 & & & & & & & 8 & & -1 & \\
& & & & & & -1 & 7 & 8 & 4 & -1 & 9 & & & & & & & 7 & -1 & 9 & \\
& & & & & & & -1 & 8 & -1 & 4 & -1 & & & & & & & 8 & 10 & 9 & \\
& & & & & & & & -1 & 9 & -1 & 4 & & & & & & & 9 & 10 & 9 & -1 \\
& & & & & & & & & & & & 4 & -1 & & -1 & & & & & -1 & \\
& & & & & & & & & & & & -1 & 4 & -1 & 13 & -1 & & & & 13 & \\
& & & & & & & & & & & & & -1 & 4 & 14 & 14 & -1 & & & 14 & \\
& & & & & & & & & & & & -1 & 13 & 14 & 4 & -1 & 15 & & & 13 & -1 \\
& & & & & & & & & & & & & -1 & 14 & -1 & 4 & -1 & & & 14 & 16 \\
& & & & & & & & & & & & & & -1 & 15 & -1 & 4 & & & 15 & 16 \\
& & -1 & 3 & 3 & 3 & -1 & 7 & 8 & 7 & 8 & 9 & & & & & & & 4 & -1 & 9 & 12 \\
& & & & & -1 & & & & -1 & 10 & 10 & & & & & & & -1 & 4 & 10 & 12 \\
& & & & & & & & & & & & -1 & 13 & 14 & 13 & 14 & 15 & 9 & 10 & 4 & -1 \\
& & & & & & & & & & & & & & & -1 & 16 & 16 & 12 & 12 & -1 & 4
\end{array}
$$

Figure 9.9: *Domain Decomposition Matrix and Fill Pattern*

The above discussion has illustrated in a very simple case the principle of domain decomposition. More generally, if we have p subdomains and separator sets that prevent unknowns in one subdomain from being connected to unknowns in any other subdomain, then the coefficient matrix will take the *block arrowhead matrix* form,

$$
A = \begin{bmatrix} A_1 & & & & B_1^T \\ & A_2 & & & B_2^T \\ & & \ddots & & \vdots \\ & & & A_p & B_p^T \\ B_1 & B_2 & \cdots & B_p & A_s \end{bmatrix}. \tag{9.2.6}
$$

Figure 9.9 (without the fill) is the special case $p = 3$. This domain decomposition process can be applied to more general partial differential equations, more general domains, and more general discretizations.

Now assume that the matrix of (9.2.6) has an LU factorization. Then the factors may be written in the form (Exercise 9.2.5)

$$L = \begin{bmatrix} L_1 & & & & \\ & L_2 & & & \\ \vdots & & \ddots & & \\ & & & L_p & \\ \hat{L}_1 & \hat{L}_2 & \cdots & \hat{L}_p & L_s \end{bmatrix} \qquad U = \begin{bmatrix} U_1 & & \cdots & & \hat{U}_1 \\ & U_2 & & & \hat{U}_2 \\ & & \ddots & & \vdots \\ & & & U_p & \hat{U}_p \\ & & & & U_s \end{bmatrix}, \tag{9.2.7}$$

where

$$L_i U_i = A_i, \qquad i = 1, \ldots, p, \tag{9.2.8a}$$

$$\hat{L}_i U_i = B_i, \qquad L_i \hat{U}_i = B_i^T, \qquad i = 1, \ldots, p, \tag{9.2.8b}$$

$$L_s U_s = A_s - \sum_{i=1}^{p} \hat{L}_i \hat{U}_i. \tag{9.2.8c}$$

Thus the factors L and U have the same block structure as A itself. All fill occurs only within these blocks and not outside them, as illustrated in Figure 9.9.

2 6 4 20 8 12 10 22 14 18 16

1 5 3 19 7 11 9 21 13 17 15

S_1 S S_1 S S_1

Figure 9.10: *Nested Dissection Reordering of Grid Points*

Nested Dissection

We return to the example of Figure 9.9. The number of fill elements shown in the figure is 76. This is to be contrasted with 182 fill elements had we used the natural ordering and applied Gaussian elimination to the resulting banded matrix (Exercise 9.2.6). We can obtain still further improvement in the amount of fill by applying the domain decomposition principle again to each subdomain of Figure 9.8. The resulting numbering of the grid points is shown in Figure 9.10 and the corresponding coefficient matrix with fill elements in Figure 9.11. The ordering of Figure 9.8 is sometimes called *one-way dissection*, and repeated use of this within the subdomains leads to *nested dissection*.

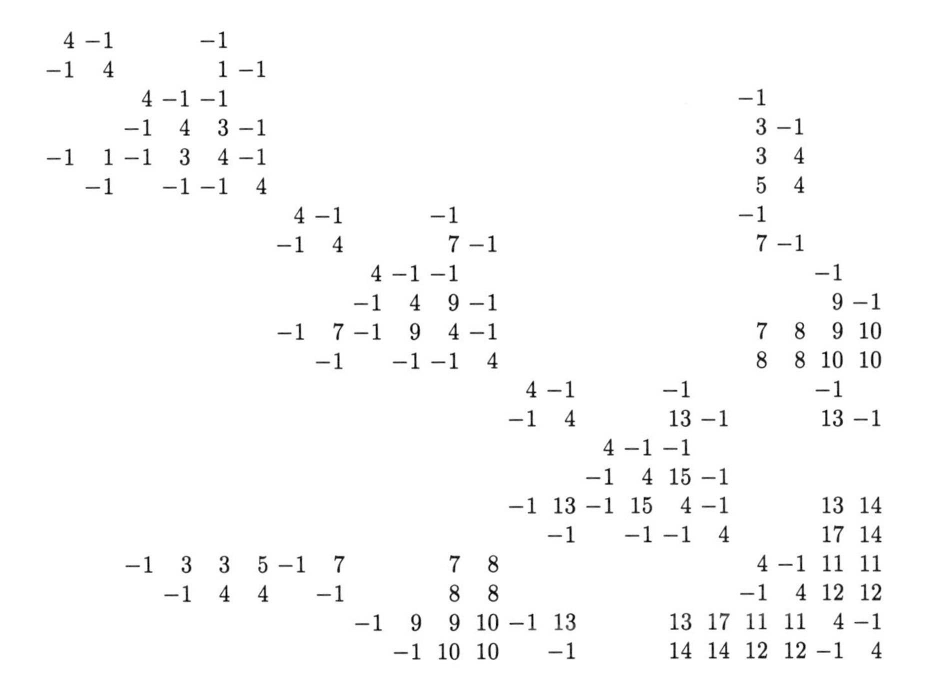

Figure 9.11: *Nested Dissection Matrix and Fill*

There are 60 fill elements in Figure 9.11, as opposed to 76 in the matrix of Figure 9.9, a relatively small but still significant saving. The main utility of these reordering techniques is, of course, for much larger problems.

In summary, the main purpose of this section has been to show that Gaussian elimination on the discrete equations of a partial differential equation leads to considerable fill, and hence extra computation, but by judicious reorderings of the equations the amount of fill can be significantly reduced.

Supplementary Discussion and References: 9.2

An excellent reference for further reading on direct methods for sparse linear systems is George and Liu [1981]. This book includes, in particular, detailed analyses of the storage and operation counts of one-way dissection and a more general treatment of nested dissection in which both horizontal and vertical separator lines are used. There also is a discussion and analysis of other reordering techniques that reduce the fill. For more general problems (i.e. not coming from Poisson's equation), one of the best such techniques is the *minimum degree algorithm.* Another good reference for sparse systems is

Duff et al. [1986].

The general idea of domain decomposition has broad application in the solution of partial differential equations, and has been a subject of intense research in the last several years because of its potential usefulness in parallel computing. For further discussion see, for example, Ortega [1988].

For general sparse linear systems, the algorithmic framework is the following:

Step 1. Use a reordering strategy to minimize fill.
Step 2. Do a symbolic factorization to determine the fill.
Set up data storage accordingly.
Step 3. Do the numerical factorization (LU or LL^T).
Step 4. Solve the corresponding triangular systems.

The symbolic factorization in Step 2 can be done surprisingly rapidly. Once this is done, the exact fill pattern is known so that the amount of storage needed for the factorization is also known. Storage can then be allowed for only the non-zero elements in the factors. Step 1 is predicated on the assumption that there is no need for interchanges to preserve numerical stability. If this is not the case, then interchanges for stability may conflict with interchanges to minimize fill. Generally, a compromise called *threshold pivoting* is used in which interchanges for stability are made only if the pivot element is too small, say less than .1 of the maximum element in its column. In this case there may be several candidates for the new pivot element, and the algorithm can choose the one that is best for maintaining sparsity.

For Poisson's equation and slight generalizations of it, there is another class of direct methods called *Fast Poisson Solvers.* There are many such methods and we will indicate only a few approaches. One is based on the fact that eigenvalues and eigenvectors of the matrix of (9.1.18) are known exactly in terms of sine functions. Thus if $A = QDQ^T$, where D is the diagonal matrix with the eigenvalues of A and Q is an orthogonal matrix whose columns are the eigenvectors of A, the solution of $A\mathbf{x} = \mathbf{b}$ is $\mathbf{x} = QD^{-1}Q^T\mathbf{b}$. The multiplications by Q^T and Q involve trigonometric sums and can be carried out by the *Fast Fourier Transform (FFT).* Other, still faster, methods are based on the *cyclic reduction* algorithm and a combination of this with the Fast Fourier Transform. For further discussion of Fast Poisson Solvers, see, for example, Hockney and Jesshope [1988].

EXERCISES 9.2

9.2.1. Let L and U be lower and upper triangular matrices for which $l_{ij} \neq 0$, $i \geq j$, and $u_{ij} \neq 0$, $i \leq j$. Show that all elements of UL are, in general, non-zero.

9.2.2. Consider the system

$$\begin{bmatrix} * & * & * & * & * \\ * & * & & & \\ * & & * & & \\ * & & & * & \\ * & & & & * \end{bmatrix} \begin{bmatrix} x_1 \\ x_2 \\ x_3 \\ x_4 \\ x_5 \end{bmatrix} = \begin{bmatrix} b_1 \\ b_2 \\ b_3 \\ b_4 \\ b_5 \end{bmatrix},$$

where $*$ denotes a non-zero element. Show that by the reordering of unknowns $x_1 \leftrightarrow x_5$, $x_2 \leftrightarrow x_4$, $x_3 \leftrightarrow x_3$ and the corresponding reordering of the equations, the system can be written as

$$\begin{bmatrix} * & & & & * \\ & * & & & * \\ & & * & & * \\ & & & * & * \\ * & * & * & * & * \end{bmatrix} \begin{bmatrix} x_5 \\ x_4 \\ x_3 \\ x_2 \\ x_1 \end{bmatrix} = \begin{bmatrix} b_5 \\ b_4 \\ b_3 \\ b_2 \\ b_1 \end{bmatrix}.$$

Generalize this to the case of the corresponding $n \times n$ system.

9.2.3. Show that the ordering of the grid points (and hence unknowns) of Figure 9.8 gives the coefficient matrix of 4's and -1's of Figure 9.9 for the system of equations (9.1.13). Next apply Gaussian elimination to this matrix and show that fill develops as indicated in Figure 9.9.

9.2.4. Show that the coefficient matrix A of Figure 9.9 (without the fill) is related to the matrix $\hat{A}$ that one would obtain from the natural ordering by $A = P\hat{A}P$, where P is a permutation matrix.

9.2.5. Verify that LU factorization of the matrix of (9.2.6) is given by (9.2.7) and (9.2.8).

9.2.6. For the grid of Figure 9.8, write out the 22×22 coefficient matrix for the natural ordering. Then verify, using the techniques that led to Figure 9.5, that the number of fill elements produced by Gaussian elimination is 182.

9.2.7. Consider the QR method for the matrix of Figure 9.5. Show that R fills in within the band.

9.2.8. Consider the special case of (9.2.6) in which

$$A = \begin{bmatrix} A_1 & B_1^T \\ B_1 & 0 \end{bmatrix},$$

and assume that A_1 is symmetric positive definite. Show that A is not positive definite but that there is a Cholesky-like factorization of the form

$$A = \begin{bmatrix} F & 0 \\ G & H \end{bmatrix} \begin{bmatrix} F^T & G^T \\ 0 & -H^T \end{bmatrix}.$$

Give an efficient algorithm for computing this factorization.

9.2.9. Consider the special case of Exercise 9.2.8 in which $A_1 = I$ and $B_1 = E^T$. Show that the solution of the normal equations $E^T E\mathbf{a} = E^T\mathbf{f}$ can be obtained by solving the system

$$\begin{bmatrix} I & E \\ E^T & 0 \end{bmatrix}\begin{bmatrix} \mathbf{r} \\ \mathbf{a} \end{bmatrix} = \begin{bmatrix} \mathbf{f} \\ 0 \end{bmatrix}.$$

Discuss the circumstances in which you might wish to solve this expanded system in place of the normal equations.

9.2.10. Use the fact that fill develops as in Figure 9.6 to show that the operation count for Gaussian elimination applied to the system (9.1.19) is $0(N^4)$.

9.3 Iterative Methods

An alternative to the direct methods discussed in the previous section is an iterative method, and we now describe some of the basic iterative methods for large sparse systems of equations.

Jacobi's Method

We consider the linear system $A\mathbf{x} = \mathbf{b}$ and make no assumptions about A at this time except that the diagonal elements are non-zero:

$$a_{ii} \neq 0, \qquad i = 1, \ldots, n. \tag{9.3.1}$$

Perhaps the simplest iterative procedure is *Jacobi's method*. Assume that an initial approximation $\mathbf{x}^0$ to the solution is chosen. Then the next iterate is given by

$$x_i^{(1)} = \frac{1}{a_{ii}}\Big(b_i - \sum_{j\neq i} a_{ij}x_j^{(0)}\Big), \qquad i = 1, \ldots, n. \tag{9.3.2}$$

It will be useful to write this in matrix-vector notation, and for this purpose, we let $D = \text{diag}(a_{11}, \ldots, a_{nn})$ and $B = D - A$. Then it is easy to verify that (9.3.2) may be written as

$$\mathbf{x}^1 = D^{-1}(\mathbf{b} + B\mathbf{x}^0),$$

and the entire sequence of Jacobi iterates is defined by

$$\mathbf{x}^{k+1} = D^{-1}(\mathbf{b} + B\mathbf{x}^k), \qquad k = 0, 1, \ldots\ . \tag{9.3.3}$$

The Gauss-Seidel Method

A closely related iteration is derived from the following observation. After $x_1^{(1)}$ is computed in (9.3.2) it is available to use in the computation of $x_2^{(1)}$, and it is natural to use this updated value rather than the original estimate $x_1^{(0)}$. If we use updated values as soon as they are available, then (9.3.2) becomes

$$x_i^{(1)} = \frac{1}{a_{ii}}\Big(b_i - \sum_{j<i} a_{ij}x_j^{(1)} - \sum_{j>i} a_{ij}x_j^{(0)}\Big), \qquad i = 1, \ldots, n, \tag{9.3.4}$$

which is the first step in the *Gauss-Seidel iteration*. To write this iteration in matrix-vector form, let $-L$ and $-U$ denote the strictly lower and upper triangular parts of A; that is, both L and U have zero main diagonals and

$$A = D - L - U. \tag{9.3.5}$$

If we multiply (9.3.4) through by a_{ii}, then it is easy to verify that the n equations in (9.3.4) can be written as

$$D\mathbf{x}^1 - L\mathbf{x}^1 = \mathbf{b} + U\mathbf{x}^0. \tag{9.3.6}$$

Since $D - L$ is a lower-triangular matrix with non-zero diagonal elements, it is nonsingular. Hence the entire sequence of Gauss-Seidel iterates is defined by

$$\mathbf{x}^{k+1} = (D - L)^{-1}[U\mathbf{x}^k + \mathbf{b}], \qquad k = 0, 1, \ldots . \tag{9.3.7}$$

The representations (9.3.3) and (9.3.7) of the Jacobi and Gauss-Seidel iterations are useful for theoretical purposes, but the actual computations would usually be done using the componentwise representations (9.3.2) and (9.3.4).

We next consider the application of these iterative methods to Laplace's equation on a square. The difference equations for this problem were given by (9.1.13) (with $f_{ij} = 0$) in the form

$$-u_{i+1,j} - u_{i-1,j} - u_{i,j+1} - u_{i,j-1} + 4u_{ij} = 0, \qquad i, j = 1, \ldots, N. \tag{9.3.8}$$

Here, the unknowns are the u_{ij}, $i, j = 1, \ldots, N$, and the remaining values of the u's are assumed known from the boundary conditions. Given initial approximations $u_{ij}^{(0)}$, a Jacobi step applied to (9.3.8) is

$$u_{ij}^{(1)} = \tfrac{1}{4}(u_{i+1,j}^{(0)} + u_{i-1,j}^{(0)} + u_{i,j+1}^{(0)} + u_{i,j-1}^{(0)}),$$

so that the new Jacobi approximation at the (i, j) grid point is simply the average of the previous approximations at the four surrounding grid points $(i \pm 1, j)$, $(i, j \pm 1)$. It is for this reason that the Jacobi method is sometimes known as the *method of simultaneous displacements*. Note that for the Jacobi method the order in which the equations are processed is immaterial. For the Gauss-Seidel method, this is not true, and each different ordering of the equations actually corresponds to a different iterative process. If we order the grid points left to right and bottom to top, as was done in Section 9.1, then a typical Gauss-Seidel step is

$$u_{ij}^{(1)} = \frac{1}{4}(u_{i-1,j}^{(1)} + u_{i,j-1}^{(1)} + u_{i+1,j}^{(0)} + u_{i,j+1}^{(0)}).$$

This new approximation at the (i, j) grid point is again an average of the approximations at the four surrounding grid points, but now using two old values and two new values. The difference between the two methods is shown schematically in Figure 9.12.

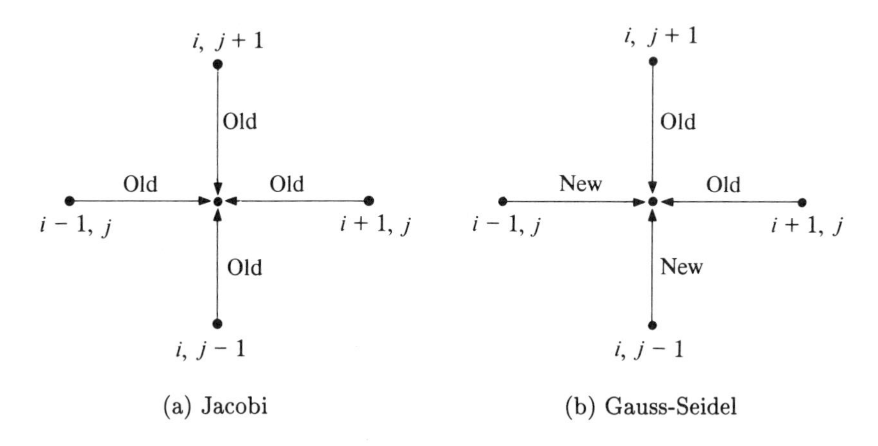

Figure 9.12: *Jacobi and Gauss-Seidel Updates*

Convergence

We consider next the question of the convergence of iterative methods. Both the Jacobi and Gauss-Seidel methods can be written in the form

$$\mathbf{x}^{k+1} = H\mathbf{x}^k + \mathbf{d}, \qquad k = 0, 1, \dots . \tag{9.3.9}$$

In particular, $H = D^{-1}B$ and $\mathbf{d} = D^{-1}\mathbf{b}$ for the Jacobi process, whereas $H = (D - L)^{-1}U$ and $\mathbf{d} = (D - L)^{-1}\mathbf{b}$ for Gauss-Seidel. Now assume that $\mathbf{x}^*$ is the exact solution of the system $A\mathbf{x} = \mathbf{b}$. For the Jacobi method we then have

$$(D - B)\mathbf{x}^* = \mathbf{b} \quad \text{or} \quad \mathbf{x}^* = D^{-1}B\mathbf{x}^* + D^{-1}\mathbf{b},$$

and for the Gauss-Seidel method

$$(D - L - U)\mathbf{x}^* = \mathbf{b} \quad \text{or} \quad \mathbf{x}^* = (D - L)^{-1}U\mathbf{x}^* + (D - L)^{-1}\mathbf{b}.$$

Thus in both cases

$$\mathbf{x}^* = H\mathbf{x}^* + \mathbf{d}. \tag{9.3.10}$$

If we subtract (9.3.10) from (9.3.9), we have

$$\mathbf{e}^{k+1} = H\mathbf{e}^k, \qquad k = 0, 1, \ldots, \tag{9.3.11}$$

where $\mathbf{e}^k = \mathbf{x}^k - \mathbf{x}^*$ is the error at the kth step.

Iterative methods of the form (9.3.9) are called *stationary one-step* methods and (9.3.10) is the *consistency condition.* Then (9.3.11) is the basic error relation for such methods. We can analyze the errors in much the same way as we analyzed the power method in Section 7.3. Assume that H has n linearly independent eigenvectors $\mathbf{v}_1, \ldots, \mathbf{v}_n$ with corresponding eigenvalues $\lambda_1, \ldots, \lambda_n$. The initial error $\mathbf{e}^0$ can then be expressed as some linear combination of the eigenvectors:

$$\mathbf{e}^0 = c_1\mathbf{v}_1 + c_2\mathbf{v}_2 + \cdots + c_n\mathbf{v}_n. \tag{9.3.12}$$

Thus,

$$\mathbf{e}^k = H^k\mathbf{e}^0 = c_1\lambda_1^k\mathbf{v}_1 + c_2\lambda_2^k\mathbf{v}_2 + \cdots + c_n\lambda_n^k\mathbf{v}_n. \tag{9.3.13}$$

In order that $\mathbf{e}^k \to 0$ as $k \to \infty$ for any $\mathbf{x}^0$ (and, hence, any c_i in (9.3.12)), we must have $|\lambda_i| < 1$, $i = 1, \ldots, n$; that is, the spectral radius, $\rho(H)$, must be less than one. This result is true also when H does not have n linearly independent eigenvectors, but the proof is more difficult (see the Supplementary Discussion). We state this basic convergence theorem as:

THEOREM 9.3.1 *If (9.3.10) holds, the iterates* (9.3.9) *converge to the solution* $\mathbf{x}^*$ *for any starting vector* $\mathbf{x}^0$ *if and only if* $\rho(H) < 1$.

Theorem 9.3.1 is the basic theoretical result for one-step iterative methods but it does not immediately tell us if a particular iterative method is convergent; we need to ascertain if the spectral radius of the iteration matrix for the method is less than 1. In general, this is a very difficult problem, for which one might have to resort to computing all the eigenvalues of the iteration matrix. But for some iterative methods and for certain classes of matrices it is relatively easy to determine that the convergence criterion is satisfied. We next give some examples of this for the Jacobi and Gauss-Seidel methods.

THEOREM 9.3.2 *Assume that the matrix A is strictly diagonally dominant:*

$$|a_{ii}| > \sum_{j \neq i} |a_{ij}|, \qquad i = 1, \ldots, n. \tag{9.3.14}$$

Then both the Jacobi and Gauss-Seidel iterations converge to the unique solution of $A\mathbf{x} = \mathbf{b}$ *for any starting vector* $\mathbf{x}^0$.

The proof of this theorem is very simple for the Jacobi method. Since $H = D^{-1}B$, the condition (9.3.14) implies that the sums of the absolute values of the elements in each row of H are less that 1. Hence $||H||_\infty < 1$, and therefore all eigenvalues of H are less that 1 in absolute value. Thus Theorem 9.3.1 applies. The proof for Gauss-Seidel is a little more complicated. Let λ be any eigenvalue of H and $\mathbf{v}$ a corresponding eigenvector. Then

$$\lambda \mathbf{v} = H\mathbf{v} = (D - L)^{-1}U\mathbf{v},$$

or

$$\lambda(D - L)\mathbf{v} = U\mathbf{v}. \tag{9.3.15}$$

Let

$$|v_k| = \max\{|v_i| \ : \ i = 1, \ldots, n\}. \tag{9.3.16}$$

The kth equation of (9.3.15) is

$$\lambda(a_{kk}v_k + \sum_{j<k} a_{kj}v_j) = -\sum_{j>k} a_{kj}v_j, \tag{9.3.17}$$

and we set

$$\alpha = \sum_{j<k} \frac{a_{kj}v_j}{a_{kk}v_k}, \qquad \beta = \sum_{j>k} \frac{a_{kj}v_j}{a_{kk}v_k}.$$

Then (9.3.17) can be written as

$$\lambda(1 + \alpha) = -\beta,$$

so that

$$|\lambda| \leq \frac{|\beta|}{|1 + \alpha|} \leq \frac{|\beta|}{1 - |\alpha|} < 1,$$

by (9.3.14) and (9.3.16). Thus we have shown that $\rho(H) < 1$ and Theorem 9.3.1 applies.

The condition of strict diagonal dominance is a rather stringent one and does not apply to the difference equations (9.3.8) for Laplace's equation: in most rows of the coefficient matrix there are four coefficients of absolute value 1 in the off-diagonal positions, so that strict inequality does not hold in (9.3.14). However, by using different techniques (see the Supplementary Discussion), it can be shown that both methods indeed converge for the difference equations (9.3.8).

The coefficient matrix of the equations (9.3.8) is symmetric [see (9.1.18)], and it can be shown that it is also positive-definite. Indeed, for many discrete analogs of elliptic partial differential equations, the coefficient matrix will be symmetric and positive-definite. In this case the Gauss-Seidel iteration will always converge, although symmetry and positive-definiteness is not, in general, sufficient for the Jacobi method to converge. We state the following theorem without proof:

THEOREM 9.3.3 *Assume that the matrix A is symmetric and positive-definite. Then the Gauss-Seidel iterates converge to the unique solution of $A\mathbf{x} = \mathbf{b}$ for any starting vector $\mathbf{x}^0$.*

Even when the Jacobi and Gauss-Seidel methods are convergent, the rate of convergence may be so slow as to preclude their usefulness; this is particularly so for discrete analogs of elliptic partial differential equations. For example, for equation (9.3.8) with $N = 44$, the error in each iteration of the Gauss-Seidel method will decrease asymptotically only by a factor of about 0.995. Moreover, the Jacobi method is about twice as slow on this problem, and the rate of convergence of both methods becomes worse as N increases.

The SOR Method

In certain cases it is possible to accelerate considerably the rate of convergence of the Gauss-Seidel method. Given the current approximation $\mathbf{x}^k$, we first compute the Gauss-Seidel iterate

$$\hat{x}_i^{(k+1)} = \frac{1}{a_{ii}}\Big(b_i - \sum_{j<i} a_{ij}x_j^{(k+1)} - \sum_{j>i} a_{ij}x_j^{(k)}\Big) \tag{9.3.18}$$

as an intermediate value, and then take the final value of the new approximation to the ith component to be

$$x_i^{(k+1)} = x_i^{(k)} + \omega(\hat{x}_i^{(k+1)} - x_i^{(k)}). \tag{9.3.19}$$

Here ω is a parameter that has been introduced to accelerate the rate of convergence.

We can rewrite (9.3.18) and (9.3.19) in the following way. First substitute (9.3.18) into (9.3.19):

$$x_i^{(k+1)} = (1-\omega)x_i^{(k)} + \frac{\omega}{a_{ii}}\Big(b_i - \sum_{j<i} a_{ij}x_j^{(k+1)} - \sum_{j>i} a_{ij}x_j^{(k)}\Big) \tag{9.3.20}$$

and then rearrange the equation into the form

$$a_{ii}x_i^{(k+1)} + \omega\sum_{j<i} a_{ij}x_i^{(k+1)} = (1-\omega)a_{ii}x_i^{(k)} - \omega\sum_{j>i} a_{ij}x_j^{(k)} + \omega b_i.$$

This relationship between the new iterates $x_i^{(k+1)}$ and the old $x_i^{(k)}$ holds for $i = 1, \ldots, n$, and using (9.3.5) we can write it in matrix-vector form as

$$D\mathbf{x}^{k+1} - \omega L\mathbf{x}^{k+1} = (1-\omega)D\mathbf{x}^k + \omega U\mathbf{x}^k + \omega\mathbf{b}.$$

Since the matrix $D - \omega L$ is again lower-triangular and, by assumption, has non-zero diagonal elements, it is nonsingular, so we may write

$$x^{k+1} = (D-\omega L)^{-1}[(1-\omega)D + \omega U]\mathbf{x}^k + \omega(D-\omega L)^{-1}\mathbf{b}. \tag{9.3.21}$$

This defines the *successive overrelaxation (SOR) method*, although, as with Gauss-Seidel, the componentwise prescription (9.3.18), (9.3.19) would usuallly be used for the actual computation. Note that if $\omega = 1$, (9.3.21) reduces to the Gauss-Seidel iteration.

We restrict ourselves to real values of the parameter ω. Then a necessary condition for the SOR iteration (9.3.21) to be convergent is that $0 < \omega < 2$ (see Exercise 9.3.15). In general, a choice of ω in this range will *not* give convergence, but in the important case that the coefficient matrix A is symmetric and positive-definite, we have the following extension of Theorem 9.3.3, which we also state without proof:

> THEOREM 9.3.4 (Ostrowski) *Assume that A is symmetric and positive-definite. Then for any $\omega \in (0, 2)$ and any starting vector $\mathbf{x}^0$, the SOR iterates* (9.3.21) *converge to the solution of $A\mathbf{x} = \mathbf{b}$.*

We would like to be able to choose the parameter ω so as to optimize the rate of convergence of the iteration (9.3.21). In general this is a very difficult problem, and we will attempt to summarize, without proofs, a few of the things that are known about its solution. For a class of matrices that are called *consistently ordered with property A*, there is a rather complete theory that relates the rate of convergence of the SOR method to that of the Jacobi method and gives important insights into how to choose the optimum value of ω. We will not define this class of matrices precisely; suffice it to say that it includes the matrix (9.1.18) of equations (9.3.8) as well as many other matrices that arise as discrete analogs of elliptic partial differential equations.

The fundamental result that holds for this class of matrices is a relationship between the eigenvalues of the SOR iteration matrix

$$H_\omega = (D - \omega L)^{-1}[(1-\omega)D + \omega U] \tag{9.3.22}$$

and the eigenvalues μ_i of the Jacobi iteration matrix $J = D^{-1}(L+U)$. Under the assumption that the μ_i are all real and less that 1 in absolute value, it can be shown that the optimum value of ω, denoted by ω_0, is given in terms of the spectral radius, $\rho(J)$, of J by

$$\omega_0 = \frac{2}{1 + \sqrt{1-\rho^2}}, \qquad \rho = \rho(J), \tag{9.3.23}$$

and is always between 1 and 2. The corresponding value of the spectral radius of H_ω is

$$\rho(H_{\omega_0}) = \omega_0 - 1, \tag{9.3.24}$$

and it is this spectral radius that governs the ultimate rate of convergence of the method. Moreover, we can ascertain the behavior of $\rho(H_\omega)$ as a function of ω, as is shown in Figure 9.13.

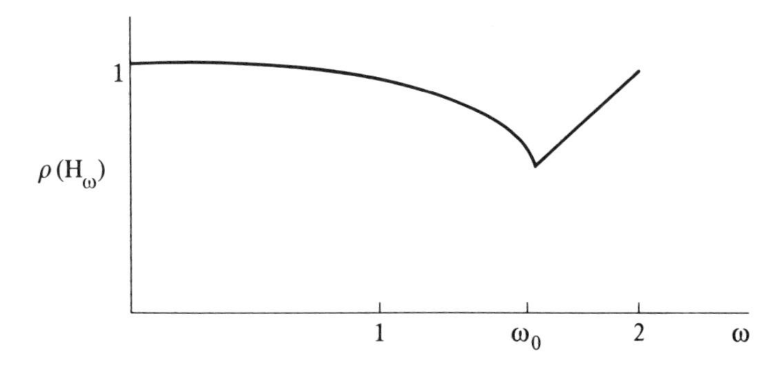

Figure 9.13: $\rho(H_\omega)$ *as a Function of* ω

We can obtain an idea of the acceleration of convergence that is possible by considering the equations (9.3.8). For this problem the eigenvalues of the Jacobi iteration matrix J can be computed explicitly, and the largest turns out to be

$$\rho(J) = \cos \pi h, \qquad h = \frac{1}{N+1}. \tag{9.3.25}$$

If we put this in (9.3.23), we obtain

$$\omega_0 = \frac{2}{1 + \sqrt{1 - \cos^2 \pi h}}, \qquad \rho(H_{\omega_0}) = \frac{1 - \sqrt{1 - \cos^2 \pi h}}{1 + \sqrt{1 - \cos^2 \pi h}}. \tag{9.3.26}$$

If, again for illustration, we take $N = 44$, then

$$p(J) \doteq 0.9976, \quad \rho(H_1) \doteq 0.995, \quad \omega_0 \doteq 1.87, \quad \rho(H_{\omega_0}) \doteq 0.87. \tag{9.3.27}$$

This shows that, asymptotically, the error in Jacobi's method will decrease by a factor of 0.9976 at each step, and that of the Gauss-Seidel method by a factor of $0.995 = (.9976)^2$, which is twice as fast. But the error in the SOR method will decrease by a factor of $0.87 = (.995)^{30}$, so that SOR is about thirty times as fast as the Gauss-Seidel method. Moreover, the improvement becomes more marked as N increases (see Exercise 9.3.8).

The preceding discussion indicates that dramatic improvements in the rate of convergence of the Gauss-Seidel method are possible. However, a number of caveats are in order. First of all, many – perhaps most – large sparse matrices that arise in practice do not enjoy being "consistently ordered with property A," and the preceding theory will not hold. It is still possible that introduction of the parameter ω into the Gauss-Seidel method will produce a substantial increase in the rate of convergence, but this will not necessarily be known in advance, nor will we know how to choose a good value of ω. Even if the coefficient matrix is "consistently ordered with property A," it still may be difficult to obtain a good estimate of ω_0. It was possible to compute explicitly the quantities of (9.3.27) only because of the very special nature of the equations (9.3.8), which allowed an exact computation of $\rho(J)$. In general this will not be possible, and to use (9.3.26) will require estimating $\rho(J)$, which is itself a difficult problem. Thus even in those cases where the preceding theory holds, it may be necessary to use an approximation process to obtain a suitable value of ω. In particular, there are "adaptive methods" that help to approximate a good value of ω as the SOR iteration proceeds (see the Supplementary Discussion).

The Conjugate Gradient Method

A large number of iterative methods for solving linear systems of equations can be derived as minimization methods. If A is symmetric and positive definite, then the quadratic function

$$q(\mathbf{x}) = \tfrac{1}{2}\mathbf{x}^T A\mathbf{x} - \mathbf{x}^T\mathbf{b} \tag{9.3.28}$$

has a unique minimizer which is the solution of $A\mathbf{x} = \mathbf{b}$ (Exercise 9.3.11). Thus methods that attempt to minimize (9.3.28) are also methods to solve $A\mathbf{x} = \mathbf{b}$. Many minimization methods for (9.3.28) can be written in the form

$$\mathbf{x}^{k+1} = \mathbf{x}^k - \alpha_k\mathbf{p}^k, \qquad k = 0, 1, \ldots . \tag{9.3.29}$$

Given the *direction vector* $\mathbf{p}^k$, one way to choose α_k is to minimize q along the line $\mathbf{x}^k - \alpha\mathbf{p}^k$; that is,

$$q(\mathbf{x}^k - \alpha_k\mathbf{p}^k) = \min_{\alpha} q(\mathbf{x}^k - \alpha\mathbf{p}^k). \tag{9.3.30}$$

For fixed $\mathbf{x}^k$ and $\mathbf{p}^k$, $q(\mathbf{x}^k - \alpha\mathbf{p}^k)$ is a quadratic function of α and may be minimized explicitly (Exercise 9.3.12) to give

$$\alpha_k = -(\mathbf{p}^k, \mathbf{r}^k)/(\mathbf{p}^k, A\mathbf{p}^k), \qquad \mathbf{r}^k = \mathbf{b} - A\mathbf{x}^k. \tag{9.3.31}$$

In (9.3.31), and henceforth, we use the notation $(\mathbf{u}, \mathbf{v})$ to denote the inner product $\mathbf{u}^T\mathbf{v}$.

Although there are many other ways to choose the α_k, we will use only (9.3.31) and concentrate on different choices of the direction vectors $\mathbf{p}^k$. One simple choice is $\mathbf{p}^k = \mathbf{r}^k$, which gives the *method of steepest descent*:

$$\mathbf{x}^{k+1} = \mathbf{x}^k - \alpha_k(\mathbf{b} - A\mathbf{x}^k), \qquad k = 0, 1, \ldots . \tag{9.3.32}$$

This is also known as *Richardson's method* and is closely related to Jacobi's method (Exercise 9.3.13). As with Jacobi's method, the convergence of (9.3.32) is usually very slow. Another simple strategy is to take $\mathbf{p}^k$ as one of the unit vectors $\mathbf{e}_i$, which has a 1 in position i and is zero elsewhere. Then, if

$\mathbf{p}^0 = \mathbf{e}_1, \mathbf{p}^1 = \mathbf{e}_2, \ldots, \mathbf{p}^{n-1} = \mathbf{e}_n$, and the α_k are chosen by (9.3.31), n steps of (9.3.29) are equivalent to one Gauss-Seidel iteration on the system $A\mathbf{x} = \mathbf{b}$ (Exercise 9.3.14).

A very interesting choice of direction vectors arises from requiring that they satisfy

$$(\mathbf{p}^i, A\mathbf{p}^j) = 0, \qquad i \neq j. \tag{9.3.33}$$

Such vectors are called *conjugate* (with respect to A). It can be shown that if $\mathbf{p}^0, \ldots, \mathbf{p}^{n-1}$ are conjugate and the α_k are chosen by (9.3.31), then the iterates $\mathbf{x}^k$ of (9.3.29) converge to the exact solution in at most n steps. This property, however, is not useful in practice because of rounding error; moreover, for large problems n is far too many iterations. For many problems, however, especially those arising from partial differential equations, methods based on conjugate directions may converge, up to a convergence criterion, in far fewer than n steps.

To use a conjugate direction method it is necessary to obtain the vectors $\mathbf{p}^k$ that satisfy (9.3.33). The *preconditioned conjugate gradient* algorithm generates these vectors as part of the overall method:

$$\text{Choose } \mathbf{x}^0. \text{ Set } \mathbf{r}^0 = \mathbf{b} - A\mathbf{x}^0. \text{ Solve } M\tilde{\mathbf{r}}^0 = \mathbf{r}^0. \text{ Set } \mathbf{p}^0 = \tilde{\mathbf{r}}^0. \tag{9.3.34a}$$

For $k = 0, 1, \ldots$

$$\alpha_k = -(\tilde{\mathbf{r}}^k, \mathbf{r}^k)/(\mathbf{p}^k, A\mathbf{p}^k) \tag{9.3.34b}$$

$$\mathbf{x}^{k+1} = \mathbf{x}^k - \alpha_k \mathbf{p}^k \tag{9.3.34c}$$

$$\mathbf{r}^{k+1} = \mathbf{r}^k + \alpha_k A\mathbf{p}^k \tag{9.3.34d}$$

$$\text{Test for convergence} \tag{9.3.34e}$$

$$\text{Solve } M\tilde{\mathbf{r}}^{k+1} = \mathbf{r}^{k+1} \tag{9.3.34f}$$

$$\beta_k = (\tilde{\mathbf{r}}^{k+1}, \mathbf{r}^{k+1})/(\tilde{\mathbf{r}}^k, \mathbf{r}^k) \tag{9.3.34g}$$

$$\mathbf{p}^{k+1} = \tilde{\mathbf{r}}^{k+1} + \beta_k \mathbf{p}^k \tag{9.3.34h}$$

We next make several comments about this algorithm. Assume for now that $M = I$ in (9.3.34a,f); then $\tilde{\mathbf{r}}^k = \mathbf{r}^k$ and (9.3.34) defines the *conjugate gradiant method.* The formula for α_k in (9.3.34b) differs from (9.3.31) because of an identity, $(\mathbf{p}^k, \mathbf{r}^k) = (\mathbf{r}^k, \mathbf{r}^k)$, in the conjugate gradient method; thus (9.3.34b,c) compute the minimum in the direction $\mathbf{p}^k$. The next step computes the new residual vector since

$$\mathbf{r}^{k+1} = \mathbf{b} - A\mathbf{x}^{k+1} = b - A(\mathbf{x}^k - \alpha_k \mathbf{p}^k) = \mathbf{r}^k + \alpha A\mathbf{p}^k.$$

Because $A\mathbf{p}^k$ is already known, this way of computing $\mathbf{r}^{k+1}$ saves the multiplication $A\mathbf{x}^{k+1}$. Next there would be a convergence test; the usual one is $(\mathbf{r}^{k+1}, \mathbf{r}^{k+1}) = ||\mathbf{r}^{k+1}||_2^2 \leq \varepsilon$. Finally, the last two steps, (9.3.34g,h), compute a new direction vector $\mathbf{p}^{k+1}$. It is not obvious that the direction vectors computed in this fashion are conjugate, but it can be proved relatively easily (Exercise 9.3.9).

Assuming still that $M = I$, the major work in each step of the conjugate gradient algorithm is the matrix-vector product $A\mathbf{p}^k$. Note that as in the Lanczos algorithm of Chapter 7, A need not be known explicitly as long as a matrix-vector product can be computed. In addition to this product, three "saxpy" operations of the form vector + scalar * vector are required, as well as inner products for $(\mathbf{p}^k, A\mathbf{p}^k)$ and $(\mathbf{r}^k, \mathbf{r}^k)$; if the latter quantity is used in the convergence test, no further work is required there except a comparison.

We now return to the assumption that $M = I$. It can be shown that if $\mathbf{x}^*$ is the exact solution, then

$$||\mathbf{x}^k - \mathbf{x}^*||_2 \leq 2\sqrt{\kappa}\alpha^k||x^0 - x^*||_2, \tag{9.3.35a}$$

where $\kappa = \text{cond}_2(A) = ||A||_2||A^{-1}||_2$ and

$$\alpha = (\sqrt{\kappa} - 1)/(\sqrt{\kappa} + 1). \tag{9.3.35b}$$

The larger the condition number, the closer α is to 1 and the slower the rate of convergence. The role of the matrix M in (9.3.34f) is to "precondition" the matrix A and reduce its condition number so as to obtain faster convergence.

Preconditioning

There are numerous ways to choose the matrix M but we discuss only two. First, consider $M = D$, the main diagonal of A. This can be very beneficial if the main diagonal elements of A have considerable variability. On the other hand, if they are constant, as for the Poisson equations (9.1.13), then this choice of M does not change the rate of convergence. Another, more generally applicable, way to choose M is known as *incomplete Cholesky* factorization:

For $j = 1, \ldots, n$

$$l_{jj} = \left(a_{jj} - \sum_{k=1}^{j-1} l_{jk}^2\right)^{1/2} \tag{9.3.36a}$$

For $i = j+1, \ldots, n$

If $a_{ij} = 0$ then $l_{ij} = 0$ else (9.3.36b)

$$l_{ij} = \frac{a_{ij} - \sum_{k=1}^{j-1} l_{ik}l_{jk}}{l_{jj}}. \tag{9.3.36c}$$

Without the statement (9.3.36b), this is the Cholesky factorization of Figure 4.5. The effect of (9.3.36b) is to bypass the computation of l_{ij} if the corresponding element of A is zero. Thus (9.3.36b) suppresses the fill that would normally occur in a Cholesky factorization, and we obtain a "no-fill" approximate Cholesky factor L. We then take $M = LL^T$. The incomplete factorization (9.3.36) is done once at the beginning of the iteration; then during each iteration (9.3.34f) is carried out by solving two triangular systems with coefficient matrices L and L^T. The hope is that the extra work of incorporating this preconditioner will be more than repaid by a reduction in the iteration count. For example, for a two dimensional Laplace equation on a 99×101 grid (9999 equations), the conjugate gradient iteration takes 266 iterations (for a given convergence criterion), whereas with incomplete Cholesky preconditioning this is reduced to 45 iterations. The extra work to incorporate the preconditioning does no more than double the time per iteration (and can be much less with certain optimizations) so that the gain in speed is at least a factor of three. Although the Cholesky factorization can always be performed if A is symmetric positive definite, the incomplete factorization of (9.3.36) may fail. However, it is possible to modify (9.3.36) in various ways so that an incomplete factorization is possible (see the Supplementary Discussion).

Summary

In this section we have given an introduction to some of the simplest iterative methods for large sparse systems of linear equations, especially those which arise from elliptic partial differential equations. There are a number of other methods, some of which are mentioned in the Supplementary Discussion, and which method to use on a given problem is usually not clear. Moreover, direct methods such as Gaussian elimination with suitable reorderings are sometimes very efficient. In fact, for elliptic equations in two dimensions, direct methods are probably to be preferred, whereas for three-dimensional problems iterative methods are probably the best. However, which method to use will depend upon several factors, including the computer, the particular equation to be solved, and the accuracy required in the solution.

Supplementary Discussion and References: 9.3

The Jacobi and Gauss-Seidel iterations are classical methods that go back to the last century. The basic theory of the SOR method was developed by Young [1950]. For a complete discussion of the Jacobi, Gauss-Seidel, and SOR methods and their many variants, see Varga [1962] and Young [1971]. For ways to compute adaptively the ω in SOR, see Hageman and Young [1981] and Young and Mai [1990].

The basic convergence Theorem 9.3.1 is really equivalent to powers of a matrix tending to zero. The condition that $H^k \mathbf{e}_0 \to 0$ as $k \to \infty$ for any $\mathbf{e}_0$ is equivalent to $H^k \to 0$ as $k \to \infty$. If $H = PJP^{-1}$, where J is the Jordan canonical form of H, then $H^k = PJ^kP^{-1}$ and $H^k \to 0$ if and only if $J^k \to 0$ as $k \to \infty$. If J is diagonal, then the result is essentially what was shown in the text. Otherwise one needs to analyze powers of a Jordan block. By considering $(\lambda I + E)^k$, where E is a matrix of 1's on the first superdiagonal, it is easy to see that for $k \geq n$

$$\begin{bmatrix} \lambda & 1 & & \\ & \lambda & \ddots & \\ & & \ddots & 1 \\ & & & \lambda \end{bmatrix}^k = \begin{bmatrix} \lambda^k & k\lambda^{k-1} & \binom{k}{2}\lambda^{k-2} & \cdots & \binom{k}{n-1}\lambda^{k-n+1} \\ & \lambda^k & k\lambda^{k-1} & \cdots & \vdots \\ & & \ddots & \ddots & \\ & & & & k\lambda^{k-1} \\ & & & & \lambda^k \end{bmatrix},$$

which shows that the powers of a Jordan block tend to zero if and only if $|\lambda| < 1$. Although $\rho(H) < 1$ is a necessary and sufficient condition that the iterates (9.3.9) converge for any $\mathbf{x}^0$, it is not necessarily the case that the relation $||\mathbf{e}^{k+1}|| < ||\mathbf{e}^k||$ for the error vectors will hold in some usual norm (see Exercise 9.3.10).

Theorem 9.3.2 can be extended to a form suitable for handling equations such as (9.3.8). If A is irreducibly diagonally dominant (see the Supplementary Discussion of Section 3.1), then the system of equations $A\mathbf{x} = \mathbf{b}$ has a unique solution $\mathbf{x}^*$, and both the Jacobi and Gauss-Seidel iterates converge to $\mathbf{x}^*$ for any starting vector $\mathbf{x}^0$. It can be shown that the coefficient matrix of (9.3.8) is irreducible and that this result then applies.

Theorem 9.3.4 is a special case of a more general convergence theorem: if A is symmetric positive definite and $A = P - Q$, where P is nonsingular and $\mathbf{x}^T(P + Q)\mathbf{x} > 0$ for all $\mathbf{x} \neq 0$, then $\rho(P^{-1}Q) < 1$. This is sometimes referred to as the Householder-John Theorem but it is actually due to Weissinger; see Ortega [1990], where it is called the P-regular splitting theorem. A converse of Theorem 9.3.4 also holds: If A is symmetric with positive diagonal elements and SOR converges for some $\omega \in (0, 2)$ and every $\mathbf{x}^0$, then A is positive definite. If this converse is added, Theorem 9.3.4 is known as the Ostrowski-Reich Theorem.

Iterative methods such as Gauss-Seidel tend to make fairly rapid progress in the early stages and then slow down. As a consequence Gauss-Seidel is rarely used by itself but it can be a key component of the *multigrid* method. Assume that a partial differential equation is discretized on a grid of points for which the approximate solution is desired. We call this the *fine* grid, and subsets of these grid points are *coarser* grids. A multigrid method takes a few Gauss-Seidel iterations on the fine grid, then stops and "restricts" information from the fine grid to a coarser grid and performs a few Gauss-Seidel iterations on this coarser grid. The process is continued by using still coarser grids. Information from these coarser grids is transmitted back to the finer grids by interpolation. There are many important details in this process and many variations on the basic multigrid idea. Properly implemented, multigrid methods are increasingly becoming the method of choice for many problems. For an introduction to the multigrid method, see Briggs [1987] and, for a more advanced treatment, Hackbrush [1985].

The conjugate gradient method was introduced by Hestenes and Stiefel [1952]. Because of its finite convergence property it is a direct method, although it was not a successful competitor against Gaussian elimination. Since the 1970's, however, it has enjoyed a resurgence as an iterative method for large sparse systems. Another sometimes useful way to view the conjugate gradient method is as an acceleration technique. It can be shown that the conjugate gradient iterates satisfy the three-term recurrence relation known as the *Rutishauser form*,

$$\mathbf{x}^{k+1} = \rho_{k+1}\{\delta_{k+1}[(I-A)\mathbf{x}^k + \mathbf{b}] + (1-\delta_{k+1})\mathbf{x}^k\} + (1-\rho_{k+1})\mathbf{x}^{k-1}, \quad (9.3.37)$$

where the scalar parameters ρ_k and δ_k are themselves computed by recurrence relations. In this form, (9.3.37) can be considered an acceleration of the basic iterative method $\mathbf{y}^{k+1} = (I-A)\mathbf{y}^k + \mathbf{b}$, which is (9.3.32) with the $\alpha_k = 1$. For further reading on the conjugate gradient method, see Golub and Van Loan [1989], Ortega [1988], and Hageman and Young [1981]. In particular, see Ortega [1988] for various ways to modify the incomplete Cholesky factorization (9.3.36) in case it should fail.

Many elliptic equations that arise in practice are nonlinear. The methods of this section do not apply immediately in this case, although extensions of them to nonlinear equations have been developed (see Ortega and Rheinboldt [1970]). On the other hand, if a method such as Newton's method is used (Section 5.3), then at each stage a large sparse linear system must be solved approximately, and iterative methods can be used for this purpose.

EXERCISES 9.3

9.3.1. Apply the Jacobi and Gauss-Seidel iterations to the system of equations $A\mathbf{x} = \mathbf{b}$ where

$$A = \begin{bmatrix} 3 & 1 & 1 \\ 1 & 3 & 1 \\ 1 & 1 & 3 \end{bmatrix}, \qquad \mathbf{b} = \begin{bmatrix} 1 \\ 2 \\ 3 \end{bmatrix}.$$

Use the starting approximation $\mathbf{x}^0 = (1,1,1)$ and take enough steps of the iterative processes for the pattern of convergence to become clear.

9.3.2. Write computer programs for the Jacobi and Gauss-Seidel methods. Test them on the problem of Exercise 9.3.1.

9.3.3. Write out in detail the Jacobi and Gauss-Seidel iterations for the equations (9.3.8) for $N = 3$.

9.3.4. Consider the elliptic equation $u_{xx} + u_{yy} + cu = 0$ with the values of u prescribed on the boundary of a square domain. Derive the difference equations corresponding to (9.3.8). If c is a negative constant, show that the resulting coefficient matrix is strictly diagonally dominant.

9.3.5. Let A be a real $n \times n$ symmetric positive definite matrix.

a. Show that the diagonal elements of A are necessarily positive.

b. If C is any real $n \times n$ nonsingular matrix, show that $C^T AC$ is also positive definite.

9.3.6. Carry out several steps of the SOR iteration for the problem of Exercise 9.3.1. Use the values $\omega = 0.6$ and $\omega = 1.4$ and compare the rates of convergence to the Gauss-Seidel iteration.

9.3.7. Write a computer program to apply the SOR iteration to (9.3.8).

9.3.8. Use the relations (9.3.25) and (9.3.26) to compute $\rho(J)$, ω_0, and $\rho(H_{\omega_0})$ for (9.3.8) for $N = 99$ and $N = 999$.

9.3.9. Prove that the relationship (9.3.33) holds for the conjugate gradient method by using the following induction argument. As the induction hypothesis, assume that $(\mathbf{p}^k, A\mathbf{p}^j) = (\mathbf{r}^k, \mathbf{r}^j) = 0$, $j = 0, \ldots, k-1$. $((\mathbf{p}^1, A\mathbf{p}^0) = 0$; $(\mathbf{r}^1, \mathbf{r}^0) = (\mathbf{r}^0 - \alpha_0 A\mathbf{p}^0, \mathbf{p}^0) = 0$ by the definition of α_0.) Then show that $(\mathbf{p}^{k+1}, A\mathbf{p}^j) = (\mathbf{r}^{k+1}, \mathbf{r}^j) = 0$ for $j = 0, 1, \ldots, k$.

9.3.10. Let $H = \begin{bmatrix} 0.5 & \alpha \\ 0 & 0 \end{bmatrix}$. Compute H^k and show that if $\mathbf{e}^0 = (0,1)^T$, then $\mathbf{e}^k = H^k \mathbf{e}^0 = (\alpha 2^{-k+1}, 0)^T$. Thus if $\alpha = 2^0$, $||e^k||_2 \geq ||e_0||_2$.

9.3.11. Show that the gradient vector for (9.3.28) is $\nabla q(\mathbf{x}) = A\mathbf{x} - \mathbf{b}$. If A is symmetric positive definite, show that the unique solution of $A\mathbf{x} = \mathbf{b}$ is the unique minimizer of $\mathbf{q}$.

9.3.12. For the function q of (9.3.28), show that

$$q(\mathbf{x} - \alpha\mathbf{p}) = \tfrac{1}{2}\mathbf{q}^T A\mathbf{p}\alpha^2 + \mathbf{p}^T(\mathbf{b} - A\mathbf{x})\alpha - \tfrac{1}{2}\mathbf{x}^T(2\mathbf{b} - A\mathbf{x}).$$

For $\mathbf{x} = \mathbf{x}^k$ and $\mathbf{p} = \mathbf{p}^k$, minimize this function of α to obtain (9.3.31).

9.3.13. If $\alpha_k \equiv 1$ and A is such that its main diagonal is I, show that (9.3.32) is Jacobi's method.

9.3.14. Let $\mathbf{p}^0 = \mathbf{e}_1, \mathbf{p}^1 = \mathbf{e}_2, \ldots, \mathbf{p}^{n-1} = \mathbf{e}_n$, where $\mathbf{e}_i$ is the vector with 1 in the ith component and zero elsewhere. Show that n steps of (9.3.29) with α_k given by (9.3.31) is equivalent to one Gauss-Seidel iteration on the system $A\mathbf{x} = \mathbf{b}$.

9.3.15. For the *SOR* iteration matrix (9.3.22), show that det $(D - \omega L)^{-1} = \det D^{-1}$ and then that det $H_\omega = (1-\omega)^n$. Using the fact that the determinant of a matrix is the product of its eigenvalues, conclude that if ω is real and $\omega \leq 0$ or $\omega \geq 2$, then at least one eigenvalue of H_ω must be greater than or equal to 1 in magnitude.

9.3.16. Let $A = \epsilon L + D - U$, where ϵ is a small parameter and $\|U\|_\infty = 1$.

a. Give an upper bound on the spectral radius of the Gauss-Seidel iteration matrix $(\epsilon L + D)^{-1}U$.

b. Now consider the "backward" Gauss-Seidel iteration

$$(D + U)\mathbf{x}^{k+1} = -\epsilon L\mathbf{x}^k + \mathbf{b}.$$

Give an upper bound on the spectral radius of the iteration matrix for this method and show its dependence on ϵ.

Appendix 1

Analysis and Differential Equations

In this appendix we review briefly and without proofs some of the basic results from calculus and ordinary differential equations that are used in the text.

> MEAN-VALUE THEOREM: *If the function f is continuously differentiable on an interval $[a, b]$, then there is a point ξ between a and b such that*

$$f(b) - f(a) = f'(\xi)(b - a). \tag{A.1.1}$$

> TAYLOR EXPANSION: *If the function f is k times continuously differentiable on an interval $[a, b]$, then for any x and x_0 between a and b, there is a ξ between x and x_0 such that*

$$\begin{aligned} f(x) &= f(x_0) + f'(x_0)(x - x_0) + \tfrac{1}{2}f''(x_0)(x - x_0)^2 + \cdots \\ &\quad + \frac{1}{(k-1)!}f^{(k-1)}(x_0)(x - x_0)^{(k-1)} + \frac{1}{k!}f^{(k)}(\xi)(x - x_0)^k. \end{aligned} \tag{A.1.2}$$

Note that the mean-value theorem can be considered a special case of the Taylor expansion for $k = 1$ and $a = x_0$, $b = x$.

> CHAIN RULE: *If f and g are two differentiable functions, then the composite function $h(x) = f(g(x))$ is differentiable, and*

$$h'(x) = f'(g(x))g'(x). \tag{A.1.3}$$

SECOND MEAN-VALUE THEOREM OF THE INTEGRAL CALCULUS: *If u and v are continuous functions on the interval $[a, b]$ and v does not change sign in $[a, b]$, then there is a $\xi \in [a, b]$ such that*

$$\int_a^b u(x)v(x)dx = u(\xi) \int_a^b v(x)dx. \tag{A.1.4}$$

The next results deal with functions of several variables $f(x_1, \ldots, x_n)$, or $f(\mathbf{x})$ where $\mathbf{x}$ is the vector with components $x_1, \ldots, x_n$. The partial derivative of f with respect to the ith variable is defined at a point $\mathbf{x}$ by

$$\frac{\partial f}{\partial x_i}(\mathbf{x}) = \lim_{h \to 0} \frac{1}{h}[f(x_1, \ldots, x_{i-1}, x_i + h, x_{i+1}, \ldots, x_n) - f(\mathbf{x})], \tag{A.1.5}$$

and similarly for partial derivatives of higher order. The *derivative* of f at a point $\mathbf{x}$ is defined by

$$f'(\mathbf{x}) = \left(\frac{\partial f}{\partial x_1}(\mathbf{x}), \ldots, \frac{\partial f}{\partial x_n}(\mathbf{x})\right), \tag{A.1.6}$$

and is considered to be a row vector. The transpose of this vector is sometimes called the *gradient* of f and is denoted by ∇f. In this context, it is often convenient to view ∇ as the vector operator of partial derivatives:

$$\nabla = \left(\frac{\partial}{\partial x_1}, \ldots, \frac{\partial}{\partial x_n}\right).$$

Then, the operator ∇^2, often denoted by Δ, is the dot product of ∇ with itself, so that

$$\Delta f = \nabla^2 f = \frac{\partial^2 f}{\partial x_1^2} + \cdots + \frac{\partial^2 f}{\partial x_n^2}.$$

This sum of second partial derivatives is very important in the study of partial differential equations (see Chapters 8 and 9).

The function f is said to be continuously differentiable in some region of n space if each (first) partial derivative of f exists and is continuous within that region. For functions of several variables the mean-value theorem again holds, as follows.

MEAN-VALUE THEOREM FOR FUNCTIONS OF SEVERAL VARIABLES: *If f is a continuously differentiable function of n variables in some region, and if $\mathbf{x}$ and $\mathbf{y}$ are two points such that the points*

$$t\mathbf{x} + (1 - t)\mathbf{y}, \qquad 0 \le t \le 1$$

are all in the region, then there is a ξ between 0 and 1 such that

$$f(\mathbf{y}) - f(\mathbf{x}) = f'(\xi\mathbf{x} + (1-\xi)\mathbf{y})(\mathbf{y} - \mathbf{x}). \tag{A.1.7}$$

We note that this mean-value theorem is simply the usual one for functions of a single variable as applied to the function

$$\hat{f}(t) = f(t\mathbf{x} + (1-t)\mathbf{y}).$$

If $f_1, \ldots, f_m$ are all functions of n variables, then we denote the vector-valued function with components $f_1, \ldots, f_m$ by $\mathbf{F}$. For such vector-valued functions, there is a natural derivative defined by

$$\mathbf{F}'(\mathbf{x}) = \left(\frac{\partial f_i(\mathbf{x})}{\partial x_j}\right), \tag{A.1.8}$$

where the notation means that $\mathbf{F}'(\mathbf{x})$ is an $m \times n$ matrix, usually called the *Jacobian matrix*, whose i, j element is the partial derivative of the ith component of $\mathbf{F}$ with respect to the jth variable. For example, if $m = n = 2$, then

$$\mathbf{F}'(\mathbf{x}) = \begin{bmatrix} \dfrac{\partial f_1(\mathbf{x})}{\partial x_1} & \dfrac{\partial f_1(\mathbf{x})}{\partial x_2} \\ \dfrac{\partial f_2(\mathbf{x})}{\partial x_1} & \dfrac{\partial f_2(\mathbf{x})}{\partial x_2} \end{bmatrix}.$$

Note that in the special case $m = 1$, $\mathbf{F}$ is simply the single function f_1, and the Jacobian matrix reduces to the row vector given by (A.1.6).

We next consider results for ordinary differential equations. If y is a function of a single variable t, then an ordinary differential equation for y is a relation of the form

$$F(t, y(t), y'(t), \ldots, y^{(n)}(t)) = 0 \tag{A.1.9}$$

for some given function F of $n+2$ variables, where the independent variable t ranges over some finite or infinite interval. Equation (A.1.9) is the most general nth-order ordinary differential equation, where the *order* is determined by the highest-order derivative of the unknown function y that appears in the equation. Usually the equation is assumed to be explicit in the highest derivative and is written as

$$y^{(n)}(t) = f(t, y, (t), y'(t), \ldots, y^{(n-1)}(t)). \tag{A.1.10}$$

If the function f is linear in y and its derivatives, then the equation is called linear and can be written in the form

$$y^{(n)}(t) = a_0(t) + a_1(t)y(t) + \cdots + a_n(t)y^{(n-1)}(t) \tag{A.1.11}$$

for given functions $a_0, \ldots, a_n$.

The equation (A.1.10) can also be considered for vector-valued functions $\mathbf{y}$ and $\mathbf{f}$, in which case we would have a system of nth-order equations. The simplest such possibility is a system of first-order equations

$$\mathbf{y}'(t) = \mathbf{f}(t, \mathbf{y}(t)), \tag{A.1.12}$$

where we assume that $\mathbf{y}$ and $\mathbf{f}$ are n vectors with components $y_1, \ldots, y_n$ and $f_1, \ldots, f_n$.

In principle, a system of first-order equations is all that we need to consider, since a single nth-order equation can be reduced to a system of n first-order equations (and, consequently, a system of m nth-order equations to a system of nm first-order equations). This reduction can be achieved, for example, as follows. Define new variables

$$y_i(t) \equiv y^{(i-1)}(t), \qquad i = 1, \ldots, n. \tag{A.1.13}$$

In terms of these variables, (A.1.10) becomes

$$y_n' = f(t, y_1, y_2, \ldots, y_{n-1}), \tag{A.1.14}$$

whereas from (A.1.13) we obtain

$$y_i' = y_{i+1}, \qquad i = 1, \ldots, n-1. \tag{A.1.15}$$

Equations (A.1.14), (A.1.15) give a first-order system of equations in the unknowns $y_1, \ldots, y_n$, where the component y_1 is the original unknown y of equation (A.1.10).

A very important special case of (A.1.12) is when $\mathbf{f}$ is linear in $\mathbf{y}$, and the equation takes the form

$$\mathbf{y}'(t) = A(t)\mathbf{y}(t) + \mathbf{b}(t), \tag{A.1.16}$$

where A is a given $n \times n$ matrix whose elements are functions of t, and $\mathbf{b}$ is a given vector function of t. An important special case of (A.1.16), in turn, is when A is independent of t, and $\mathbf{b} = 0$, so that the equation is

$$\mathbf{y}' = A\mathbf{y}. \tag{A.1.17}$$

Such a *linear homogeneous* system with *constant coefficients* can, in principle, be solved explicitly by the series expansion

$$\mathbf{y}(t) = (I + At + \tfrac{1}{2}A^2t^2 + \cdots)\mathbf{c}, \tag{A.1.18}$$

where $\mathbf{c}$ is an arbitrary constant vector. The series expansion is simply that of the exponential of a matrix, and (A.1.18) can be written in the compact form

$$\mathbf{y}(t) = e^{At}\mathbf{c}. \tag{A.1.19}$$

Equation (A.1.19) shows that the general solution of (A.1.17) depends on n arbitrary constants – the n components of the vector $\mathbf{c}$. Thus, to obtain a unique solution of the system (A.1.17) n additional conditions must be specified, and these are usually given in terms of *initial* or *boundary conditions.* For example, suppose that we desire a solution of (A.1.17) for $t \geq 0$ such that at $t = 0$ the solution takes on the initial condition $\mathbf{y}_0$. The solution is then given by (A.1.19) as $\mathbf{y}(t) = e^{At}\mathbf{y}_0$.

For more complicated equations the initial condition will not be represented in the solution in such a straightforward fashion. Indeed, it is not immediately obvious under what conditions the general *initial-value problem*

$$\mathbf{y}'(t) = \mathbf{f}(t, \mathbf{y}(t)), \qquad \mathbf{y}(0) = \mathbf{y}_0 \tag{A.1.20}$$

will even have a unique solution, but a number of basic theorems in this regard are known and may be found in any book on ordinary differential equations.

Appendix 2

Linear Algebra

The most important tool in many areas of scientific computing is linear algebra, and we review here some of the basic results that will be used.

If $A = (a_{ij})$ is a real $n \times n$ matrix, we denote the inverse of A by A^{-1} and the determinant by $\det A$. If the inverse of A exists, then A is *nonsingular*. The following basic result gives various other ways of stating this.

THEOREM A.2.1 *The following are equivalent:*

1. *A is nonsingular.*

2. $\det A \neq 0$.

3. *The linear system $A\mathbf{x} = 0$ has only the solution $\mathbf{x} = 0$.*

4. *For any vector $\mathbf{b}$, the linear system $A\mathbf{x} = \mathbf{b}$ has a unique solution.*

5. *The columns (rows) of A are linearly independent; that is, if $\mathbf{a}_1, \ldots, \mathbf{a}_n$ are the columns of A and $\alpha_1\mathbf{a}_1 + \cdots + \alpha_n\mathbf{a}_n = 0$, then the scalars α_i are necessarily zero.*

The last condition may be rephrased to say that A has rank n where, in general, the *rank* is defined as the number of linearly independent columns (or rows) of the matrix.

The *transpose* of $A = (a_{ij})$ is $A^T = (a_{ji})$. A basic fact about determinants is that $\det A^T = \det A$. Thus, by **2.** of Theorem A.2.1, A^T is nonsingular if and only if A is nonsingular. A particularly important type of matrix satisfies $A^T = A$ and is called *symmetric*. If, in addition, $\mathbf{x}^T A\mathbf{x} > 0$ for $\mathbf{x} \neq 0$, then A is *positive definite*. By **3.** of Theorem A.2.1, a positive definite matrix is nonsingular.

A *submatrix* of A is obtained by deleting rows and columns of A. A *principle* submatrix results from deleting corresponding rows and columns; in particular, a *leading* principle submatrix of size k is obtained by deleting rows and columns $k+1, k+2, \ldots$. An important fact is that any principle submatrix of a symmetric positive definite matrix is also symmetric positive definite.

Section 7.1 is devoted to a review of eigenvalues and we give in this appendix only the basic facts. A (real or complex) scalar λ and a vector $\mathbf{x} \neq 0$ are an *eigenvalue* and *eigenvector*, respectively, of the matrix A if

$$A\mathbf{x} = \lambda\mathbf{x}. \tag{A.2.1}$$

By Theorem A.2.1, it follows that λ is an eigenvalue if and only if

$$\det(A - \lambda I) = 0. \tag{A.2.2}$$

This is the *characteristic equation* of A and is a polynomial of degree n in λ. (Here, as always, I is the identity matrix.) Consequently, A has precisely n (not necessarily distinct) eigenvalues – the n roots of (A.2.2). The collection of these n eigenvalues $\lambda_1, \cdots, \lambda_n$ is called the *spectrum* of A, and

$$\rho(A) = \max_{1 \leq i \leq n} |\lambda_i| \tag{A.2.3}$$

is the *spectral radius* of A. Even if the matrix A is real, the eigenvalues of A may be complex. If A is symmetric, however, then its eigenvalues are necessarily real. Moreover, if A is also positive-definite, then its eigenvalues are also positive. The converse also holds; that is, if all the eigenvalues of a symmetric matrix are positive, then the matrix is positive-definite.

Eigenvalues are generally difficult to compute, but there is an important class of matrices in which they are available by inspection. These are upper- or lower- *triangular matrices*

$$A = \begin{bmatrix} a_{11} & \cdots & a_{1n} \\ & \ddots & \vdots \\ & & a_{nn} \end{bmatrix}, \qquad A = \begin{bmatrix} a_{11} & & \\ \vdots & \ddots & \\ a_{n1} & \cdots & a_{nn} \end{bmatrix},$$

for which the eigenvalues are simply the main diagonal elements. An important special case of triangular matrices are *diagonal matrices*

$$D = \begin{bmatrix} d_1 & & \\ & \ddots & \\ & & d_n \end{bmatrix},$$

which we will usually denote by $D = \operatorname{diag}(d_1, \ldots, d_n)$.

The Euclidean length of a vector $\mathbf{x}$ is defined by

$$||\mathbf{x}||_2 = \left(\sum_{i=1}^{n} x_i^2\right)^{1/2}. \tag{A.2.4}$$

This is a special case of a *vector norm*, which is a real-valued function that satisfies the following distance-like properties:

$$\begin{aligned} &1.\ ||\mathbf{x}|| \geq 0 \text{ for any vector } \mathbf{x} \text{ and } ||\mathbf{x}|| = 0 \text{ only if } \mathbf{x} = 0.\\ &2.\ ||\alpha\mathbf{x}|| = |\alpha|\ ||\mathbf{x}|| \text{ for any scalar } \alpha.\\ &3.\ ||\mathbf{x}+\mathbf{y}|| \leq ||\mathbf{x}|| + ||\mathbf{y}|| \text{ for all vectors } \mathbf{x} \text{ and } \mathbf{y}.\end{aligned} \tag{A.2.5}$$

Property 3 is known as the *triangle inequality.*

The Euclidean length (A.2.4) satisfies these properties and is usually called the Euclidean norm, or l_2 norm. Other commonly used norms are defined by

$$||\mathbf{x}||_1 = \sum_{i=1}^{n} |x_i|, \qquad ||\mathbf{x}||_\infty = \max_{1\leq i\leq n} |x_i|, \tag{A.2.6}$$

which are known as the l_1 norm, and the l_∞ or max norm, respectively. The three norms (A.2.4) and (A.2.6) are special cases of the general class of l_p norms

$$||\mathbf{x}||_p = \left(\sum_{i=1}^{n} |x_i^p|\right)^{1/p}, \tag{A.2.7}$$

defined for any real number $p \in [1, \infty)$. The l_∞ norm is the limiting case of (A.2.7) as $p \to \infty$. Another important class of norms consists of the *elliptic norms* defined by

$$||\mathbf{x}|| = (\mathbf{x}^T B\mathbf{x})^{1/2}$$

for some given symmetric positive-definite matrix B; the Euclidean norm is the special case $B = I$.

These various norms can be visualized geometrically in terms of the set of vectors $\{\mathbf{x} : ||\mathbf{x}|| = 1\}$, which is known as the *unit sphere.* These are shown in Figure A.2.1 for vectors in the plane. Note that only for the Euclidean norm are the unit vectors on the circle of radius 1.

The elliptic norms play a particularly central role in matrix theory because they arise in terms of an *inner product*, which in turn defines orthogonality of vectors. An inner product is a real-valued function of two vector variables that satisfies the following conditions (stated only for real vectors):

$$\begin{aligned} &1.\ (\mathbf{x},\mathbf{x}) \geq 0 \text{ for all vectors } x;\ (\mathbf{x},\mathbf{x}) = 0 \text{ only if } \mathbf{x} = 0.\\ &2.\ (\alpha\mathbf{x},\mathbf{y}) = \alpha(\mathbf{x},\mathbf{y}) \text{ for all vectors } \mathbf{x} \text{ and } \mathbf{y} \text{ and scalars } \alpha.\\ &3.\ (\mathbf{x},\mathbf{y}) = (\mathbf{y},\mathbf{x}) \text{ for all vectors } \mathbf{x} \text{ and } \mathbf{y}.\\ &4.\ (\mathbf{x}+\mathbf{z},\mathbf{y}) = (\mathbf{x},\mathbf{y}) + (\mathbf{z},\mathbf{y}) \text{ for all vectors } \mathbf{x},\ \mathbf{y}, \text{ and } \mathbf{z}.\end{aligned} \tag{A.2.8}$$

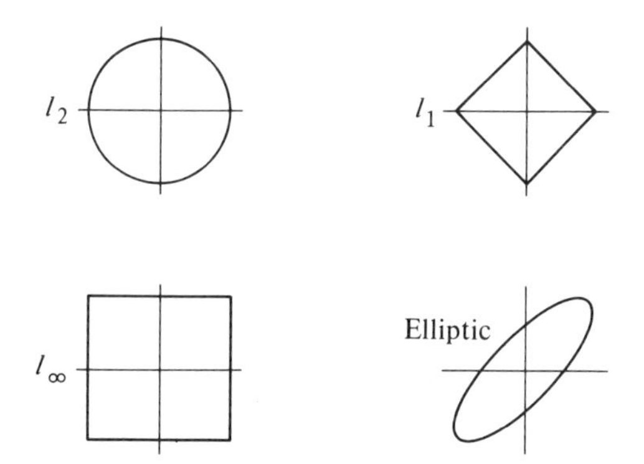

Figure A.2.1: *Unit Spheres of Several Norms*

For any inner product a norm may be defined by

$$||\mathbf{x}|| = (\mathbf{x}, \mathbf{x})^{1/2},$$

and the elliptic norms then derive from the inner product

$$(\mathbf{x}, \mathbf{y}) \equiv \mathbf{x}^T B \mathbf{y}. \tag{A.2.9}$$

Two nonzero vectors $\mathbf{x}$ and $\mathbf{y}$ are *orthogonal* with respect to some inner product if

$$(\mathbf{x}, \mathbf{y}) = 0.$$

If the inner product is the Euclidean one defined by (A.2.9) with $B = I$, then this gives the usual and intuitive concept of orthogonality. A set of nonzero vectors $\mathbf{x}_1, \ldots, \mathbf{x}_m$ is *orthogonal* if

$$(\mathbf{x}_i, \mathbf{x}_j) = 0, \qquad i \neq j.$$

A set of orthogonal vectors is necessarily linearly independent, and a set of n such vectors is said to be an *orthogonal basis*. If, in addition, $||\mathbf{x}_i||_2 = 1, \quad i = 1, \cdots, n$, the vectors are *orthonormal.*

If the columns of a matrix A are orthonormal in the inner product $\mathbf{x}^T\mathbf{y}$,then $A^T A = I$ and the matrix is *orthogonal.* Orthogonal matrices have the important property that they preserve the length of a vector; that is, $||A\mathbf{x}||_2 = ||\mathbf{x}||_2$.

Convergence of a sequence of vectors $\{\mathbf{x}^k\}$ to a limit vector $\mathbf{x}$ is defined in terms of a norm by

$$||\mathbf{x}^k - \mathbf{x}|| \to 0 \qquad \text{as} \qquad k \to \infty.$$

It is natural to suppose that a sequence might converge in one norm but not in another. Surprisingly, this cannot happen.

THEOREM A.2.2: *The following are equivalent:*

1. *The sequence* $\{\mathbf{x}^k\}$ *converges to* $\mathbf{x}$ *in some norm.*
2. *The sequence* $\{\mathbf{x}^k\}$ *converges to* $\mathbf{x}$ *in every norm.*
3. *The components of the sequence* $\{\mathbf{x}^k\}$ *all converge to the corresponding components of* $\mathbf{x}$*; that is,* $x_i^k \to x_i$ *as* $k \to \infty$ *for* $i = 1, \ldots, n$.

As a consequence of this result – sometimes known as the *norm equivalence theorem* – when we speak of the convergence of a sequence of vectors, it is immaterial whether or not we specify the norm.

Any vector norm gives rise to a corresponding matrix norm by means of the definition

$$||A|| = \max_{\mathbf{x}\neq 0} \frac{||A\mathbf{x}||}{||\mathbf{x}||} = \max_{||\mathbf{x}||=1} ||A\mathbf{x}||. \tag{A.2.10}$$

The properties (A.2.5) also hold for a matrix norm; in addition, there is the multiplicative property $||AB|| \leq ||A||\ ||B||$. The geometric interpretation of a matrix norm is that $||A||$ is the maximum length of a unit vector after transformation by A; this is depicted in Figure A.2.2 for the l_2 norm.

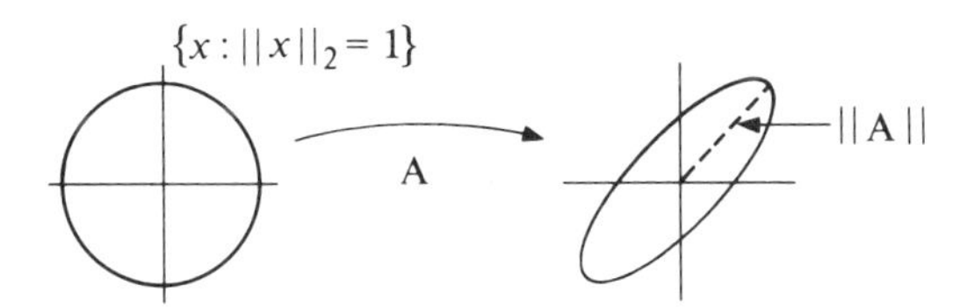

Figure A.2.2: *The* l_2 *Norm*

As with vectors, the convergence of a sequence of matrices may be defined component-wise or, equivalently, in terms of any matrix norm. That is, we write $A_k \to A$ as $k \to \infty$ if in some norm $||A_k - A|| \to 0$ as $k \to \infty$. Again, convergence in some norm implies convergence in any norm.

The matrix norms corresponding to the l_1 and l_∞ vector norms are easily computed by

$$||A||_1 = \max_{1\leq j\leq n} \sum_{i=1}^{n} |a_{ij}|, \qquad ||A||_\infty = \max_{1\leq i\leq n} \sum_{j=1}^{n} |a_{ij}|. \tag{A.2.11}$$

That is, $||A||_1$ is the maximum absolute value column sum of the elements of A, and $||A||_\infty$ is the maximum row sum. The Euclidean matrix norm is given in terms of the spectral radius of A^TA by

$$||A||_2 = [\rho(A^TA)]^{1/2} \tag{A.2.12}$$

and is much more difficult to compute. If A is symmetric, (A.2.12) reduces to

$$||A||_2 = \rho(A), \tag{A.2.13}$$

which is still difficult to compute but is more directly related to the matrix A.

We note that it follows immediately from (A.2.10) that if λ is any eigenvalue of A and $\mathbf{x}$ a corresponding eigenvector, then

$$|\lambda|\ ||\mathbf{x}|| = ||\lambda\mathbf{x}|| = ||A\mathbf{x}|| \le ||A||\ ||\mathbf{x}||,$$

so that $|\lambda| \le ||A||$; thus, any norm of the matrix A gives a bound on all eigenvalues of A. In general, however, the property (A.2.13) of the l_2 norm will not hold.

Bibliography

A. Aho, R. Sethi, and J. Ullman [1988]. *Compilers.* Addison-Wesley, Reading, MA.

E. Allgower and K. Georg (eds.) [1990]. *Computational Solution of Nonlinear Systems of Equations.* American Mathematical Society, Providence, RI.

E. Allgower and K. Georg [1990]. *Numerical Continuation Methods.* Springer-Verlag, New York.

W. Ames [1977]. *Numerical Methods for Partial Differential Equations.* Academic Press, New York.

M. Arioli, J. Demmel, and I. Duff [1989]. Solving Sparse Linear Systems with Sparse Backward Error. *SIAM J. Matrix Anal. Appl.*, 10: pp. 165–190.

U. Ascher, R. Mattheij, and R. Russell [1988]. *Numerical Solution of Boundary Value Problems for Ordinary Differential Equations.* Prentice-Hall, Englewood Cliffs, NJ.

O. Axelsson and V. Barker [1984]. *Finite Element Solution of Boundary Value Problems.* Academic Press, Orlando.

E. Becker, G. Carey, and J. Oden [1981]. *Finite Elements, An Introduction.* Prentice-Hall, Englewood Cliffs, NJ.

J. Bramble and B. Hubbard [1964]. New Monotone Type Approximations for Elliptic Problems. *Math. Comp.*, 18: pp. 349–367.

K. Brenan, S. Campbell, and L. Petzold [1989]. *Numerical Solution of Initial-Value Problems in Differential-Algebraic Equations.* American Elsevier, New York.

W. Briggs [1987]. *A Multigrid Tutorial.* SIAM, Philadelphia.

P. Brown, G. Byrne, and A. Hindmarsh [1989]. VODE: A Variable Coefficient ODE Solver. *SIAM J. Sci. Stat. Comp.*, 10: pp. 1038–1051.

J. Butcher [1987]. *The Numerical Analysis of Ordinary Differential Equations.* John Wiley & Sons, New York.

G. Carey and J. Oden [1984]. *Finite Elements, Computational Aspects.* Prentice-Hall, Englewood Cliffs, NJ.

B. Char, K. Geddes, G. Gonnet, and S. Watt [1985]. *First Leaves: A Tutorial Introduction to MAPLE, in MAPLE User's Guide.* Watcom Publications Ltd., Waterloo, Ontario.

R. Courant and D. Hilbert [1953 , 1962]. *Methods of Mathematical Physics,* volume 1 and 2. Interscience, New York.

J. Daniel and R. Moore [1970]. *Computation and Theory in Ordinary Differential Equations.* Freeman, San Francisco.

P. Davis and P. Rabinowitz [1984]. *Methods of Numerical Integration.* Academic Press, New York.

C. de Boor [1978]. *A Practical Guide to Splines.* Springer-Verlag, New York.

J. Dennis and J. Moré [1977]. Quasi-Newton Methods: Motivation and Theory. *SIAM Review,* 19: pp. 46–89.

J. Dennis and R. Schnabel [1983]. *Numerical Methods for Unconstrained Optimization and Nonlinear Equations.* Prentice-Hall, Englewood Cliffs, NJ.

J. Dongarra, E. Anderson, Z. Bai, A. Greenbaum, A. McKenney, J. DuCroz, S. Hammerling, J. Demmel, C. Bischof, and D. Sorensen [1990]. *LAPACK: A Portable Linear Algebra Library for High Performance Computers.* Proc. Supercomputing '90, IEEE Computer Society Press, Washington, DC.

J. Dongarra, J. Bunch, C. Moler, and G. Stewart [1979]. *LINPACK Users' Guide.* SIAM, Philadelphia.

J. Dongarra, I. Duff, D. Sorensen, and H. van der Vorst [1990]. *Solving Linear Systems on Vector and Shared Memory Computers.* SIAM, Philadelphia.

I. Duff, A. Erisman, and J. Reid [1986]. *Direct Methods for Sparse Matrices.* Oxford University Press, Oxford.

A. Edelman and M. Ohlrich [1991]. *Editors Note.* SIAM J. Mat. Anal. Appl. 12, no.3.

R. Elmasri and S. Navathe [1989]. *Fundamentals of Database Systems.* Benjamin/Cummings Publishing Co., Menlo Park, CA.

C. Fischer and R. LeBlanc Jr. [1988]. *Crafting a Compiler.* Benjamin/Cummings Publishing Co., Menlo Park, CA.

G. Forsythe and W. Wasow [1960]. *Finite Difference Methods for Partial Differential Equations.* Wiley, New York.

J. Francis [1961, 1962]. The QR Transformation, Parts I & II. *Computer Journal,* 4: pp. 265–271, 332–345.

R. Friedhoff and W. Benzon [1989]. *Visualization.* Henry N. Abrams, Inc., New York.

P. Garabedian [1986]. *Partial Differential Equations; 2nd Edition.* Chelsea, New York.

B. Garbow, J. Boyle, J. Dongarra, and C. Moler [1977]. *Matrix Eigensystem Routines - EISPACK Guide Extension,* volume 51. Springer-Verlag, New York.

C. W. Gear [1971]. *Numerical Initial Value Problems in Ordinary Differential Equations.* Prentice-Hall, Englewood Cliffs, NJ.

A. George and J. Liu [1981]. *Computer Solution of Large Sparse Positive Definite Systems.* Prentice-Hall, Englewood Cliffs, NJ.

G. Golub and C. Van Loan [1989]. *Matrix Computations, Second Edition.* Johns Hopkins Press, Baltimore.

N. Gould [1991]. On Growth in Gaussian Elimination with Complete Pivoting. *SIAM J. Mat. Anal. Appl.,* 12: pp. 354–361.

A. Griewank [1989]. *On Automatic Differentiation.* In "Mathematical Programming: Recent Developments and Applications," M. Iri and K. Tanake (eds.), Klewer Academic Publishers, Norwell, MA.

A. Griewank [1990]. *Direct Calculation of Newton Steps without Accumulating Jacobians.* In "Large Scale Optimization," T. Coleman and G. Li (eds.), SIAM, Philadelphia.

R. Haberman [1983]. *Elementary Applied Partial Differential Equations.* Prentice-Hall, Englewood Cliffs, NJ.

W. Hackbush [1985]. *Multigrid Methods with Applications.* Springer-Verlag, New York.

L. Hageman and D. Young [1981]. *Applied Iterative Methods.* Academic Press, New York.

W. Hager [1989]. Updating the Inverse of a Matrix. *SIAM Review,* 31: pp. 221–239.

E. Hairer, C. Lubich, and M. Roche [1989]. *The Solution of Differential-Algebraic Systems by Runge-Kutta Methods.* Springer-Verlag, New York.

E. Hairer, S. Norsett, and G. Wanner [1987]. *Solving Ordinary Differential Equations: I. Non-Stiff Problems.* Springer-Verlag, New York.

C. Hall and T. Porsching [1990]. *Numerical Analysis of Partial Differential Equations.* Prentice-Hall, Englewood Cliffs, NJ.

J. Hennessy and J. Patterson [1990]. *Computer Architecture: A Quantitative Approach.* Morgan Kaufman Publishers, Inc., San Mateo, CA.

P. Henrici [1962]. *Discrete Variable Methods in Ordinary Differential Equations.* Wiley, New York.

M. Hestenes and E. Stiefel [1952]. Methods of Conjugate Gradients for Solving Linear Systems. *Journal of Research of the National Bureau of Standards,* 49: pp. 409–436.

N. Higham [1990]. Exploiting Fast Matrix Multiplication within the Level 3 BLAS. *ACM Trans. Math Softw.*, 16: pp. 352–368.

A. Hindmarsh [1983]. *ODEPACK, A Systematized Collection of ODE Solvers.* In "Scientific Computing", R. Stepleman (ed.), North Holland, Amsterdam.

R. Hockney and C. Jesshope [1988]. *Parallel Computers 2.* Adam Hilger, Bristol and Philadelphia.

A. Householder [1964]. *The Theory of Matrices in Numerical Analysis.* Ginn (Blaisdell), Boston.

E. Isaacson and H. Keller [1966]. *Analysis of Numerical Methods.* Wiley, New York.

J. Keener [1988]. *Principles of Applied Mathematics.* Addison-Wesley Publishing Co., Reading, MA.

H. Keller [1968]. *Numerical Methods for Two-Point Boundary Value Problems.* Ginn (Blaisdell), New York.

V. Kublanovskaya [1961]. On Some Algorithms for the Solution of the Complete Eigenvalue Problem. *Zh. Vych. Mat.*, 1: pp. 555–570.

J. Lambert [1973]. *Computational Methods in Ordinary Differential Equations.* Wiley, New York.

J. LaSalle and S. Lefschetz [1961]. *Stability by Liapunov's Direct Method.* Academic Press, New York.

C. Lawson and R. Hanson [1974]. *Solving Least Squares Problems.* Prentice-Hall, Englewood Cliffs, NJ.

R. Mendez (ed.) [1990]. *Visualization in Supercomputing.* Springer Verlag, New York.

G. Meyer [1973]. *Initial Value Methods for Boundary Value Problems; Theory and Application of Invariant Imbedding.* Academic Press, New York.

C. Moler and G. Stewart [1973]. An Algorithm for Generalized Matrix Eigenvalue Problems. *SIAM J. Numer. Anal.*, 10: pp. 241–256.

W. Newman and R. Sproul [1979]. *Principles of Interactive Computer Graphics (2nd Edition).* McGraw-Hill, New York.

J. Ortega [1987]. *Matrix Theory: A Second Course.* Plenum Press, New York.

J. Ortega [1988]. *Introduction to Parallel and Vecter Solution of Linear Systems.* Plenum Press, New York.

J. Ortega [1990]. *Numerical Analysis: A Second Course.* Reprint of 1972 original. SIAM, Philadelphia.

J. Ortega and W. Rheinboldt [1970]. *Iterative Solution of Nonlinear Equations in Several Variables.* Academic Press, New York.

B. Parlett [1980]. *The Symmetric Eigenvalue Problem.* Prentice-Hall, Englewood Cliffs, NJ.

J. Peterson and J. Silberschatz [1985]. *Operating Systems Concepts, 2nd Edition.* Addison-Wesley Publishing Co., Reading, MA.

T. Pratt [1984]. *Programming Languages.* Prentice-Hall, Englewood Cliffs, NJ.

P. Prenter [1975]. *Splines and Variational Methods.* Wiley, New York.

G. Rayna [1987]. *Software for Algebraic Computation.* Springer-Verlag, New York.

R. Richtmyer and K. Morton [1967]. *Difference Methods for Initial Value Problems.* Interscience-Wiley, New York.

P. Roache [1972]. *Computational Fluid Dynamics.* Hermosa Publishers, Albuquerque, NM.

S. Roberts and J. Shipman [1972]. *Two-Point Boundary Value Problems: Shooting Methods.* American Elsevier, New York.

J. Rosser, R. Newton, and G. Gross [1974]. *The Mathematical Theory of Rocket Flight.* McGraw-Hill, New York.

S. Rubinow [1975]. *Introduction to Mathematical Biology.* Wiley-Interscience, New York.

W. Schiesser [1991]. *Numerical Method of Lines.* Academic Press, New York.

R. Sethi [1989]. *Programming Languages.* Addison-Wesley, Reading, MA.

L. Shampine and M. Gordon [1976]. *Computer Solutions of Ordinary Differential Equations: The Initial Value Problem.* W. H. Freeman, San Francisco.

L. Shampine, H. Watts, and S. Davenport [1976]. Solving Nonstiff Ordinary Differential Equations - The State of the Art. *SIAM Review*, 18: pp.376–411.

J. Shoosmith [1973]. *A Study of Monotone Matrices with an Application to the High-Order Finite Difference Solution of a Linear Two-Point Boundary Value Problem.* Ph.D. Thesis, Applied Mathematics, University of Virginia.

R. Skeel [1980]. Iterative Refinement Implies Numerical Stability for Gaussian Elimination. *Math. Comp.*, 35: pp. 817–832.

G. W. Stewart [1973]. *Introduction to Matrix Computations.* Academic Press, New York.

G. W. Stewart [1974]. Modifying Pivot Elements in Gaussian Elimination. *Math. Comp.*, 28: pp. 527–542.

G. W. Stewart and J.-G. Sun [1990]. *Matrix Perturbation Theory.* Academic Press, New York.

G. Strang and G. Fix [1973]. *An Analysis of the Finite Element Method.* Prentice-Hall, Englewood Cliffs, NJ.

V. Strassen [1969]. Gaussian Elimination Is Not Optimal. *Numerische Mathematik*, 13: pp. 354–356.

A. Stroud [1971]. *Approximate Calculation of Multiple Integrals.* Prentice-Hall, Englewood Cliffs, NJ.

Symbolics [1987]. *Macsyma User's Guide.* Symbolics, Inc., Cambridge, MA.

J. Traub [1964]. *Iterative Methods for the Solution of Equations.* Prentice-Hall, Englewood Cliffs, NJ.

R. Varga [1962]. *Matrix Iterative Analysis.* Prentice-Hall, Englewood Cliffs, NJ.

J. Wilkinson [1961]. Error Analysis of Direct Methods of Matrix Inversion. *J. Assoc. Comput. Mach.*, 10: pp. 281–330.

J. Wilkinson [1963]. *Rounding Errors in Algebraic Processes.* Prentice-Hall, Englewood Cliffs, NJ.

J. Wilkinson [1965]. *The Algebraic Eignevalue Problem.* Oxford University Press (Clarendon), London and New York.

S. Wolfram [1988]. *A System for Doing Mathematics by Computer.* Addison-Wesley, Redwood City, CA.

E. Yip [1986]. A Note on the Stability of Solving a Rank p Modification of a Linear System by the Sherman-Morrison-Woodbury Formula. *SIAM J. Sci. Stat. Comput.*, 7: pp. 507–513.

D. Young [1950]. *Iterative Methods for Solving Partial Difference Equations of Elliptic Type.* Ph.D. Thesis, Harvard University.

D. Young [1971]. *Iterative Solution of Large Linear Systems.* Academic Press, New York.

D. Young and R. Gregory [1990]. *A Survey of Numerical Mathematics*, volume 1 and 2. Chelsea Publishing Co., New York.

D. Young and T.-Z. Mai [1990]. *The Search for Omega.* In "Iterative Methods for Large Linear Systems," D. Kincaid and L. Hayes (eds.), Academic Press, New York.

Author Index

Subject Index

A

B

C

D

E

F

G

H

I

J

L

M

N

ISBN 0-12-289255-0

90040

9 780122 892554